64th Annual Safety in Ammonia Plants and Related Facilities Symposium 2019

AIChE Technical Manual 2019 Volume 60

San Francisco, California, USA
8 - 12 September 2019

ISBN: 979-8-3313-0306-8

Printed from e-media with permission by:

Curran Associates, Inc.
57 Morehouse Lane
Red Hook, NY 12571

AIChE
120 Wall Street, FL 23
New York, NY 10005-4020

Phone: (800) 242-4363
Fax: (203) 775-5177

www.aiche.org

Additional copies of this publication are available from:

Curran Associates, Inc.
57 Morehouse Lane
Red Hook, NY 12571 USA
Phone: 845-758-0400
Fax: 845-758-2633
Email: curran@proceedings.com
Web: www.proceedings.com

TABLE OF CONTENTS

Syngas Compressor Oil Tank Explosion in Ammonia Plant

On April 27th, 2017, 02:43 am, a power outage tripped the ammonia plant in Porsgrunn, Norway. 30-minutes later the syngas compressor oil tank exploded. A thorough investigation identified two independent root-causes that led to an unexpected and dangerous situation.

Svein Hansen
Yara Norge A.S. (Porsgrunn)

Introduction

The Porsgrunn Ammonia Plant was commissioned in 1968 and produces 1450 MTPD. The syngas compressor has four stages, a circulator, an electrical motor and a steam turbine to increase the pressure from 20 to 280 bars (290-4060 psi). The syngas compressor is located in a building with a gross volume of 24.000 m^3 (850.000 ft^3). The process air compressor is in the same building. There are two separate oil systems, one for the process air and one for the syngas compressor train. The oil system for the syngas compressor contains an amount of 35 m^3(1235 ft^3), covering lube, seal and regulator oil.

Safeguarding the Building and Machines

Thirteen explosive gas detectors are located in the inside of the roof, close to the vents in the building. The 3rd and 4th stage, circulator driven by the steam turbine are equipped with a water mist system, which is activated automatically by flame detectors. These detectors react on fire containing hydrocarbons, they thereby target fire originating from oil. A camera monitors the exact same area and is visible on a screen in the control room. There is also a foam system installed in the building designed to fill the ground floor with fire suffocating foam within 5 minutes. This system is activated manually by the operator.

Incident Summary

On the 24th of April 2017, at 02:43 am, a trip occurred in the ammonia plant in Porsgrunn due to a major external power supply fluctuation. The electrical driven process air compressor stopped, a shut-down sequence was initiated, and the syngas compressor tripped.

After stabilizing the ammonia plant's front-end, the shift operators were preparing for a restart of the process air compressor. Suddenly, an explosion occurred in the main syngas oil tank (T-401), located at the ground floor of the compressor hall. The explosion occurred at 03:13, 30 minutes after the trip of the back-end of the ammonia plant. An intense fire developed rapidly inside the building.

Two field operators were present in the compressor building at that moment, both were able to escape. The panel operator immediately initiated the emergency shutdown of the entire ammonia plant, contacted the emergency response team and activated the mitigation device i.e. the foam system.

The Emergency Response Team quickly arrived on site and after identifying the main fire source and setting up their equipment, they were able to control the fire. No one from the emergency response team and operating shift was physically injured.

The material damage was concentrated on the electrical/instrumentation cabling on the west side of the compressor building. Civil damage to the building was extensive but impact on the structural integrity was limited.

Figure 1. Syngas compressor (3rd and 4th stage) and turbine after explosion and fire.

The turbine and the 3rd and 4th stage of the syngas compressor located near the fire source had been protected by the water mist system. In figure 1 this is clearly visible by the yellow paint being intact on the compressor piping. The process air compressor - located on the opposite side of the fire in the building - had damages limited to dirt and debris. Due to a lot of water and foam in the building, the oil trap (the ground floor itself) overflowed to the outside of the building. Minor environmental pollution was therefore reported because oil had reached the nearby fjord via the sewer system.

Figure 2. Distortion of oil tank T-401

What Had Happened?

The reality was clear: the oil tank ruptured or exploded (figures 2 and 3) and caught fire. In order to generate such a rupture of the oil tank several scenarios were proposed. We ended up exploring the possibility of an explosion in the vapor phase of T-401.

We found that the air separation unit (ASU) also had a partial trip due to the external power fluctuation at 02:43. This unit supplies the site including the ammonia plant with nitrogen. In the ammonia plant this nitrogen is mainly used for blanketing of several units, among which T-401.

Figure 3. Top view of ruptured tank T-401

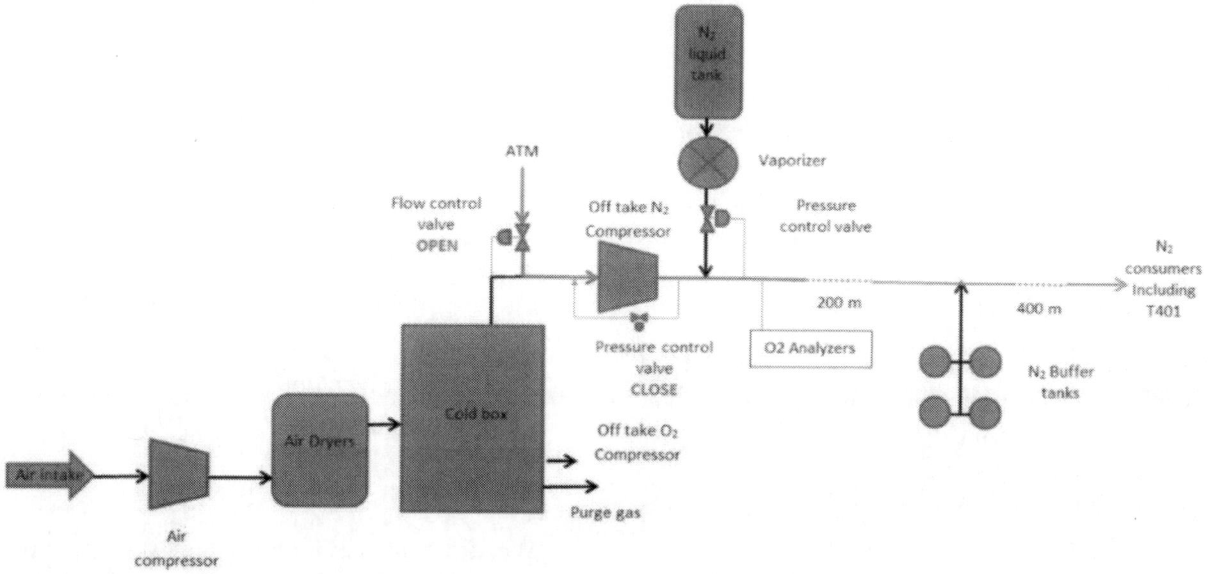

Figure 4. ASU scheme in the status with an air compressor shutdown and the nitrogen compressor still running. The flow control valve is forced to open, and air intake occurred via this valve (red).

The partial trip of the ASU evolved as follows:

- The air compressor supplying the ASU stopped, but the nitrogen compressor kept on running. The flow scheme of the ASU is illustrated in figure 4.
- The nitrogen compressor started to take in ambient air via the flow control valve as the pressure from the ASU decreased. In addition, the ambient air was enriched by oxygen from the ASU's coldbox. The time delay for the analyzer to respond to an oxygen contamination in the nitrogen was found to be approximately 10 minutes.

The analyzer detect oxygen just before 03:00. Given the uncertainties around the precise event timing and residence time evaluation, this is consistent with the assumption that air was introduced in the nitrogen grid at around 02:48. At 03:08 the nitrogen compressor was manually stopped, and the nitrogen supply was thereafter 100% from the main back up supply (liquid nitrogen tank).

Preliminary Conclusion

We concluded that we had all the components needed for an explosion in the tank. Oil vapor probably containing hydrogen degassed from oil to a limited extent, ignited as oxygen enriched ambient air entered the T-401 tank.

The nitrogen compressor did not stop because the pressure switch on the suction side was removed. It proved itself difficult to determine why, or even when the pressure switch was removed. It probably happened when we did a PLC upgrade in 2010. There was no updated HAZID or PHA done for the ASU.

When It Rains It Pours...

There are several explosive gas detectors in the building where the fire occurred. If two out of three detectors exceed a defined limit, a vent system operates to dilute the combustible gases by adding air. Three fans go to a maximum vent mode forcing 170 m^3/s of air into the building. The building has a net volume of 21000 m^3 (741,600 ft^3).

Our explosive gas detectors are designed to be "fail safe". So, when they were destroyed by the fire and failed, they indicated high level of combustible gas in the building; the extensive venting was initiated, and we blew air into the fire! This unintended fire boost caused some large doors to open on the ground floor, allowing the foam to escape to the outside instead of filling the ground floor with fire suffocating foam.

Repair and Restart

Damages were repaired within seven months. All recommended actions were implemented, and the plant was restarted.

However, as there was a possibility for a hydrogen "leak" from the compressor seals to the tank (T-401), we scheduled for a gas sample to be taken from the tank as soon as the plant was up and running again. Surprisingly, we found a large amount of hydrogen in the gas phase of the tank as shown in table 1. All possible resources were now allocated to one single task: where is this hydrogen coming from?

	Nov 2017	Jun 2018
O_2	1.5 %	0.9 %
H_2	16 %	< 0.5 %
N_2	82 %	98.6 %
CH_4	0.1 %	0.01 %
NH_3	traces	-

Table 1. Gas phase composition of T-401 after the start-up in 2017 and in June 2018

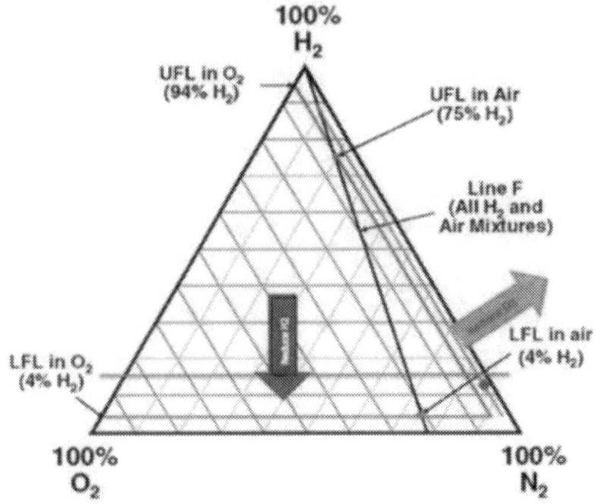

Figure 5. Lower/upper flammable limit for oxygen and hydrogen mix

The Suspect

We did a massive amount of studies. Since the plant was in operation, it gave some limitations with regards to investigations, but it also gave opportunities, for example we could verify the impact of imposed changes immediately.

We ended up with a new theory: there was a leak in the 4th stage compressor end lid. There are oil channels in the end lid and these are routed in angles to get desired oil flow as shown in figure 6. The excess holes were plugged after drilling. We suspected that one of these plugs was corroded and started leaking. This led to a connection between lube oil and process gas. The lube oil return is routed straight back to the tank. Our suspicion was so strong that we even ordered a new end lid.

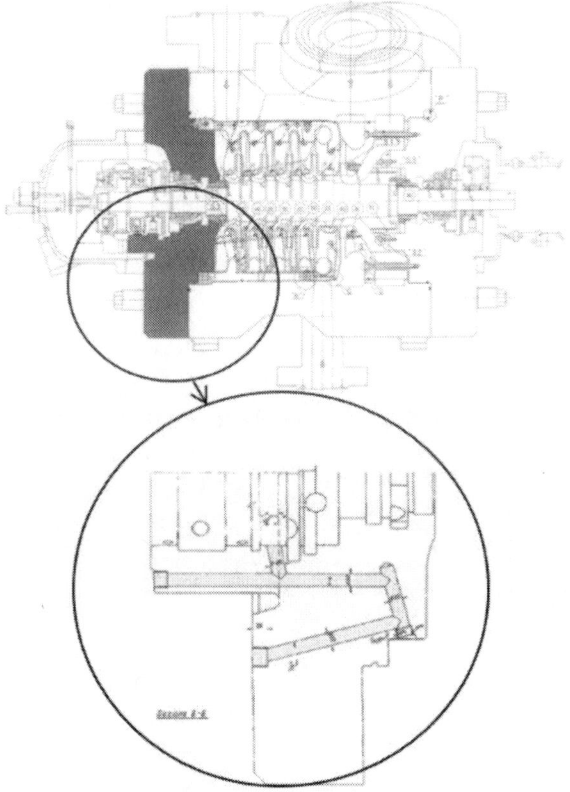

Figure 6. End lid oil channels highlighted.

In June 2018 we had to stop the plant due to operational issues. We took the opportunity to dismantle and check the 4th stage end lid. The leak we found was significant. We re-plugged the leakage, as the new lid hadn't arrived yet. The plant was restarted, and new samples were taken from the gas phase of T-401 as shown in the last column of table 1. In figure 7 it is also clear that since the repair in June 2018 the gas phase is significantly less explosive.

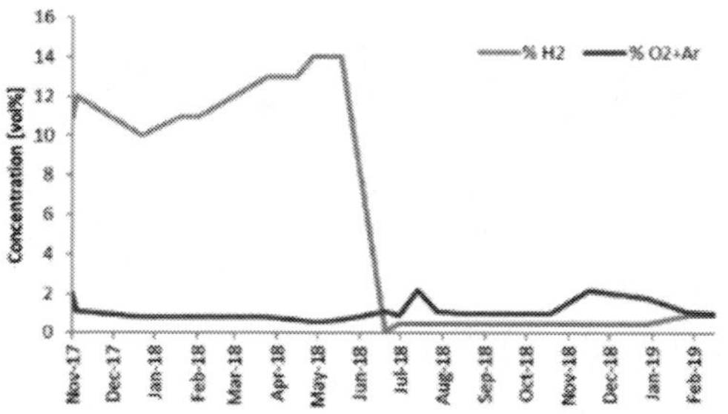

Figure 7. Development of T-401 gas phase composition. The June 2018 plant stop, and repair was a success!

Discussion

The incident shows the complexity of identifying when two errors occur simultaneously creating "the perfect storm": The missing pressure switch trip on the nitrogen compressor and the leaking end lid. It is very unlikely to identify such a scenario through a HAZOP, where double jeopardy isn't usually considered.

Nevertheless, the combination of potential errors may lead to dangerous situations and expensive repairs. All unwanted incidents, like a nitrogen compressor not stopping when it should, must have two independent technical barriers. In our case the water mist system limited the damages.

Another fact is that during the 30 minutes between the plant trip by the power outage and the fire, the control room operator got 900 alarms from the DCS. In addition; the nitrogen compressor that did not stop gave no alarm.

Finally, after the repairs in 2017, a photo was taken from the compressor hall as in figure 8 below.

Figure 8. Compressor building after repair.

Advances in High Temperature Hydrogen Attack (HTHA) Inspection and Fitness-for-Service Relating to Ammonia Producers

Assessments of HTHA risks, Inspection, and Fitness-for-Service (FFS) are progressing in API and will directly impact Ammonia Producers. This paper summarizes the ongoing changes in the industry and discusses methods for modeling damage progression and developments in non-destructive testing, which are being considered for inclusion in applicable codes and recommended practices in the coming years. These approaches have been presented to other International Regulatory agencies and are rapidly evolving. An overview of these developments and how an owner-user may position themselves to better manage the risk associated with operating equipment prone to HTHA is presented. Furthermore, based on advanced analysis Integrity Operating Windows (IOWs) can be established to maintain safe operating limits for affected equipment. This paper offers insight into identifying HTHA risks, predicting damage progression over time, and implementing risk mitigation strategies.

Phillip E. Prueter, P.E.
The Equity Engineering Group, Inc.

Introduction

High temperature hydrogen attack (HTHA) has long been an issue since the initial production of ammonia over 100 years ago. Carl Bosch experienced HTHA during his experiments to synthesize ammonia with what later became known as the Haber-Bosch process. This confrontation with HTHA in his laboratory scale reactor led to the utilization of one of the first known internal liners to protect the pressure boundary from high hydrogen partial pressures. In conjunction with "Bosch Holes", (through-wall holes to allow hydrogen to escape) this system was successful in preventing HTHA from adversely affecting the mechanical integrity of the system (see Figure 1). A historical high-pressure steel reactor ammonia production via the Haber process is displayed at an office complex in Germany (see Figure 2). While it is no longer advisable to in-stall through-wall holes in the pressure boundary of reactors, the industry has learned a lot about this mechanism over the decades and awareness has increased tremendously along with improved design and material selection practices.

In this paper, a case study by John Zink Hamworthy Combustion (JZHC) of a reformer optimization will be presented using JZHC's Smart Combustion solutions. Although reformers consist of multiple combustion zones throughout the process, the approach described in this paper will only focus on the primary reformer radiant (arch burners) section.

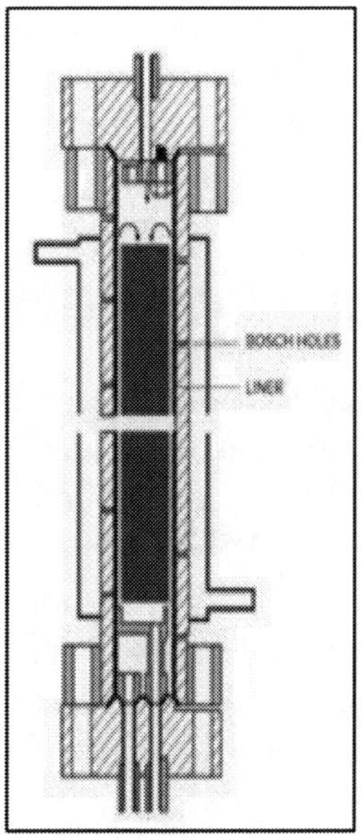

Figure 1. Internal liner / Bosch Holes.

Figure 2. Historical Reactor on display (circa 1921).

Nevertheless, there are still many unknowns, and HTHA has proven to be a challenging damage mechanism to detect and analyze, becoming more difficult to manage due to changing recommendations, industrial guidance, and "best practice". However, through an international joint industry project (JIP) spanning over half a decade, the introduction of reliable non-destructive examination (NDE) techniques have allowed for the creation of new fitness for service (FFS) rules to evaluate HTHA damage.

Changes in API 941 Guidance

Several C-0.5Mo equipment items, some of which are in Ammonia Production, currently operate above what is now the recommended Nelson Curve for C-0.5Mo; that is the Post Weld Heat Treated (PWHT'd) Carbon Steel (CS). The most recent shift of the Nelson Curves came in the 8th Edition of API RP 941 (2016), where CS was delineated into separate PWHT and non-PWHT curves.

Following a catastrophic rupture due to HTHA at the Tesoro Anacortes Refinery that included seven fatalities in 2010, a significant amount of industry resources became focused on HTHA. The ensuing Chemical Safety and Hazard Board (CSB) investigation, along with other HTHA incidents in liquid filled hydrotreating lines, prompted the introduction of a new non-PWHT CS Nelson Curve, that lies approximately 50°F (28°C) below the original CS Nelson curve (now PWHT'd CS). The original and current curves are shown in Figures 3 and 4, respectively.

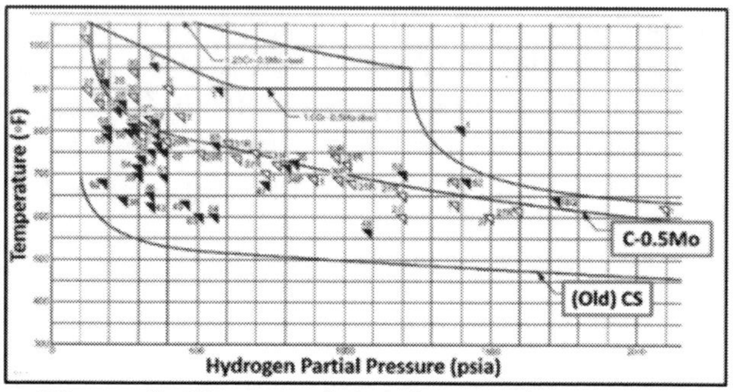

Figure 3. Historical C-0.5Mo Nelson Curve.

(FMC/TFM) are complimentary methods, although PAUT (used previously but with less array element and processing capability) has been further refined for HTHA. These techniques have been gaining widespread acceptance by Owner-Users, NDE Service Providers and regulatory agencies such as the Health and Safety Executive (HSE) in the UK to inspect for HTHA.

When using these techniques to inspect for HTHA, it is always recommended that the inspector has earned their certification from an internationally acknowledged HTHA training course. This certification was developed to make field technicians more familiar with the various NDE modalities, as well as to optimize the inspection techniques to repeatedly and accurately locate HTHA damage, if present. Figure 5 and Figure 6 (upper and lower images show an example of two different ToFD scans of the same location, one scan was performed using the proper parameters (probe spacing, probe angle, gain, etc.) and one scan was performed using improper parameters. The scan setup with improper parameters fails to locate the HTHA damage, (the yellow circle depicts regions of HTHA damage) or presents a false negative.

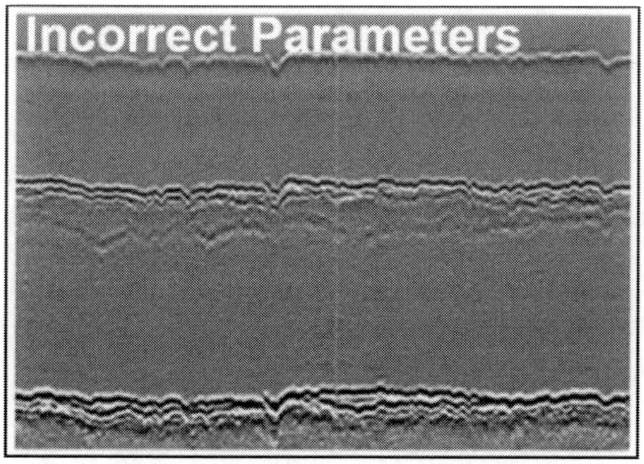

Figure 5. Incorrectly set parameters fail to reveal regions of HTHA.

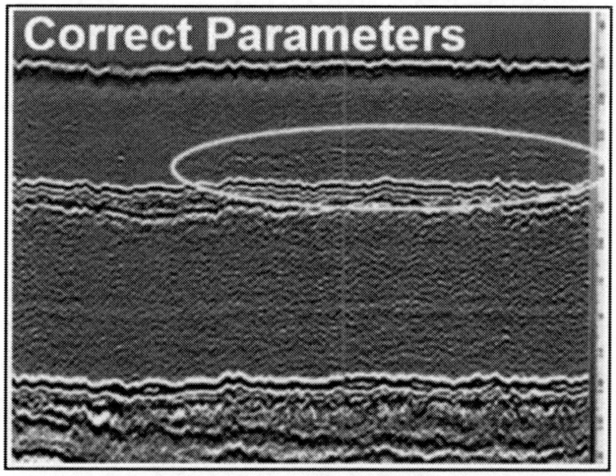

Figure 6. Same weld from Figure 5 using optimized parameters.

Fitness-for-Service Developments

In general, API 579-1/ASME FFS-1 Fitness-For-Service (API 579) Fitness-for-Service (FFS) assessment rules are dependent on accurate in-puts to achieve meaningful and reliable results. For example, some parts of API 597 rely on understanding and definition of the geometry of the local metal loss, and other parts depend on the sizing of a given crack like flaw. The requirement of accurate location and sizing for FFS (damage characterization) is what has created difficulty in the past for HTHA evaluation.

Prior to the introduction of the new NDE techniques (TFD, PAUT, FMC/TFM) there was minimal confidence in accurate sizing HTHA damage and therefore no usefulness in creating FFS rules. Now that experience has shown HTHA damage can repeatedly and reliably be characterized and sized using NDE, this allows the introduction of FFS rules for HTHA evaluation.

The second task of the ongoing HTHA JIP was to develop FFS rules for components in HTHA service to assist owner-users in the assessment of equipment. This methodology has successfully been benchmarked against the data points on the Nelson Curves, ensuring accurate predictions.

Additionally, this model has been applied to numerous assessments of individual components, pressure vessels, heat exchangers, and piping.

Further Advances in HTHA Time-Based Design Curves

The API 941 task group has approved a project to produce time-based (progression over 1-100 years) HTHA screening curves from a partnership of two major international engineering firms. These draft curves are remarkably similar to the experienced-based Nelson Curves, currently in API 941. Based on the preliminary results, two different design life methodologies will be compared to the empirical data. The API 579 FFS task group also has a draft set of rules for HTHA. Additionally, elevated temperature material testing in hydrogen is being carried out to validate and provide crack growth rate models. Some examples of applications of these methods, with attention to Ammonia Plant and C-1/2 Mo equipment are discussed in this paper.

FFS HTHA Case Studies

This paper provides a summary of two case studies: an in-service high temperature ammonia shift converter and an in-service hydrotreater reactor that received targeted inspection for HTHA damage. Components in the ammonia industry commonly exposed to hot hydrogen service include methanators, high temperature shift converters, shift converter feed/effluent exchangers, and waste heat boilers. It is noted that cold-walled (refractory-lined) components that generally operate at temperatures below the HTHA regime may be subject to damage if internal liners or refractory are compromised, resulting in hot spots.

High Temperature Shift Convertor Case Study

The evaluation of a C-0.5Mo high temperature shift converter due to historic operation above the PWHT'd CS Nelson Curve (recommended for C-0.5Mo equipment) is discussed below. The original C-0.5Mo Nelson Curve still existed around the time that many of the ammonia plants were constructed, leading to numerous C-0.5Mo high temperature shift converters still in service today. This vessel operated at a maxi-mum of approximately 850°F (454°C) and 230 H2 psia (1585 kPa) over the life of the plant (Figure 4 red circle on curve). Using the recently developed mechanistic HTHA damage model (from an ongoing JIP), this vessel was evaluated with regard to current estimated HTHA dam-age, flaw sensitivity analysis, brittle fracture concerns, and recommended replacement time-line.

Using approximately 10 years of past operating history (temperature and hydrogen partial pressure), calculated stress state, and year of construction as an input to the model, the vessel was predicted to contain a moderate to high level of HTHA damage (approximately 65%), which can be correlated to methane formation and degradation of material properties. This was determined using hydrogen diffusion and me-thane reaction rates at operating conditions, along with stress states local to structural dis-continuities. Due to the moderately high level of damage predicted, an MPT (Minimum Pressurization Temperature) envelope was provided to minimize the risk of brittle fracture during unit start-ups and shutdowns.

Example MPT envelopes for an ammonia shift converter are shown in Figure 7 (including curves for different flaw sizes remote from and near structural discontinuities). The individual points on this plot indicate historical operating pressure-temperature points. These curves aided in determining a critical flaw size, or in other words the smallest flaw that must be detected to ensure operation in the acceptable region of the MPT envelope occurs. To inspect for this critical flaw size a HTHA JIP inspection was conducted to confirm model predictions regarding HTHA damage, and to inspect for the critical flaw size.

Based on predicted HTHA crack growth rates, the largest flaw found during the inspection, and required MPT guidance, a recommended end of life and inspection interval was established. In this case, sufficient remaining life was demonstrated to allow replacement of the vessel at the next turnaround, instead of the current turnaround. In the end, this saved the owner-user time and money by not having to delay start-up until replacement of the shift converter. This methodology added value to the owner-user by providing a more refined understanding of the current HTHA risk and being able to justify prolonged operation until the next turnaround through sound engineering judgement.

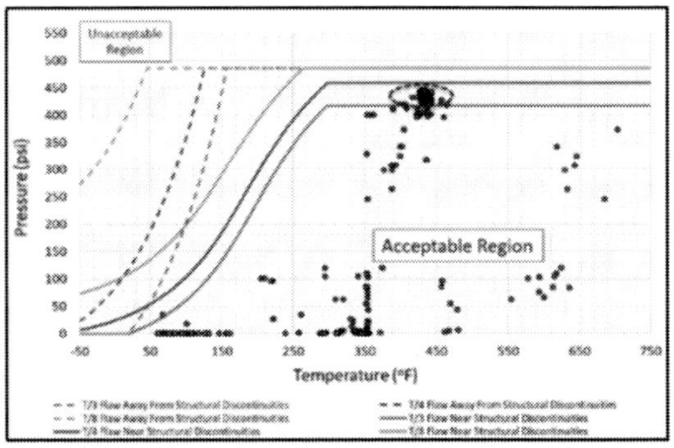

Figure 7. Example of MPT Envelope for multiple reference flaw sizes (for example only)

Hydrotreater Reactor Case Study

A second case study is on an in-service hydrotreater reactor (SA-204 Gr B, 1975 Vintage DT 775°F, DP 626 psig) that received targeted inspection for HTHA damage. The reactor inspection map (including a roll- out of the cylindrical shell) is provided in Figure 8 to show the targeted weld locations that were examined. Longitudinal-to-circumferential weld T-junctions were targeted in addition to nozzle and manway-to-shell junctions. The inspection identified possible incipient HTHA damage and a few damaged regions near welds consistent with crack-like flaws (potentially originating as fabrication defects). Even though this example applies to the refining industry, the overall methodology can be directly applied to ammonia production equipment.

HTHA damage calculations are carried out based on the previously mentioned FFS and JIP efforts. Additionally, two different future operating temperatures are considered: 625°F and 650°F (329°C and 343°.C). For both assumed temperatures, crack growth calculations are per-formed based on different flaw sizes (see Figure 9 for flaw growth predictions for different flaw depths and different flaw length-to-depth ratios).

The concept of these crack growth estimates is to determine if a current crack-like flaw is likely to propagate to critical dimensions (determined using the fracture mechanics approach described below) before planned reactor replacement. The reactor replacement is the recommended long-term strategy given the vintage of the reactor and the identified potential HTHA damage. In the short-term, risk mitigation strategies including establishing technically-based pressurization approaches to avoid brittle fracture during start-up or shut-down and understanding the risk of crack-growth due to HTHA are employed.

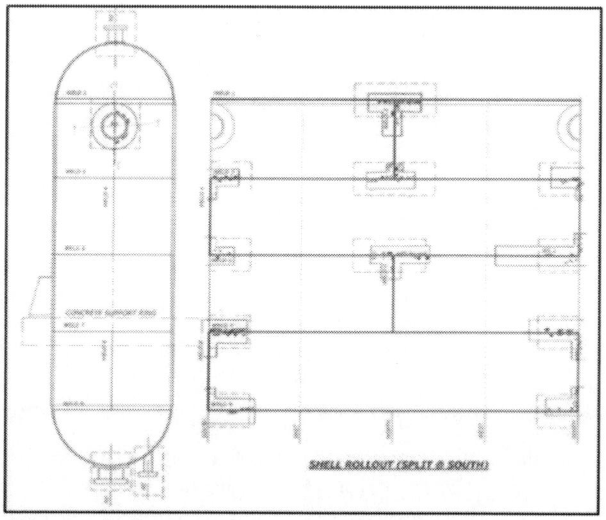

Figure 8. Reactor Inspection Map with Roll-Out of Cylindrical Shell.

Following completion of the inspection summarized above, the most limiting defects were evaluated (actual measured flaw dimensions) using the same MPT assessment approach and buffers or margins of 25°F and 50°F were ap-plied to the most limiting MPT envelope. Figure 10 shows a plot of these two MPT curves in addition to the operating data points representative of typical reactor start-ups and shut-downs.

The reason for invoking some margin or buffer on the most limiting MPT curve is to attempt to account for analysis unknowns such as fracture toughness and potentially, the presence of a larger defect that was not identified during the targeted inspection (that is, because 100 percent coverage of all pressure boundary welds was not carried out). Figure 10 highlights that historical pressurization procedures technically violate these proposed MPT envelopes. The MPT curve with a 50°F buffer indicates that essentially no pressurization is recommended for temperatures below 150°F. While this may seem overly limiting, the extent of the identified damage coupled with potentially conservative fracture toughness assumptions and the crack driving forces associated with weld and cladding residual stress results in limiting pressurization curves, even at little-to-no internal pressure.

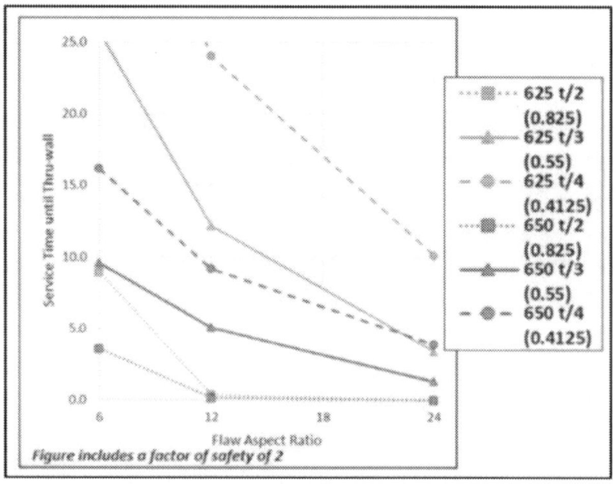

Figure 9. Predicted Crack Growth Rates in the Cylindrical Shell.

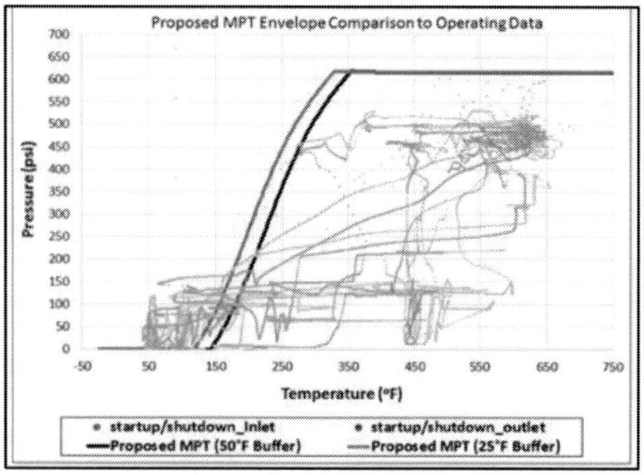

Figures 10. Example of MPT Envelope for multiple reference flaw sizes.

Integrity Operating Windows (IOWs)

Integrity Operating Windows (IOWs) are established limits for process variables or parameters that can affect the integrity of equipment if the process operation deviates from the established limits for a predetermined amount of time. Furthermore, IOWs are generally classified into different levels, distinguished by risk, in order to set priorities on notifications (including alarms, alerts, and/or other notifications) and timing of actions to be implemented when IOWs are exceeded. In general, it is essential that a Management of Change (MOC) process is applied whenever critical or standard IOW variables are revised, updated, or created utilizing the same types of experienced Subject Matter Experts (SMEs) and engineering analysis that were used to generate the original IOW list for a given process unit. Furthermore, the MOC work process must be closely integrated with the IOW work process for any changes, additions, or deletions to be made to the IOW list.

Summary and Conclusions

This paper offers background information on the mechanism of HTHA, the resistance of C-0.5Mo material to HTHA damage and fracture, and recent developments in NDE technology to accurately identify HTHA damage. Additionally, a case study of an in-service high temperature ammonia shift converter is discussed and a second case study of a C-0.5Mo hydrotreater reactor is summarized. For the latter, modern HTHA inspection techniques such as TOFD and PAUT (performed by a properly certified inspector) identified incipient HTHA and the presence of some crack-like flaws in pressure boundary welds. Furthermore, a significant portion of historical operating points fall above the current PWHT carbon steel Nelson Curve (currently recommended for C-0.5Mo in API 941).

For the reactor, fracture mechanics-based MPT envelopes are generated based on this identified damage to guide pressurization procedures that mitigate the risk for brittle fracture, particularly during start-up and shut-down. Also, historical operating hydrogen partial pressure and temperature histograms are established to try and bound past operating data, and HTHA damage is evaluated based on methods developed through an ongoing JIP. Crack growth predictions are also employed to gain confidence in safe short-term operation, until planned reactor replacement can feasibly take place (an approximately two-year time-line).

Ultimately, new IOWs were established to mitigate the risk for both brittle fracture and time-dependent HTHA crack growth. External steam tracing was installed to achieve a bulk pressure boundary metal temperature of 200°F prior to reactor pressurization to avoid brittle fracture. These IOWs ensure that the proposed MPT envelope with a 50°F shift or buffer relative to the most limiting calculated MPT curve is always adhered to and that maximum operating reactor metal temperatures are limited to 625°F to minimize short-term HTHA crack growth. If an unexpected unit trip (shut-down) occurs, targeted follow-up inspection is recommended to benchmark any damage progression or crack-growth at some of the most severely (known) damaged locations.

While the long-term recommendation is to replace the reactor, the analysis methods described herein and implementation of new IOWs, coupled with modern NDE techniques carried out by a fully-trained inspector, offer a means to manage the risks associated with operating this 1970s vintage hydrotreater reactor (with possible HTHA damage) in the short-term. These recent developments in inspection and analysis techniques offer owner-users the ability to quantify the risk of HTHA damage like never before.

References

[1] Bosen, C., 1933. "Attack of Steel Under Conditions of Ammonia Synthesis". Die Chem. Fabrik, p. 127.

[2] Evans, T., 1948. "Hydrogen Attack on Carbon Steels". Mechanical Engineering, 70, pp. 414–416.

[3] API, 2016. Recommended Practice 941: Steels for Hydrogen Service at Elevated Temperatures and Pressures in Petroleum Refineries and Petrochemical Plants, 8th ed. American Petroleum Institute, Washington, DC.

[4] Lundin, C., Bharadwaj, M., Barlow, W., Barnard, B., and White, S., 2014. Welding Re-search Council Bulletin 548: Effect of Hydro-gen Attack on the Mechanical Properties of a C-1/2Mo Steel. Welding Research Council, Inc., Shaker Heights, OH.

[5] Nelson, G., 1949. "Hydrogenation Plant Steels". In 1949 Proceedings API. Vol. 29M, pp. 163-174.

[6] API, 1967. API Publication 940, Steel Deterioration in Hydrogen (Out of Print). American Petroleum Institute, Washington, DC.

[7] API, 1975. API Publication 945, A Study of the Effects of High-temperature, High-pressure Hydrogen on Low-alloy Steels (Out of Print). American Petroleum Institute, Washington, DC.

[8] Fletcher, E., and Elsea, A., 1964. DMIC Report 202: The Effects of high-Pressure, High-Temperature Hydrogen on Steel. Defense Metals Information Center - Battelle Memorial Institute, Columbus, OH.

[9]Panzarella, C., and Cochran, J., 2015. E2G Technical Re- port 48: Report on the HTHA Models Used for FFS Assessments and RBI Implementations HTHA JIP Phase 1 Final Re-port. The Equity Engineering Group, Inc., Shaker Heights, OH.

[10] Panzarella, C., and Cochran, J., 2016. E2G Technical Re- port 60: Technical Background for the E2G HTHA Model. The Equity Engineering Group, Inc., Shaker Heights, OH.

[11] Osage, D., Panzarella, C., and Cochran, J., 2017. E2G Technical Report 94: E2G HTHA JIP FFS Rules for API 579- 1/ASME FFS-1, Part 15 Version 7. The Equity Engineering Group, Inc., Shaker Heights, OH.

[12] Nugent, M., Silfies, T., Dobis, J., and Armitt, T., 2017. "High Temperature Hydrogen

Attack (HTHA) Modeling Prediction and Non-Intrusive Inspection Review".

[13] Prueter, P., and Nugent, M., 2019. "Fitness-for-Service Case Studies in HTHA Equipment". In 23rd International Pressure Equipment Integrity Association (IPEIA) Conference.

[14] Kowalski, P., Prueter, P., and Nugent, M., 2018. "Fitness- for-Service Case Studies in HTHA Equipment". In Paper 17-3D: AIChE Ammonia Safety Symposium.

[15] API, 2016. API 579-1/ASME FFS-1 Fitness-For-Service. The American Petroleum Institute, Washington D.C.

[16] Saba, B., 2003. Evaluation of mechanical fitness for service of high temperature hydrogen attacked steels. Louisiana State University Master's Thesis, Baton Rouge, Louisiana.

[17] Silfies, T., 2016. "Looking Behind the Curtain of API 941 HTHA Data". NACE International, Corrosion Conference and Expo.

[18] Martin, M., Dadfarnia, M., Orwig, S., Moore, D., and Sofronis, P., 2017. "A microstructure-based mechanism of cracking in high temperature hydrogen attack". Acta Materialia, 140, pp. 300–304.

[19] Tang, S., Guo, T., and Cheng, L., 2011. "Modeling hydrogen attack effect on creep fracture toughness". International Journal of Solids and Structures, 48, pp. 2909–2919.

[20] Nageswaran, C., 2018. Maintaining the integrity of process plant susceptible to high temperature hydrogen attack. Part 1: analysis of non-destructive testing techniques. Health and Safety Executive (HSE), London.

[21] Rothwell, J., 2018. Maintaining the integrity of process plant susceptible to high temperature hydrogen attack. Part 2: factors affecting carbon steels. Research Report RR1134 - Health and Safety Executive (HSE), London.

[22] EEMUA, 2018. Publication 240 - Factors to Consider when Managing High Temperature Hydrogen Attack on Petroleum and Chemical Facilities. The Engineering Equipment and Materials Users Association (EEMUA), London.

[23] McLaughlin, J., Krynicki, J., and Bruno, T., 2010. "Cracking of Non-PWHTd Carbon Steel Operating at Conditions Immediately Below the Nelson Curve". In ASME Pressure Vessels and Piping Conference. PVP2010-25455.

[24] API, 1977. Recommended Practice 941: Steels for Hydrogen Service at Elevated Temperatures and Pressures in Petroleum Refineries and Petrochemical Plants, 2nd ed. American Petroleum Institute.

[25] API, 1990. Recommended Practice 941: Steels for Hydrogen Service at Elevated Temperatures and Pressures in Petroleum Refineries and Petrochemical Plants, 4th ed. American Petroleum Institute.

[26] Nugent, M., and Armitt, T., 2018. "HTHA Inspection Detection Confidence Improves Basis for New API Recommendations".

[27] ASME, 1974 Addenda. Section VIII Division 1: Rules for Construction of Pressure Vessels. American Society of Mechanical Engineers Boiler and Pressure Vessel Committee, New York, NY.

[28] Shewmon, P., and Xue, Y., 1991. "Effect of High-Pressure Hydrogen on Crack Growth in Carbon Steel". Metallurgical Transactions, 22A, pp. 2703–2707.

[29] Osage, D., Spring, D., Anderson, T., Kummari, S., Prueter, P., and Wallin, K., 2017.

Welding Research Council Bulletin 562: Recommendations for Establishing the Minimum Pressurization Temperature (MPT) for Equipment. Welding Research Council, Shaker Heights, OH.

[30] Prueter, P., Kummari, S., and Osage, D., 2017. "A Fracture Mechanics-Based Approach to Establish Minimum Pressurization Temperature Envelopes". In ASME Pressure Vessels and Piping Conference. PVP2017-65805.

[31] Scott, P., Anderson, T., and Osage, D., 1998. Welding Re- search Council Bulletin 430: A Review of Existing Fitness- For-Service Criteria for Crack-Like Flaws. Welding Re- search Council, New York.

[32] Anderson, T., 2005. Fracture Mechanics: Fundamentals and Applications, 3rd ed.

[33] Anderson, T., 2000. "A Compressive Fitness-For-Service Guide". International Journal of Pressure Vessels and Piping, 77, pp. 953–963.

[34] Wallin, K., 2011. Fracture Toughness of Engineering Materials. EMAS Publishing, United Kingdom.

[35] ASME, 2017. Boiler and Pressure Vessel Code, Section VIII Rules for Construction of Pressure Vessels, Division 2 Alternative Rules. The American Society of Mechanical Engineers, New York.

[36] API, 2017. API Recommended Practice 970: Corrosion Control Documents. The American Petroleum Institute, Washington D.C.

[37] API, 2014. API Recommended Practice 584: Integrity Operating Windows. The American Petroleum Institute, Washington D.C.

[38] Prueter, P.E., Jones, J.R., Hess, J. and DeLuca, J., "Managing the Risks Associated with Operating a Hydrotreater Reactor with Possible High-Temperature Hydrogen Attack Damage," PVP2019-93533, Proceedings of 2019 ASME Pressure Vessels and Piping Conference, San Antonio, TX, July 14-19, 2019.

Mechanical Integrity of Process Installations: An Assessment Based on Bow-Ties

A special Safety Research project was carried out in an ammonia plant of OCI Nitrogen, located at the Chemelot site in Geleen, The Netherlands. This research focused on the development of a method to monitor process safety accident processes in which loss of mechanical integrity is the root cause. A significant part of the mechanical integrity failure scenarios originates from material and corrosion mechanisms which may develop over a relatively long-time period, possibly taking months, years or even longer before turning into an event. Based on two examples, failure mechanisms are worked out and visualized using a bow-tie. The research object shows that the monitoring of 'early warnings' and barriers can provide information about the current probability of scenarios in order to intervene prematurely. In addition, early warnings can be used to initiate inspections if there is a demonstrable probability that the failure mechanism and the scenario may be taking place. Considering the historical development of breakdown maintenance to preventive and predictive maintenance and risk-based inspection (RBI), this method could be a new step in the field of maintenance efficiency.

Peter Schmitz
OCI Nitrogen, The Netherlands

Introduction

Located at the Chemelot site in Geleen in The Netherlands, OCI Nitrogen operates two ammonia plants, each with a nameplate capacity of 1360 MTPD of ammonia. One of these plants was designed and constructed by M.W. Kellogg based on Kellogg's reduced energy ammonia technology for steam reforming of natural gas. The plant was commissioned in 1984 and has a current capacity is 1550 MTPD. This plant has been subjected to a mechanical integrity study.

In the past years, there has been a number of major process safety incidents at OCI Nitrogen's two ammonia plants. According to an internal investigation, these incidents were mainly caused by incorrect material choice or unexpected mechanical failure mechanisms. The latter in particular played an important role. The incidents were not detected in previous safety studies, nor were the related phenomena looked at during regular inspections. These 'leak before break' incidents were always unexpected and occurred without prior warning. Due to these incidents, the ammonia plant had to be stopped. The question was raised to what extent aging played a role. After all, OCI Nitrogen's two ammonia plants are almost 50 and 35 years old, respectively.

This sub-study is part of a larger study aimed at preventing major process safety incidents by taking timely measures. The research question of this sub-study is: *How can accident processes caused by mechanical failure of static process units be influenced at OCI Nitrogen's ammonia plants?* This paper describes a method for investigating mechanical failure mechanisms that can affect the integrity of static process units of ammonia plants in such a way that it can lead to catastrophic failure. This method is based on a mechanical integrity assessment of one of OCI Nitrogen's ammonia plants.

Although aging is not explicitly mentioned in the research question, some incidents seem to be related to it. Research shows that aging is a factor in a significant proportion of major accidents [10, 23, 24]. The deteriorating state of chemical installations due to failure mechanisms related to aging is becoming an important issue on a global scale [3]. Literature research shows that the technical aspects of aging play a role in almost all industries, possibly with the exception of the IT industry, and that they receive special attention during maintenance, inspection and service [24]. Professional literature provides all kinds of guidelines with programs for asset management, asset integrity or risk management, whether or not aimed at preventing aging of (process) installations [4, 8, 9, 10, 11, 14]. Scientific literature is more model-based and provides proposals for risk-based asset integrity indicators [6], condition monitoring [26] or an integrity operating window [13] which can foresee increased risks and thus promote timely action.

Method and theory

Definitions

Bow-ties are user-friendly and suitable for mapping scenarios. They are a metaphor for an accident process and show the cause of a scenario, the consequences and the barriers that can stop the scenario [21]. The simple, sequential set-up of bow-ties is strongly reminiscent of the Swiss cheese model by Reason [17], who used cheese slices to represent the barriers. The organizational aspects in the bow-tie however are shown as 'delivery systems', the management actions to guarantee the barriers' quality [22].

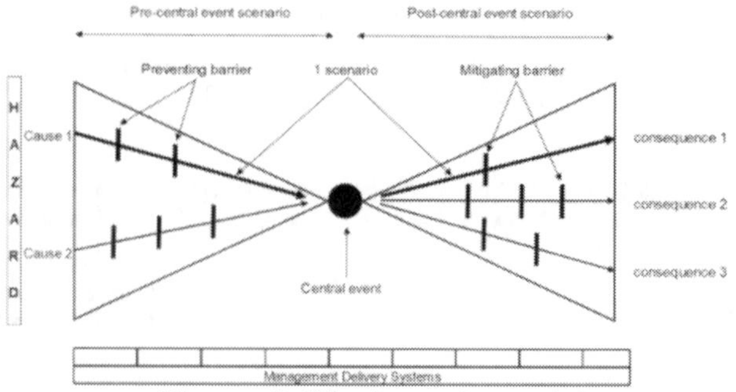

Figure 1. The bow-tie model

Figure 1 shows an example of a bow-tie in which the central event is at the center of the bow-tie. The central event in (petro) chemical installations is often characterized by an undesirable and uncontrolled release of a hazardous substance and/or energy from the plant. As a result, a situation with an uncontrollable hazard arises. A hazard has the intrinsic ability to cause material damage, casualties and injuries and consists of the substance and energy of a process unit. An ammonia plant contains inflammable gases such as hydrogen and natural gas, toxic ammonia in gas and liquid form and steam.

A barrier can be defined as anything that can prevent a cause from developing into a consequence, including preventing the cause itself [2]. Safety barriers can be physical and/or non-physical means to prevent, control or reduce undesired events or accidents [20]. When developing scenarios, the installed (preventive) barriers must function at all times so as to prevent a central event from taking place. If these are broken or not present, a scenario may develop into a central event.
'Early warnings' are used as a synonym for 'leading process safety indicators'. They can be seen as an observable collection of data which can indicate the faults and threats of a system in a timely manner [5, 12, 15, 16, 26]. Based on these indicators, it is examined here to what extent scenarios can be influenced and actions triggered to timely stop the occurrence and development of (mechanical) failure mechanisms.

Root causes

Deterioration of mechanical integrity – often referred to as aging – is a major cause of incidents in the process industry in many cases. This is also true at the Chemelot site, a major chemical industry site in the south of the Netherlands where OCI Nitrogen's ammonia plants are located: during the period 2011 – 2015 approximately 50% of the 'loss of containment' incidents were due to material deterioration and corrosion [7]. A majority of these incidents were not foreseen as the failure mechanisms were not known to the plant organization. The technical causes have been investigated based on 89 mechanical integrity incidents. Five categories have been identified:

1) Corrosion under insulation;
2) Contact with aggressive chemicals;
3) Continuous presence of vibrations;
4) Extreme process conditions including frequent starting / stopping and heating / cooling;
5) Mechanical stress in the material.

Based on this, there were four root causes for mechanical failures identified:

1) External conditions, such as the weather, the environment and (plant) emissions;
2) Internal process conditions due to (aggressive) chemicals;
3) Maintenance activities, for example, assembly under stress or wrong material selection;
4) Process conditions like vibrations, pressure peaks, extreme temperatures, rapid temperature changes.

Method

An assessment has been conducted on one of OCI Nitrogen's ammonia plants. The assessment focused on scenarios related to mechanical integrity. A multi-disciplinary team has been formed that was able to assess the construction and material choice, understand the mechanical failure mechanisms, and explain the deviations in the operation of the various process units (such as starting or stopping). The expertise of the team consisted of knowledge of the process, construction and structure of the process units, materials and corrosion, and the performance of inspections. During the preparation phase, it was investigated which incidents occurred at OCI Nitrogen and other ammonia manufacturers in order to get an overview of the probability of the different failure mechanisms. In addition, there have been extensive discussions with the control room operators about operational management, especially the starting and stopping of process units. An important issue was to understand why and to what extent they deviate from the normal operating window. Increased temperatures and temperature gradients have proven to be important input parameters for the assessment.

Figure 2 shows the checklist that was used to assess the process units, which have been selected in a preliminary sub-study [18]. The checklist aims to include mechanical failure mechanisms for both normal and deviating operational modes such as starting and stopping but also, for example, catalyst reductions.

In step 4 of figure 2, attention should be paid to barriers that have already proven their effectiveness in preventing the failure mechanism from developing into a central event. This can give the image of an inherently safe design or a very unlikely scenario. Therefore, the existing barriers should be considered as non-existent. This means that it is necessary to check the influence of the process conditions and the process medium on the materials as if there was no protection at all.

The probability is divided into four groups:

• Very probable. The failure mechanism has already occurred in the concerned process unit (mostly without catastrophic failure);
• Probable. The failure mechanism has not occurred yet, but it seems probable based on the current process conditions or in case of a minor deviation from current operation. External casuistry can also indicate the probability of the failure mechanism;
• Improbable. The failure mechanism will probably not occur but cannot be excluded. The

failure mechanism will only occur in case of major (process) deviations.

• Very improbable. The failure mechanism will not occur and is excluded from further consideration.

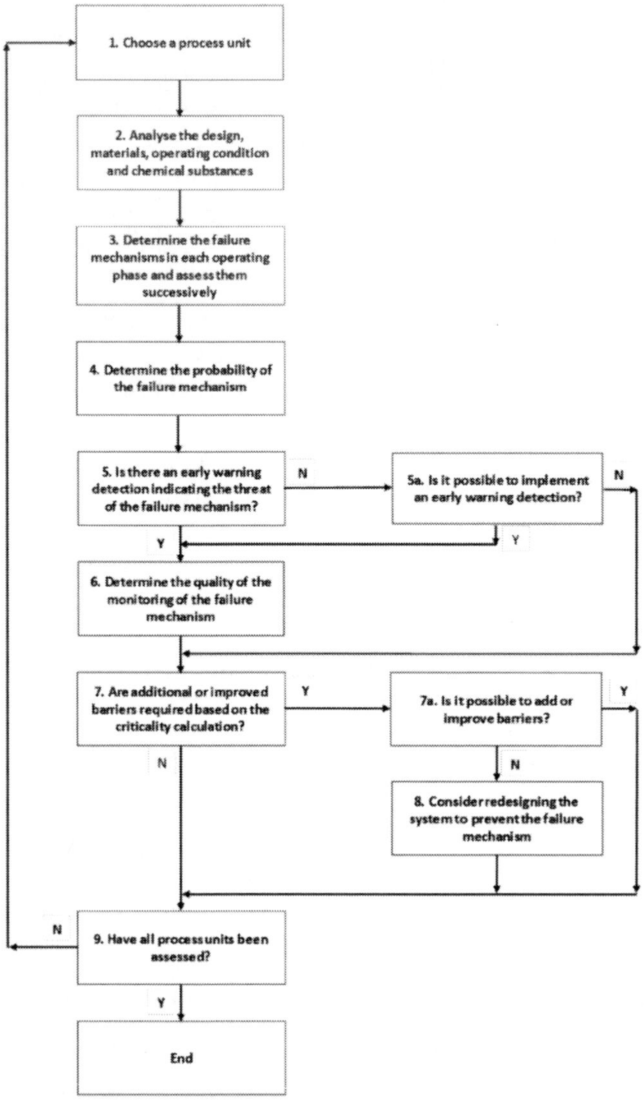

Figure 2. Process unit assessment checklist

Early warning detection (steps 5 and 5a) provides an indication of when the failure mechanism will occur. The combination of an early warning detection and a high-quality monitoring of the failure mechanism (step 6) can be considered a full barrier if there is a proper procedure in place which includes follow-up analysis to investigate the severity. It is evident that this is preferred over a (latent) failure mechanism without an early warning detection.

In step 7, a criticality calculation is used to assess whether additional barriers are required, or existing barriers need improvement. The criticality C is determined by the probability P of the failure mechanism, the quality D of the detection and monitoring of the failure mechanism, and the reliability R of the existing barriers using the formula: C = P x D x R. Table 1 shows the numerical values for P, D, and R. If the criticality is 16 or lower, there are no additional or improved barriers needed and the current situation can be considered as safe. The threshold of 16 is somewhat conservative, i.e. most companies opt for ALARP (as low as reasonably practicable) in their risk assessment.

	1	2	3	4
P	Very improbable	Improbable	Probable	Very probable
D	Very good	Good	Reasonable	Unreasonable or none
R	Very good	Good	Reasonable	Unreasonable or none

Table 1. Numerical value of P, D and R

If the calculated criticality is higher than 16, table 2 can be used to determine how many barriers are needed and what quality and reliability they should have. An early warning can be considered a barrier if it is connected to high quality monitoring of the failure mechanism. The detection of early warnings often consists of measurements on the basis of which further investigation follows, all of which is managed by a procedure or work process.

Probability	1 barrier	2 barriers	3 barriers
4: Very probable	1 very good barrier	2 good barriers	1 good and 2 reasonable barriers
3: Probable	1 very good barrier	2 good barriers	1 good and 2 reasonable barriers
2: Improbable	1 good barrier	2 reasonable barriers	
1: Very improbable	No barriers required		

Table 2. Necessary barriers as a function of the probability

Table 3 is a non-exhaustive list with examples of the quality or reliability of barriers and detections applied at OCI Nitrogen. The table provides direction for barriers and detections that are already in place or which can be implemented. It is the team's responsibility to determine how good the quality or reliability of the barriers and detections are. This should be assessed on a case-by-case basis.

If the failure mechanism does not have sufficient barriers and the implementation of additional barriers is not possible, a redesign should be considered in accordance with step 8 of figure 2. A redesign could be accomplished by a different material choice, a change in process conditions or altered start-up or shutdown procedures in order to prevent catastrophic failure of the process unit caused by the assessed mechanical failure mechanism.

Quality of the barrier or detection	Value of R or D	Examples
Very good	1	• Covered with sheet steel (cladding, a uniform binding of a protective metal cover) • Distribution pipe to prevent erosion caused by intruding gas • SIL2 instrumental protection
Good	2	• Covered with sheet steel (lining, a local binding of a protective metal cover) • Covered with heat-resistant stone and/or plaster (refractory) • SIL 1 instrumental protection • Safety critical work instruction or procedure, procedural safeguards, alarm with management attention
Reasonable	3	• Coating, preservation • Non-SIL instrumental protection • Normal work instruction or procedure, alarm with operator attention
Unreasonable or none	4	

Table 3. Quality indication of barriers and detections

Finally, the mechanical failure mechanisms are listed in the left part of the bow-tie together with the existing and implemented barriers, which should prevent the occurrence of a loss of containment in the central event.

Results

Two scenarios of a process unit have been worked out and schematically shown as the (preventive) left part of the bow-tie. It has been assessed whether the scenarios have sufficient barriers, if there is an early warning or if this can be implemented and where improvements can be made in case there are insufficient barriers.

Scenario 1. Start-up heater, low cycle fatigue (LCF)

During start-up, the start-up heater is used to heat synthesis gas, a mixture of hydrogen and nitrogen in a ratio of 3:1, from approx. 270 °C (518 F) to approx. 400 °C (752 F). The process flow for the start-up heater and synthesis reactor is shown schematically in figure 3. The synthesis gas is provided by the synthesis gas compressor and has a pressure of approx. 200 bar (2900 PSI). When the supply to the start-up heater (via MEV3001) is opened, the temperature in the supply line to the start-up heater rises quickly due to the hot synthesis gas at 270 °C (518 F). When there is sufficient flow, the start-up heater is ignited to further heat the synthesis gas until the synthesis reaction is activated. When the synthesis reactor is generating enough heat from its reaction to heat up the supplied gas, the burner in the start-up heater will be switched off and the start-up heater taken out of line by closing valve MEV3001.

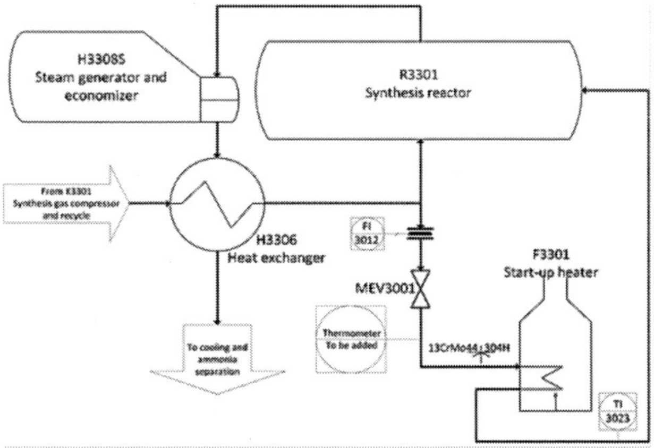

Figure 3. Start-up heater and synthesis reactor

The supply line of the start-up heater is made of Cr-Mo steel (13CrMo44), whereas the coils are made of austenitic steel (304H). This means that there is a dissimilar welding joint in the transition of the supply line to the inlet header. This welding joint is not designed for high levels of stress caused by large temperature gradients, which can lead to low cycle fatigue (LCF), i.e. fatigue involving cyclic, plastic deformation and eventually cracking. A trend for temperature measurement in the outlet of the start-up heater (TI3023) shows that during the start-up, the temperature rises approx. 200 °C (392 F) per hour after opening the supply valve during more than half an hour. The temperature of the dissimilar welding joint experiences quite a similar temperature gradient in this operating phase. This creates unacceptable material stress in the mentioned welding joint.

The inlet and outlet header, made of the same material as the coils (304H), contain little cracks which are, among other things, associated with LCF and are due to the design. There is external casuistry, but this is not related to the welding joint described above.

In the current situation:

- The work instruction only mentions a temperature rise for the synthesis reactor of 50 °C (122 F) per hour. It does not mention anything about the start-up heater, the supply and discharge pipes or headers;
- There is no temperature point in the supply of the start-up heater located at the welding joint;
- Supply valve MEV3001 cannot be operated from the control room. This makes it difficult to control the heating up of the supply and discharge pipes and headers.
- No LCF has been detected at the dissimilar welding joint of the supply line so far. Due to the present high temperature gradients the failure mechanism is classified as probable.

Given the fact that there is no early warning in the current situation, it is checked, in accordance with step 5a of the checklist, whether it can be implemented. This seems possible by installing a temperature point in the supply line of the start-up heater, which generates an alarm when the temperature rises over 50 °C (122 F) per hour. On the basis of this alarm, a 'fitness for service' (FFS) analysis should be initiated (according to the procedure). If deemed necessary, an inspection may determine whether a repair is necessary. If the procedure receives management attention in accordance with table 3, it may be classified as good (step 6). In order to receive this classification, the procedure should be described and included in the safety management system and have a certain degree of management involvement during its use. The alarm will not only be visible in the control room but could also be sent to those responsible for integrity and asset management. In addition, this alarm could be discussed in the daily operation meeting.

In step 7, the criticality must be determined to check whether additional or improved barriers are needed. It is assumed that, in the current situation, there is no barrier for the failure mechanism LCF (so R=4). With a probable failure mechanism (P=3) and good detection/monitoring (D=2), a temperature gradient alarm, the criticality is 24 (C = PxDxR = 3x2x4). The scenario is currently insufficiently safeguarded and requires additional barriers (step 7a). According to table 2, this (probable) scenario should be provided with 1 good or 2 reasonable barriers in addition to the proposed detection/monitoring. This can be achieved by controlling supply valve MEV3001. Although automatic control is preferred, MEV3001 can also be manually operated from the field in accordance with the instructions of the operator in the control room (as indicated in fig. 4). To achieve a *good* barrier, the current operating instruction should be classified as safety critical. This provides the scenario with 2 good barriers (a temperature gradient alarm and a safety critical procedure for opening MEV3001), which means that the criticality of the scenario is sufficiently low (C = PxDxR = 3x2x2 = 12).

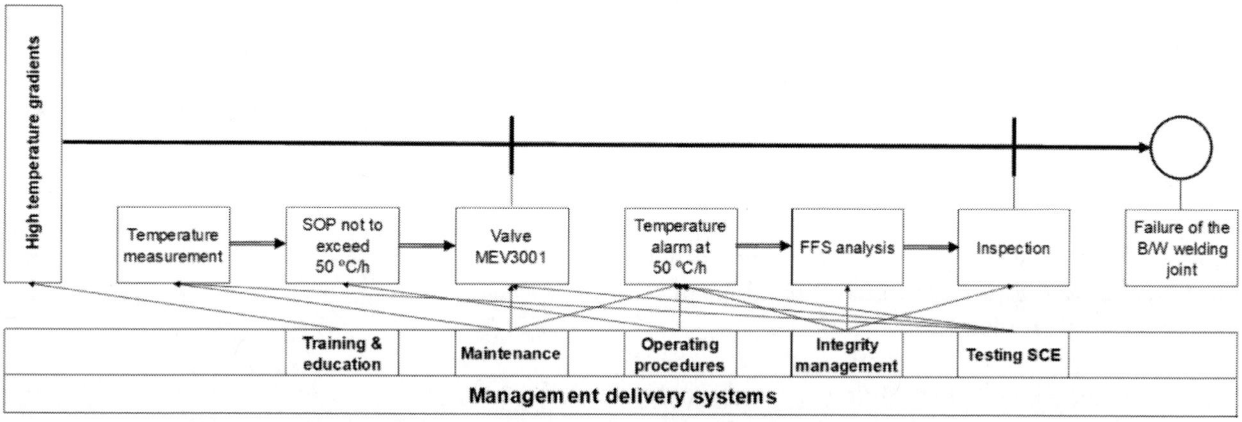

Figure 4. Bow-tie of low cycle fatigue in the start-up heater

Figure 4 shows a bow-tie for the scenario with the new barriers, which work out as a 1-out-of-2 system. Both barriers are (mostly) non-technical and based on an instruction or procedure. The barriers are constructed as a 3-out-of-3 system, in other words: all three elements must work in order to ensure the availability of the barrier.

Scenario 2. Steam superheater A, creep and Nelson hydrogen attack

The steam superheater overheats high pressure steam of 125 bar (1813 PSI) using process gas which consists of approx. 35% hydrogen. The temperature of the process gas drops from 600 °C (1112 F) to 475 °C (887 F). Figure 5 shows the flow of the process gas that goes upwards via the internal heat exchanger and returns along the wall after which it exits on the right. The process gas that leaves the internal heat exchanger on the top, passes the internal refractory that protects the outer wall against high temperatures. When the refractory is not intact, the outer wall is exposed to high temperatures whereby the metal is burdened by creep (slow, plastic deformation caused by stress and temperature) and Nelson hydrogen attack (damage mechanism in which diffused H-atoms react with carbides to methane and whereby the larger, trapped methane may cause cracks when exceeding the yield limit). A major damage of the internal refractory exposes a large part of the wall to hot process gas, which can cause the wall to weaken and eventually lead to failure of the steam superheater. Since a major damage is always preceded by small, hard-to-see defects, the scenario is focused on the latter.

Unlike the waste heat boiler, which is located upstream the steam superheater, very few external incidents have been reported with regard to this process unit. Singh reports pipe leakages due to under deposit corrosion by phosphate [19]. Given the construction, it seems impossible that a steam leakage might attack the refractory. Own casuistry shows that small defects in the refractory have been found, but they have not led to local overheating of the wall. However, larger defects to the wall, which can lead to hot spots, cannot be excluded. In view of experiences with other internally refracted process units, this scenario is considered to be probable.

During inspections, the internal refractory generally looks good and is qualified as a good barrier. There are currently no indications that the refractory is not intact. During operation this can only be seen by hot spots on the outer wall. These hot spots can be visually observed by the operator during an inspection round, but are easily overlooked. This means that there is no reliable early warning in the current situation. In accordance with the checklist in figure 2, step 5a, it has been checked whether another early warning can be implemented. This seems possible by providing the outer wall of the steam superheater with a layer of indicator paint that discolours due to the higher surface temperature where the internal refractory is damaged. This indicator paint decreases the possibility of overlooking hot spots, especially during the day time. It is also recommended to provide the outer wall with a few temperature points that trigger an alarm when the temperature gets too high. If the indicator paint is discoloured and/or one or more temperature alarms are activated, the vessel can be inspected with an IR camera to determine the size of the hot spot and whether a fitness for service analysis is needed. This will confirm whether an inspection is necessary. The need for replacement or repair will ensue from the inspection. If this procedure gets management attention, it can be qualified as a good barrier (step 6).

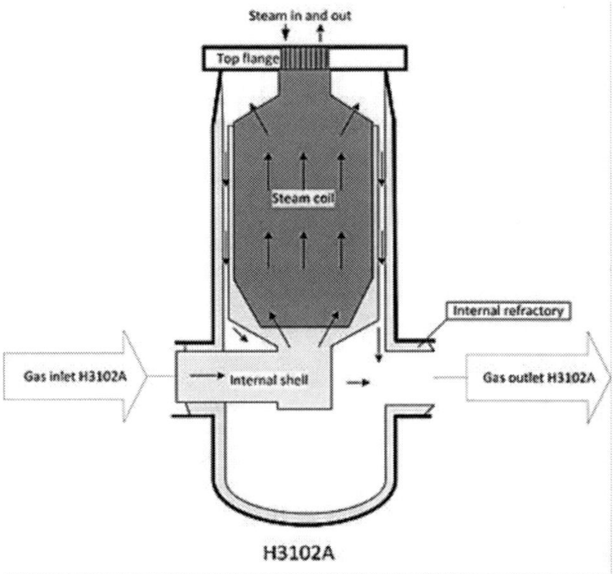

Figure 5. Steam superheater A

In step 7, the criticality must be determined to decide whether additional barriers are required. It is assumed that the failure mechanism is probable, the proposed detection and monitoring is of good quality and that the refractory is a good barrier. The criticality is 12 ($C = PxDxR = 3x2x2$) which means that there are no additional or improved barriers needed if detection and monitoring are implemented as described above.

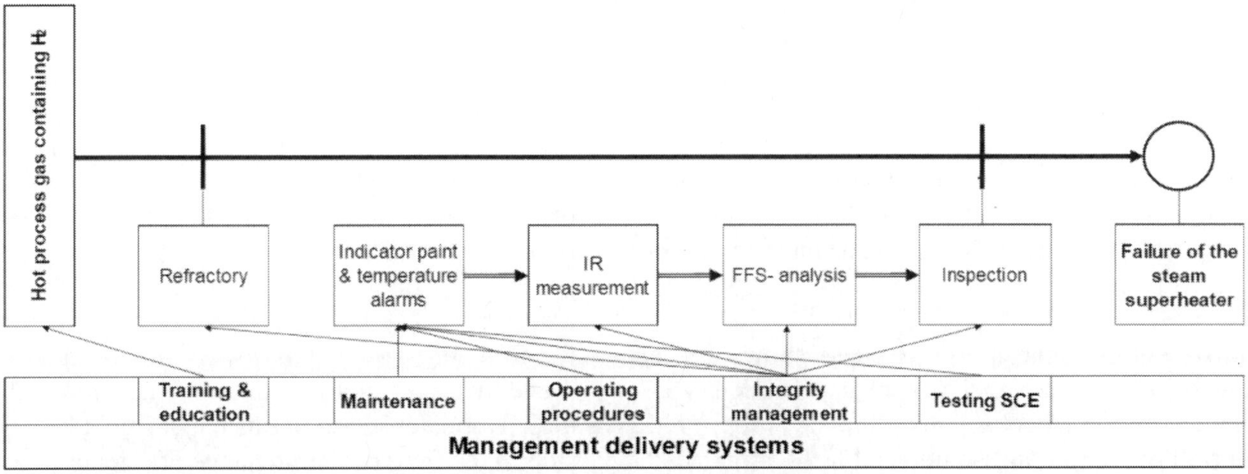

Figure 6. Bow-tie of creep and Nelson hydrogen attack for steam superheater A

Figure 6 shows a bow-tie of the scenario with the new detection and monitoring. The scenario in the new situation consists of two barriers that can be considered a 1-out-of-2 system. The existing internal refractory is a technical barrier, while the second is a non-technical barrier, consisting of detection through indicator paint, temperature alarms and IR measurement, after which a procedure with management attention initiates an FFS analysis and inspection.

Discussion

This sub-study on mechanical failure mechanisms has three 'high-level' results:

1. Important, missing information about design, material choice and inspection methods, but also potential incidents have been revealed;
2. Additional scenarios have been found by, in contrast with previous assessments, looking at all operational modes;
3. Part of the failure mechanisms have now been judged as probable by not considering the barriers.

This sub-study does not take into account poor design, incorrect assembly or repair and incomplete or bad inspections. The time dependency of mechanical failure mechanisms is not included either. However, the operation of the plant outside the 'operating window' was considered intensively, especially in start-up and shut-down situations. Many mechanical failure mechanisms are susceptible to these deviating operations, which are often not taken into account in the design. The expected (mechanical) lifespan can be considerably shortened when the process is operated outside the operating window, which is referred to as the integrity operating window [13].

Process indicators in the form of early warnings are not always present but can be implemented in most cases. When a scenario develops latently, it has to be stopped by one or more (available) barriers. To review the quality of the existing barriers as a function of the reliability and availability, process indicators can be introduced which are then monitored. This means that process indicators can be used to indicate whether a mechanical failure mechanism has been initiated, but also to assess the quality of the barriers. In this way process indicators provide an indication of the momentary probability of the occurrence of the central event.

Inspections represent an integral part of the condition monitoring of process equipment [25]. The bow-ties show that they can be carried out initiated by early warnings. These process indicators report that the mechanical failure mechanism concerned may be taking place.

Whether this is actually the case, must be proven by a more detailed consideration, a fitness for service analysis, in which the need for an inspection is determined. If an inspection is considered urgent and cannot be performed during operation, the installation has to be shut down. The speed at which and the extent to which the failure mechanism is taking place, depends on a number of factors which can be hard to oversee. A further elaborated consideration is not included in this sub-study.

Two examples have been used to show how a mechanical failure mechanism can be detected. Some failure mechanisms can be monitored during operation and some need to be monitored during regular or interim inspections. The quality of the inspection has a significant impact on the reliability of the results. The results of the inspections determine to what extent the failure mechanism has already developed. Not only (the execution of) inspections can be used as process indicators [6], but also the results of the inspections.

Conclusions

To date, (near) incidents occur due to previously unknown mechanical failure mechanisms or at previously unknown locations. A lot of knowledge has been built up because the failure mechanisms occurred. In this sub-study, scenarios based on mechanical failure mechanisms have been visualized using bow-ties with the existing barriers and early warnings. The monitoring of barriers and early warnings can provide us with information in the form of process indicators. Process indicators indicate the momentary probability of a scenario, thus allowing for quick action. The relationship between process indicators and barriers or early warnings and their effect on the development of scenarios will be investigated further in a follow-up study. This will also include the effect of management delivery systems.

The method used in this sub-study is somewhat reminiscent of the model for risk-based inspections (RBI): inspections are carried out when it appears necessary based on a risk assessment. However, this is only partly true. The difference is that in this method inspections are not necessary until there is a demonstrable probability that the failure mechanism and thus the scenario is taking place. On the contrary, RBI is a systematic method in which an inspection program is established beforehand based on a risk assessment [1]. In the light of the historical development of breakdown maintenance to preventive and predictive maintenance and RBI, the method described in this paper might be a new step in the field of maintenance efficiency.

References

1. API. (2016). *Risk-Based Inspection*. API RP 580, third edition.
2. Bellamy, L.J., Ale, B.J.M., Geyer, T.A.W., Goossens, L.H.J., Hale, A.R., Oh, J., Mud, M., Bloemhof, A., Papazoglou, I.A., & Whiston, J.Y. (2007). Storybuilder – A tool for the analysis of accident reports. *Reliability Engineering and System Safety, 92*, 735-744.
3. Candreva, F., & Houari, M. (2013). Plant Screening for Ageing Impact in the Process Industry. *Chemical Engineering Transactions, 31*, 253-258.
4. DNV. (1996). *Risk management of ageing process plants*. DNV Research Report 96-2001.
5. Dokas, M., Feehan, J., & Syed, I. (2013). EWaSAP: An early warning sign identification approach based on a systemic hazard analysis. *Safety Science, 58*, 11-26.
6. Hassan, J. & Khan, F. (2012). Risk-based asset integrity indicators. *Journal of Loss Prevention in the Process Industries, 25*, 544-554.
7. Hoedemakers, G. (2016). *Mechanische integriteitsrisico's in chemische procesinstallaties: Hoe houden we de tijger in zijn kooi?* (Thesis MoSHE course TU Delft 2016).

8. HSE. (2006). *Plant ageing*. RR509. Retrieved from http://www.hse.gov.uk/research/rrpdf/rr509.pdf
9. HSE. (2007). *Key Programme 3, Asset Integrity Programme*. Retrieved from http://www.hse.gov.uk/offshore/kp3.pdf
10. HSE. (2010). *Managing Ageing Plant, A Summary Guide*. Retrieved from http://www.hse.gov.uk/research/rrpdf/rr823-summary-guide.pdf
11. IAEA. (2009). *Ageing Management for Nuclear Power Plants*. Safety Guide No. NS-G-2.12. Retrieved from https://www-pub.iaea.org/MTCD/Publications/PDF/P1738_web.pdf
12. Knegtering, B., & Pasman, H. (2013). The safety barometer, how safe is my plant today? Is instantaneously measuring safety level utopia or realizable? *Journal of Loss Prevention in the Process Industries, 26*, 821-829.
13. Lagad, V., & Zaman, V. (2015). Utilizing Integrity Operating Windows (IOWs) for enhanced plant reliability & safety. *Journal of Loss Prevention in the Process Industries, 35*, 352-356.
14. OGP. (2008). *Asset integrity – the key to managing major incident risks*. Report No. 415. Retrieved from https://www.scribd.com/document/391778284/OGP-Report-415-Asset-integrity-the-key-to-managing-major-incident-risks-December-2008-pdf
15. Øien, K., Utne, I., & Herrera, I. (2011a). Building safety indicators, I theoretical foundations. *Safety Science, 49*, 148-161.
16. Øien, K., Utne, I., Tinmannsvik, R., & Massaiu, S. (2011b). Building safety indicators II applications. *Safety Science, 49*, 162-171.
17. Reason, J. (1990). *Human error*. Cambridge, UK: University Press.
18. Schmitz, P., Swuste, P., Theunissen, J., Reniers, G., Decramer, G., & Uijterlinde, P. (2018). Een aanpak voor het bepalen van een realistische ranking van de gevaarlijkste procesonderdelen van het ammoniakproductieproces. *Tijdschrift voor toegepaste Arbowetenschap, 2018 2*, 42-56.
19. Singh, J., Varma, S.L., & Patel, B.M. (2003). Failures in Secondary Waste Heat Boilers. *AIChE Technical Manual 2003, Safety in ammonia plants and related facilities symposium, volume 44*, 109-116.
20. Sklet, S. (2006). Safety barriers: Definition, classification, and performance. *Journal of Loss Prevention in the Process Industries, 19*, 494-506.
21. Swuste, P., Theunissen, J., Schmitz, P., Reniers, G., & Blokland, P. (2016). Process safety indicators, a review of literature. *Journal of Loss Prevention in the Process Industries, 40*, 162-173.
22. Swuste, P., van Gulijk, C., Zwaard, W., Lemkowitz, S., Oostendorp, Y., & Groeneweg, J. (2019). *Van veiligheid naar veiligheidskunde*. Alphen aan den Rijn, Nederland: B + B Vakmedianet.
23. SZW. (2016). *Ageing bij Brzo-bedrijven*. Retrieved from https://zoek.officielebekendmakingen.nl/blg-813397.pdf
24. TNO. (2015). *Literatuuronderzoek naar veroudering van installaties*. Retrieved from https://publications.tno.nl/publication/34626341/jB13ci/TNO-2015-R10878.pdf.
25. Utne, I.B., Brurok, T., & Rødseth, H. (2012). A structured approach to improved condition monitoring. *Journal of Loss Prevention in the Process Industries, 25*, 478-488.
26. Vinnem, J.E. (2010). Risk indicators for major hazards on offshore installations. *Safety Science, 48*, 770-787.

Uptime Improvement Journey via the Systematic Root Cause Analysis

Since 2016, Dyno Nobel has been operating a 2300 MTPD ammonia plant based on KBR Purifier Technology at its Waggaman site in the state of Louisiana (USA). After 72 hours of successful performance test completion, the plant was handed over to Dyno Nobel's operation in October 2016. The plant met the design capacity and specific energy consumption during the test run. This paper describes the first-year operating experience where it had more than 20 trip events for multiple reasons. Plant trip, repairs and start-ups result in overall significant higher safety risks, both people safety and process safety. The paper elaborates the systematic approach of the organization to improve the uptime of the plant and shares the key lessons learned. Since the commissioning, several challenges have contributed to the production unit's downtime. Some of the key issues were steam system management during transient conditions, process control issues, critical third-party package integration, electrical infrastructure and not, but the least, the human factor. It outlines the charter of each of the in-house champions-led uptime task forces supported by subject matter experts including process licensors and original equipment manufacturers (OEMs). The purpose, methodology, summary of findings and actions implemented are explained and it is detailed how the significant improvements in plant uptime were made.

Ashutosh Shukla
Dyno Nobel Inc.

Marc Hermus,
Incitec Pivot Ltd

Venkat Pattabathula
Incitec Pivot Ltd

Introduction

AT Waggaman, Louisiana (WALA) Dyno Nobel operates an ammonia plant of 2300 MTPD name plate capacity based on Kellogg, Brown & Root(KBR) Purifier Technology. The plant started production in October 2016 and experienced several challenges in its first year of operation which caused plant downtime due to complex steam management, process control issues, electrical infrastructure and human factors. These multiple downtime events during the first year of operation exceeded a total of 20 and showed that it was by no means a reliable operation. Additionally, each plant trip, isolation, decontamination, recommissioning and start-up results in significantly higher process safety and people safety risks. Thus, the journey began to improve uptime and safety through a process of systematic data analyses, Root Cause analyses (RCA's) and engineering studies, followed by improvement plans.

Analysis of downtime

About 20 events in the first year of operation resulted in 100 days of loss in production. (see Fig 1 and 2)

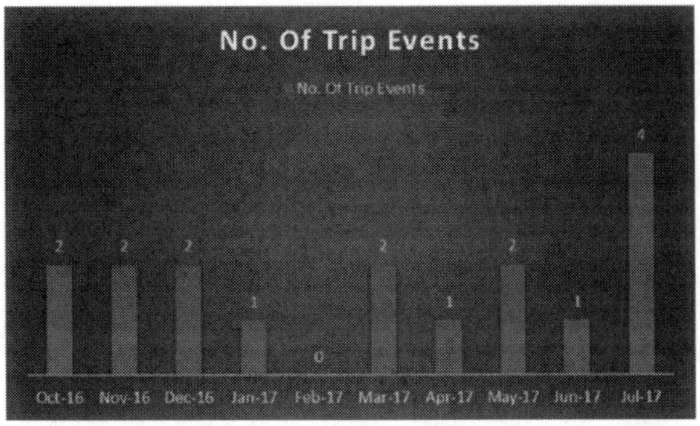

Figure 1. Number of Trip events post commissioning

These events have resulted in plant downtime ranging from at least 1 day to a maximum of 34 days.

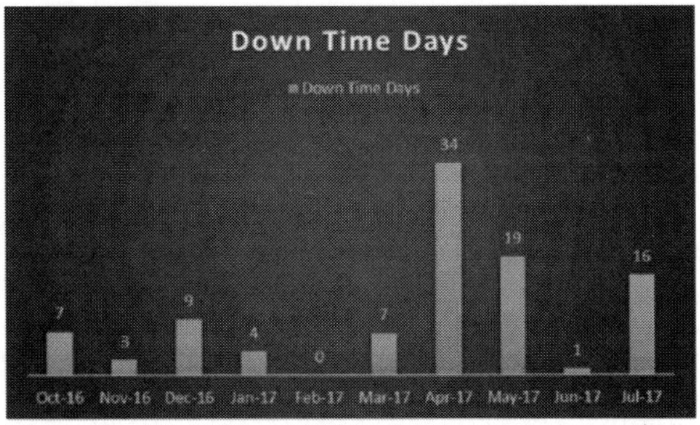

Figure 2. Downtime days post commissioning

Many of the trips did initiate in the back-end but cascaded into front-end trips (see Fig. 3), resulting in much more downtime and more unproductive use of gas.

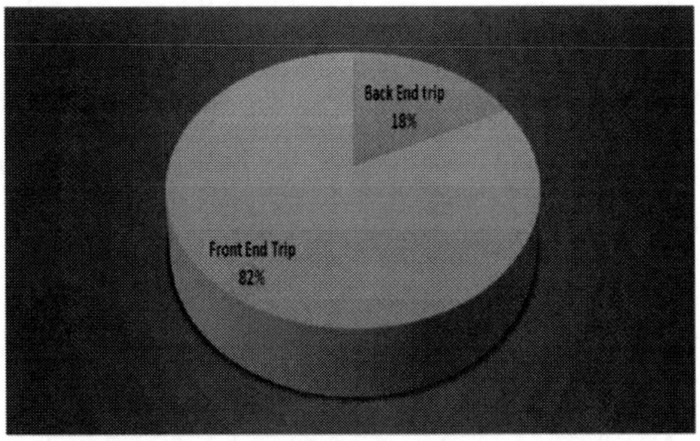

Figure 3. Front End and Back end trips

Figure 4 indicates analysis of generic causes.

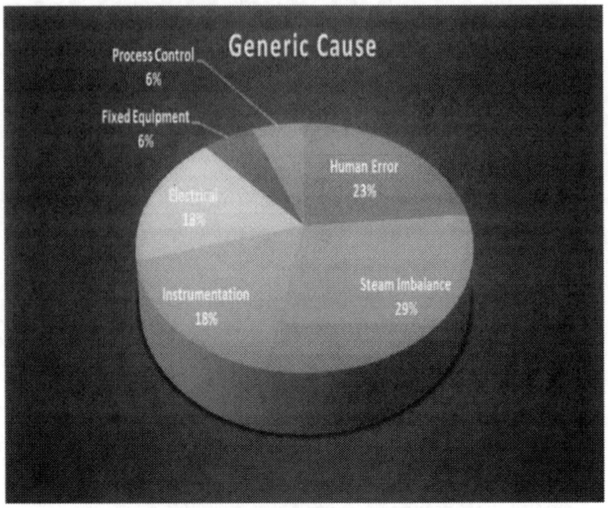

Figure 4. Generic Causes

Steam system imbalances followed by human error were the main contributing factors to the loss of production. Electrical, instrumentation and process control related issues were the remaining main causes of down time.

Improvement Strategy

The uptime was approximately 70% during the first year. This performance was well below company target of 90 %. The uptime measures the amount of time plant operates at maximum demonstrated rates. It is calculated after deducting the scheduled losses, unscheduled losses, process related rate reduction, hence it provides measure of time available to meet market demand.

$$\%\boldsymbol{UPTIME} = \frac{\boldsymbol{Maximum\ Sustained\ rate - Losses}}{\boldsymbol{Maximum\ Sustained\ Rate}}$$

The site reliability team recognized the following:

- The plant was still in the "Infant Mortality" phase and more unexpected plant trips were expected. They were called the "Unknown Unknowns".

- A certain number of defects had already surfaced and needed definition and implementation of solutions. They were called the "Known Knowns".
- Areas of concern could be identified by extrapolations of the Known Knowns and by bench marking with other similar plants. They were called the "Known Unknowns".

This simple premise was used by the reliability team to build a strategy for improving the uptime.

1. The Known Knowns need identification and implementation of solutions. Additional resources are required to guarantee progress while plant personal is focusing on the other two categories, and short-term mitigations have to be identified to allow working on long term sustainable solutions.
2. The Unknown Unknowns requires high quality, timely and structured Root Cause Analyses each time a trip occurs. This is required to quickly drive to root causes in order to prevent repetitive trip events.
3. The Known Unknows require detail engineering reviews using structured methodologies, driven by the Reliability Engineering team but supported by contract resources. Hidden defects (future plant trips) need to be brought to the surface. Specific uptime Task Forces had to be defined in order to achieve a targeted approach.

A benchmarking with plants achieving consistent high uptime (> 95 percent over multiple years) also revealed that the plant should have "inherent reliable design", "high level of automation" and "high level of redundancy." This insight further influenced the selection and definition of the Specific uptime Task Forces.

The two most significant "Known Knowns" issues were related to steam and human error. Steam issues were related to:

- Dynamics of steam balance during transient conditions
- Lack of availability of Start-Up steam due to the overall unreliable steam network of third-party supplier aggravating the amount of loss due to plant trips.

Two key initiatives were undertaken to address these high priority Known Knowns.

Steam Dynamic Study

Many plant trips were initiated by back-end trips and frequently cascading into front-end trips. A low steam - carbon ratio was the main reason for tripping the primary reformer. Multiple RCA's done by the WALA site team indicated issues with valve tuning, trip settings, control logic, overall response of the steam system due to its integration with entire manufacturing complex. The team was concerned to implement identified changes due to the complex nature of controls especially during transient conditions and the inability to fully test improvement proposals off-line first.

The reliability team reached out to KBR with their findings and requested KBR to assist with the task of performing a dynamic steam study. The study's main purpose was to identify any changes required in the control system, logic / tuning parameters, valves, etc. to prevent front-end trips resulting from the plant's back-end upset / conditions. The goal was also to see if operator intervention could further be minimized after the back-end upsets to achieve the safe and steady front-end operation.

KBR conducted a dynamic simulation study for different scenarios like Methanator trip, Synthesis Gas compressor trip, refrigeration compressor trip etc. and subsequent cascading frontend trips.

Main recommendations were:

i. Changes to HP to MP steam letdown control logic.

ii.Reducing back pressure on loss of methanator and syngas compressor.
iii.Automatic reduction of plant rate at faster ramp rate on loss of methanator and syngas compressor.

All the recommended changes were successfully and safely implemented by using the management of change (MOC) process of Dyno Nobel and Risk Assessments prior to each control system change.

The lack of Start-Up steam is mainly caused by an overall low reliability of steam generation of third-party supplier due to aging infrastructure and extremely tight integration of multiple operating units within the complex. During commissioning Dyno operation was supported with four temporary boilers so never surfaced as constraint with regard to availability of steam. Dyno Nobel imports steam from third party in order to bring the NH3 plant online and then exports to medium pressure steam network once fully online. There have been lost uptime waiting for availability of steam for start-up.

An investment decision has been made to install new start-up boilers. One additional advantage is that these boilers will be available to provide steam for ammonia plant at all transient scenarios, thus improving uptime.

Recognizing that the improvements to process controls have resulted in a significant reduction in trip events is very satisfying.

Operator Training Simulator (OTS)

The DCS operators were given training using the simulator supplied by KBR together with classroom training during the pre-commissioning phase. This arrangement formed part of the client's contract with the EPC company to impart training using OTS. But post commissioning this set up was discontinued as it wasn't part of agreement to own the OTS infrastructure by the client.

It was recognized through RCA findings that there was a need to arm our operators with intimate plant operational knowledge in various transient conditions. Furthermore, it was acknowledged that response to upset conditions is a skill which is best to be acquired either during real plant upset conditions or by simulation. Exposure to plant upset conditions by all four crews will become very limited once higher Uptime numbers are being achieved.

Dyno's management decided to invest in training simulator. A dedicated resource was engaged to ensure that that the new OTS reflects the actual plant operations in every aspect in order to meet the following objectives:

- Maintaining and enhancing operator skills through refresher training.
- The need for increased flexibility of operator teams by offering cross- training opportunities across different areas of the plant.
- To have standardized form of training.

The OTS is being used for intensive training purposes as well as testing the Standard Operating Procedures (SOPs) for plant startup. At this stage, it is difficult to quantify any tangible benefits but we have seen much improved response during transient conditions amongst the board operators due to exposure on OTS. The significant achievement is that Dyno Nobel has succeeded in bringing a few field operators to regular board operators by proper training and certification process utilizing OTS as training tool.

Setting up of Uptime Task forces

After addressing steam dynamic issues and developing a plan to improve operator skills, Dyno Nobel management approved and funded the establishment of uptime Task Forces (UTF) to systematically address the reliability issues in the category of the "Known Unknowns".

A total of five task forces were defined, based on analysis of historical data, internal learnings from Dyno Nobel other manufacturing plants and interviews with multiple stakeholders.

1) UTF-1: Critical Machine Control.
2) UTF-2: Single Point of Failure.
3) UTF-3: Review of Cause and Effect.
4) UTF-4: Alarm Management.
5) UTF-5: Electrical Infrastructure.

Each of the UTFs was led by a Dyno Nobel Reliability Engineer and complimented by other team members. The teams worked under strict time constraint because solutions requiring an outage needed to be funded, detailed and be ready for implementation well before the planned turnaround scheduled for 2019.

UTF-1: Critical Machine Control

Dyno Nobel constituted the Task Force 1 team to review critical machines control. Multiple plant trips have taken place since the plant was transferred from EPC to Dyno Nobel in 2016. The underlying root cause identified was related to the integration of various critical machine control systems and the application of various standards.

The philosophy of designing, carrying out controls and verification had been outsourced to a large extent to third parties. Knowledge by site engineers was at the time limited.

Purpose

- Review all critical machine control systems, including design philosophy, codes and standards, narratives, actual configured logic, and alarms.
- Identify reliability risks and quantify the risk for each finding.
- Provide recommended solutions for risks rank resulting in production loss or risks with a low pain/gain ratio.

Systems Analyzed

Only critical machines without redundancy have been analyzed in order to narrow down the scope. The cold box expander was not added to this initial list since the plant could run in conventional mode and the fact that spares were available.
Machines List

- 101-J Process Air Compressor
- 102-J Feed Gas Compressor
- 103-J Synthesis Gas Compressor
- 105-J Ammonia Refrigerant Compressor
- 5202-JAT Cooling Water Pump Turbine
- ID/FD fans

Methodology

The Task Force 1 Team reviewed each machine's control logic as configured. This included but was not limited to Integrated Turbine and Compressor Controls (ITCC), governor System etc. Also, devices related to these systems have been reviewed for instrumentation and actuation. Much of the work was done remotely, followed by site visits of third parties to validate the findings.

Potential points of failure as well as gaps between the actual configuration and the specification documents for the controls were identified. For each finding, a risk ranking was assigned. Management of Change (MOCs) has been written for those with a higher risk to correct the deficiencies. MOCs were also written for lower-risk items that were considered easy to implement at low cost.

Summary of Results

The task forces identified 45 MOC recommendations.
Significant improvements made in reliability against spurious machine trips were as follows:

- Radial vibration trips voting changed to from 1oo2 to 2oo2.
- Removal of bearing temperatures trips

- Common SIS shut down signal -Introduce 0.5 s (addition of de-bounce timer) for 1oo1 signal.
- ITCC HMI emergency shut down push button- Introduce 0.5 s for 1oo1 signal
- Change Anti-Surge Valve solenoid to dual solenoid with by-pass functionality.
- Increase time delay on high seal gas vent flow trip (to avoid spurious trips due to flare header back pressure fluctuation).
- Assigning fall back transmitter into Anti surge flow controller.
- Removed "Probes NOT OK" as trip initiator in Bently Nevada (BN) system for both radial and axial ones 101-J process air compressor.
- Enable Anti Surge trip to 102-J, 103-J and 105-J.

UTF-2: Single Point of failure

The task force 2 team was formed to identify single point of failure with in control systems.

Purpose

- Identify all single point of failures in control system (initiator, logic solver, final elements including power supply).
- Assess reliability risk and quantify the risk for each finding.
- Provide recommended solutions for risks rank resulting in production loss or risks with a low pain/gain ratio.

Systems Analyzed

The control systems were evaluated by task force 2 team from input sources to final elements. A total of 3255 devices, including transmitters, IO modules, cabinet power supplies and final elements were analysed. The list of systems analysed are as follows:

Plant Systems

- Distributed Control System (DCS)
- Burner Management System (BMS)
- Safety Instrumented System (SIS)
- Fire and Gas System (FGS)

Water Treatment Systems

- Raw water Clarifier Unit
- Mixed Bed Polisher (MBP)
- Reverse Osmosis (RO #1 / #2 / #3)

Network

- Networking Equipment
- Human Machine Interface (HMI)
- MODBUS Motor Control

Methodology

Failure Mode and Effects Analysis (FMEA) was used to determine the potential impact of device failures. A total of 10 workshops were organized. The team analysed the systems organized by cabinet. Each component within each cabinet was evaluated for potential single point of failure, as well as each field device connected to each cabinet. Each single point of failure identified was then risk ranked using the risk matrix of Dyno Nobel to determine where there was a risk of reliability within the plant.

The failure points with the highest risk rankings were then developed into projects to be implemented to eliminate or mitigate the identified risks. These were reviewed by DNLA management; approved projects were then implemented through the MOC process or distributed as appropriate to the plant project group.

Summary of Results

A large number of single point of failures were identified, resulting in 23 recommendations. Many recommendations covered multiple pieces equipment. To implement all approved recommendations, a total of 38 MOCs were required.

Some of the findings and proposed improvements are:

- **Single DCS transmitter for control with 2oo3 Median back up** – There are several process controls using single transmitter and 2oo3 SIS trips. If the single transmitter fails, the process is likely to be unstable and 2oo3 trip would trigger. The site has developed functional specification for this reason in which DCS will use the median value of 2oo3 trip transmitters when primary DCS control transmitter fails. For modification, a total of 18 primary DCS control transmitters were identified.

- **Install redundant Analog Output (AO) cards** – There are several control valves with potential single point of failure. Hence, it is suggested that redundant AO cards be installed.
- **Redundant Transmitters on the Same AI Card** – In several cases where transmitters are measuring the same process but are on the same analog input card. This makes the card a single point of failure. Moving one of the two redundant transmitters to another AI card is proposed. Additionally, the redundant transmitters can be configured to automatically switch control to another if failure is detected from one of the inputs. 14 such transmitters are identified for changes.
- **ID and FD Fan Solenoid System** – The ID and FD fans are installed with a single solenoid to the trip valve. These solenoids are single points of failure. It was recommended that the fans be installed with redundant trip solenoids.

UTF-3: Cause and Effect Rationalization

The task force 3 team was set up to identify opportunities to create a larger operating window with in the trip system without jeopardizing the safety functionality. This is to reduce the likelihood for spurious trips.

Purpose

- Review and rationalize all C&E registers (trip set points, time delays, voting, auto start functionality). Gain full understanding of the design reasoning and explore options to eliminate trips, increase/decrease trip set points, introduce time delays, improve voting, auto starts etc. Changes should allow the plant to ride through upset conditions without too conservatively tripping the plant.
- Create detail MOCs (containing new narratives, updated documentation and a risk assessment) to allow for implementation online.

Systems Analyzed

The Task Force 3 Team evaluated the WALA C&E logic and documentation for the following systems:

Ammonia Plant

- 101-J Process Air Compressor
- 102-J Feed Gas Compressor
- 103-J Syn Gas Compressor
- 105-J Ammonia Refrigerant Compressor
- 131-J Purifier Expander
- Balance of Ammonia Process plant

Utilities

- Ammonia Storage
- Water Treatment Plant
- Cooling Water
- Ammonia Shipping (Truck and Barge Loading, Injection)

Methodology

Multiple workshops were organized to review the Cause and Effect (C & E) matrix. These workshops were attended by process engineers and operations staff. Discussions were held with engineers at other Dyno Nobel plants.

Summary of Results

A total of 28 MOCs were created.

The main recommendations were:

1) Creating additional logic for automated action.

2) Increasing the time delay for transient upset.
3) Changing the set point for increased response time.
4) Addition of some automated action on reaching pre-trip limit.
5) Creating stage trip set point (on first trip set point longer time delay, on subsequent trip very minimal time delay)

UTF-4: Alarm Management and rationalization

Task Force 4 team was created to improve the performance of alarm management. The RCAs of most plant trips identified poor alarm management as a contributing factor. This was acknowledged by the fact that no sound alarm management philosophy was applied during the design and commissioning of the plant. To improve operability, software and maintenance processes have been implemented to manage alarms.

Purpose

- Review the existing Site Alarm Philosophy against industry standard ISA 18.2 and make update.
- Identify a tool for alarm rationalization and ongoing alarm management.
- Rationalization of all alarms by applying the updated alarm philosophy including all alarms from third party packages. Meet the performance standard as defined in ISA 18.2.
- Establish a process for ongoing maintenance of alarms (standing alarms, shelved alarms, system alarms, alarm flooding, alarm masking etc.)
- Review trip group's philosophy to enable alarm shelving to prevent alarm flooding.
- Train operational and engineering staff in the tools for alarm management.

Systems Analyzed

Task Force 4 Team reviewed all configurable alarms in the WALA control systems. More than 9400 parameters could be reviewed during a total of 8 workshops.
Plant Systems

- Distributed Control System (DCS)
- Burner Management System (BMS)
- Safety Instrumented System (SIS)
- Fire and Gas System (FGS)

Methodology

In a series of workshops, the Task Force 4 Team met to review alarms, each workshop being dedicated to a process area. MOCs were written between workshop sessions to change the configuration of the alarms.

Initial workshops focused on reviewing the philosophy of the alarm and developing a communication plan to inform operations about any changes, along with training on the software Plant State Suite (PSS). PSS is used for importing and managing DCS alarm data and generating system based KPIs to monitor progress.

Attendance of control room operators during the workshops turned out to be essential for success of the Task Force Team.

Summary of Results

New servers and software for collecting and managing alarm settings were installed. This software is also used to generate KPI reports to monitor alarm statistics for changes. Operator training sessions were held to ensure that this system is owned by Operations.

To reduce nuisance alarms around several pieces of equipment, dynamic alarm suppression has been implemented. More than 7000 parameters have been set to Logging or Reference priority that are used by engineers but not required by Operations. As a result, the number of standing alarms was significantly reduced.

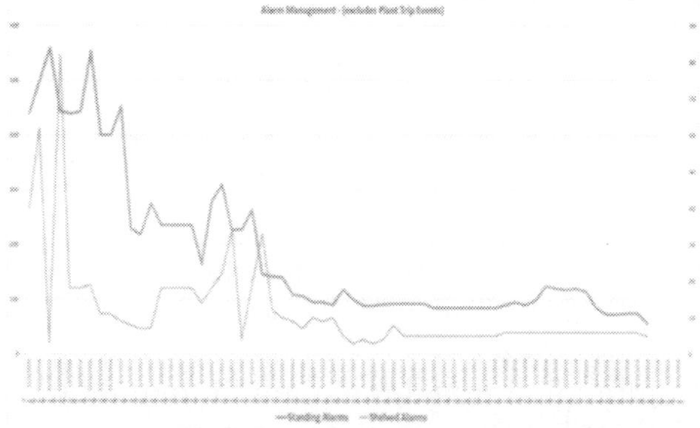

Figure 5. Reduction in standing and Shelved Alarms

Alarm priorities have been reviewed with an ISA 18.2 defined target distribution; this target distribution has been met.

Coordination of alarms was made consistent on SIS and DCS transmitters in the same service. Implementation of alarms for 2oo3 transmitters was standardized, including deviation alarms.

The site alarm philosophy document has been updated to include the results of the workshops of the task force.

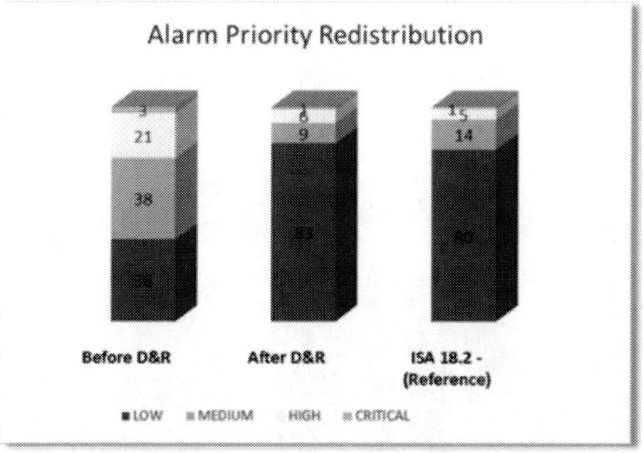

Figure 6. Alarm Priority Distribution

The following table shows the improvement in alarm metrics as a result of the initial workshops:

Alarm Performance Metrics (Week Averages)						
Alarm Metric						
Annunciated Alarms / Operating Position	Target Value	Before Rationalization	After Week 1&2 D&R	After Week 3 D&R	Trend	
% Hours > 30 alarms	~ < 1%	0%	0.00%	0.61%	→	✓
Hourly Alarm Rate	6	8.73	3.89	3.61	↓	✓
Standing Alarms (>24 Hours Old)	<5	124	66	39	↓	✓
Alarm Flood Periods	< 1% of operation	2.39%	0.51%	0.58%	→	✓
Alarm instances		1466	618	609	↓	✓
Frequent Alarms Top 10 % of Total Alarm instances	< 5%	37.20%	34.5%	39.4%	→	!
Alarm Shelving Records – Per Shift	< 20 Alarms	201	127	100	↓	✓

Figure 7. Alarm Performance Matrices

UTF-5: Electrical Infrastructure

Task Force 5 was established to investigate the electrical distribution system.

Purpose

- Survey existing electrical equipment and design to identify reliability risks.
- Perform a lightning protection study.
- Assess reliability risk and quantify the risk for each finding.

System Analyzed

Power Supply Distribution System

- Create a master equipment list and identify critical equipment
- Review one-line diagrams and panel schedules for redundancy requirements
- Field validation of critical design assumptions

Transient Conditions related System

- Review under voltage / overvoltage trip devices (relays) for motors and switchgear. Consider appropriate action for each.
- Review options to survive short under voltage events, including power management system

Grounding, Bonding, Shielding System

- Design validation of design drawings.
- Field validation of grounding and bonding installation
- Design isolated ground system testing equipment
- Inspect and test existing wiring cabinets to confirm ground system integrity
- Inspect field instrumentation and control panels to verify shielding and isolation

Lightning Protection System

- Perform plant Lightning Protection Study and
- Consider additional lightning protection methods
- Review/design additional grounding connections to bring plant into compliance with NFPA 780

Backup Power System

- Review existing plant UPS system and make recommendations for reliability improvements.
- Evaluate options for regular load tests of emergency generator (temporary load bank hook up and testing)

Summary of Results

The task force identified several areas for improvement that do not require project activity, including establishing communications between existing site (Third party) and Dyno Nobel for power-related issues, and disabling overvoltage trip on overload relays. The task force determined that the current grounding system meets NEC code but is not industry standard.
Because of Task Force 5 findings, several projects have been approved for implementation, listed below:

- Implementation of improved lightning protection (includes grounding).
- Installation of transfer switch for emergency generator testing by using a load bank.
- Installation of UPS and battery monitoring system.
- Installation of UPS power to the truck loading and barge loading Programmable logic controller (PLC) panels.
- Addition of second power feed to mixed bed polisher PLC panel.

Conclusion

By implementing majority of low hanging items WALA site has been able to reduce the duration/impact of the "Infant Mortality" phase. (See Fig.8)

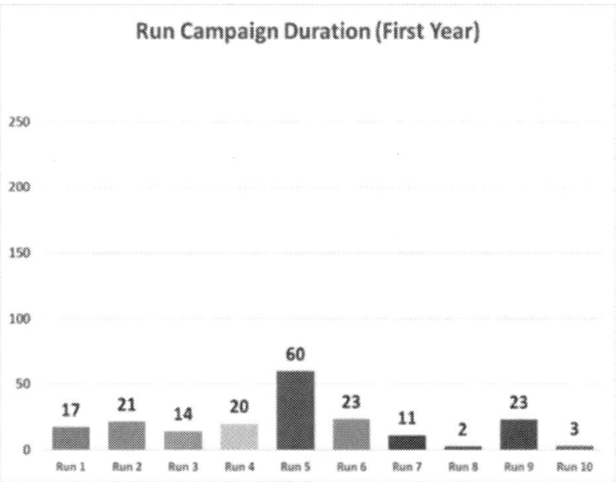

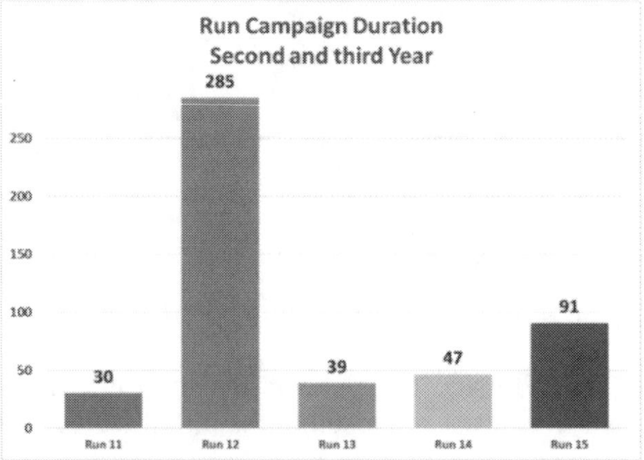

Figure 8. Run Campaigns in days First & Second year

By systematically creating a focussed approach, the site has been able to address the "known-knowns" in addition to bringing the "known-unknowns" to the surface prior to resulting in plant trips.

It is still expected that "unknown-unknowns" will randomly surface, typically following a normal distribution of bath tub like design flaws, construction defects, operability issues which should not greatly impact the uptime.

The comparative uptime performance is shown below in Fig.9

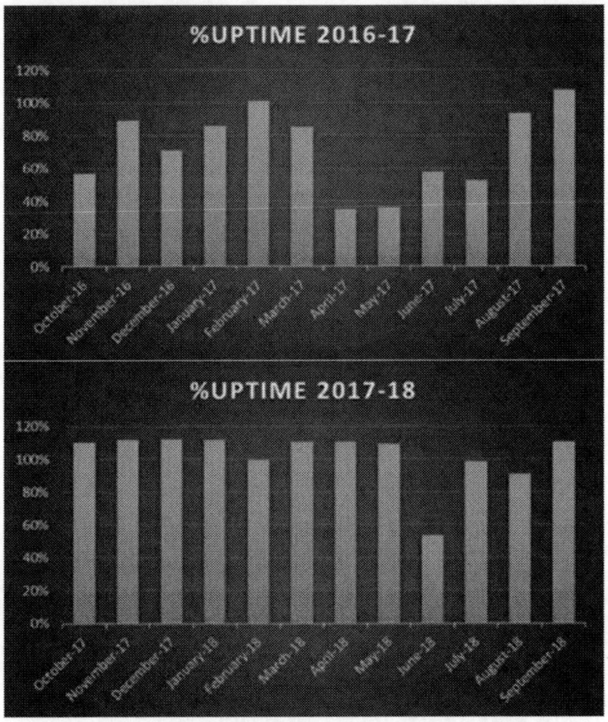

Figure 9. Uptime comparison First & Second year

The authors would like to acknowledge Dyno leadership team support and encouragement throughout the process. We would also acknowledge KBR technical team for conducting comprehensive analysis on steam dynamic study.

Causes for Process Gas Cooler Tubesheet Thermal Disintegration and Mitigating Actions

At Borealis site Linz a new process gas cooler (PGC) was installed in 2013. Over time several incidents with hot spots occurred. After removal of the tubesheet refractory in 2018 the metal surface was found to have severe damages. A failure investigation was started and several causes promoting an overheating of the tubesheet were found. By means of a fitness for service analysis the remaining life of the vessel was determined. With strong support from the manufacturer Bertsch, internal and external laboratories, notified bodies and the plant operation and maintenance team a way was found to bring the plant back into safe operation.

Daniel Sponseiler
Borealis Agrolinz Melamine GmbH, Linz, Austria

Daniel Sparr
Bertsch Energy GmbH & Co KG, Bludenz, Austria

Introduction

Borealis operates two (2) ammonia plants of different designs and name plate capacities in Linz / Austria. One plant has a nameplate capacity of approx. 850MTPD (935 STPD) and was commissioned in 1973. The plant conforms to Uhde's "Single train" design.

Operational Parameters

In a first step, the gas received (H_2 ~37vol% | CO ~7vol%) from the secondary reformer (K201) is cooled in a horizontal process gas cooler (PGC; Tag Nr. W201) and afterwards further cooled to the final receiving temperature of the high temperature shift reactor (HTSR) by a feedwater preheater (W203). The heat energy of the gas is used to produce saturated steam with approx. 120 bar (1740 psi)

Both the PGC and the feedwater preheater are of fire-tube design. An overview of the operational parameters is shown in Table 1. The original PGC, delivered by Borsig (Germany), was replaced in 1990 due to several tube failures. The replacement was of similar design, delivered by Steinmueller, Germany. It was in operation until 2013 when it was replaced by a new boiler delivered by Bertsch (Austria) with the goal of having a lower gas outlet temperature and a higher net recirculation ratio.

Parameter	Unit	Value
Name plate plant capacity	MTPD	850 (operated up to 1080)
	STPD	935 (operated up to 1188)
Gas/ steam pressure	barg	34,00 / 119,00
	psig	493 / 1726
Gas/ steam flow	kg/hr.	148.500 / 128.000
Gas/ steam temperature	°C	975 / 322
	°F	1787 / 612
Heat Flow	MW	~52
	MMBTU	~117,4

Table 1. Operational parameters (@1000MTPD)

The new PGC receives the process gas at around 975°C (1787°F) from the secondary reformer and cools it to around 430°C (806°F). An internal bypass allows to control the outlet temperature as well as allowing a bigger operational margin in case of fouling. Via riser pipes, the produced steam is delivered to a steam drum which operates at around 120 bara (1740 psia). Feedwater returns through downcomer pipes allowing the system to be operated by natural circulation. The steam amount produced by the PGC is around 128ton/h. The drum also receives steam produced in the NH3 synthesis waste heat boiler.

The PGC contains 420 fire-tubes welded to the tubesheets together with a 16" bypass pipe in the center. At the inlet the tubesheet is refractory lined and protected by ferrules of 460mm (18,1") length made of Alloy 800H material (as former PGC). The bypass tube is equipped with an Alloy 800H inliner. The overall dimensions of the PGC are a total vessel length of about 12500mm (492") and a shell OD of around 2070mm (81,5"). The tubesheet and tube as well as the in- and outlet chamber material is 1Cr-0,5Mo steel (13CrMo45) whilst the material of the water side shell is 15NiCuMoNb5-6-4.

The inlet chamber has a two layer refractory lining with a flange connection between the conical inlet section and the boiler tubesheet. The tubesheet itself has one layer of refractory lining. The outside shell is painted with temperature sensitive paint.

Water Quality and Blow Down Ratio

The water is supplied from a water treatment unit and is further degassed and stabilized with ammonia (1,8mg/l) and hydrazine (0,9mg/l). Around 1% of the steam production in the drum is purged. Further the rules of VGB-R 450 L apply.

Challenges of a Footprint Exchange

During a lifetime extension project in 2012 it was decided to replace the existing PGC (W201) to meet the process requirements related to the increase of production and energy efficiency as well as to overcome reliability concerns. The scope below was defined by the project team:

- Keep the existing steam drum (B201) without any modifications
- Keep the existing boiler feedwater preheater (W203) without any modifications
- Keep the existing secondary reformer (K201) connection without any modification
- Increase the heating surface while ensuring that natural circulation is still intact
- Keep the gas outlet temperature below 570°C (1058°F) even in high fouling conditions.

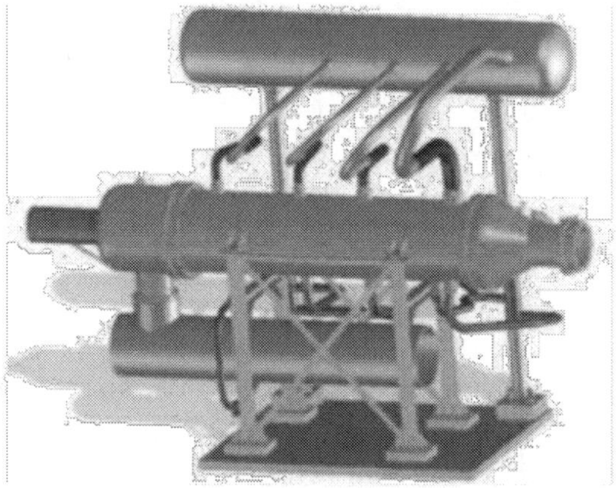

Figure 1. Overview of scope of the footprint exchange

As a consequence of the challenges of a footprint exchange only a limited number of companies were considered to be qualified. Finally, the company Bertsch (Austria) was awarded with the contract.

A footprint exchange of a PGC requires the development of a thermodynamic model to optimize the performance of the new boiler. Operation at different plant loads as well as startup and shutdown conditions need to be taken into account. In addition, the natural circulation must be ensured in all possible states. The optimization was performed by Bertsch.

The design optimization yielded a PGC with increased length resulting in longer tubes. The use of the existing drum was checked carefully since the natural circulation ratio decreased to a critical ratio of 7.7. Slightly bigger downcomer nozzles were added to account for the higher steam production, especially close to the inlet area to ensure that the natural circulation ratio is kept in the acceptable range. Hold up time in the steam drum (B201) was considered to be reasonable with a minimum hold up time from low water level to drum empty of 3.5 minutes and demister loading was checked. A sectional view of the installed PGC is shown in Figure 2.

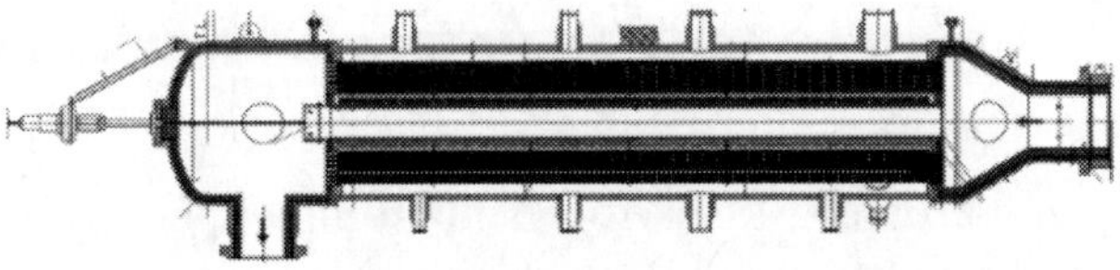

Figure 2. PGC section view (inlet on the right side)

The possibility to increase the heating surface of the PGC by increasing the shell diameter was not an option as it would decrease the gas velocity and increase fouling issues which have been observed during operational periods with low load which is a downside of horizontal PGC designs. Target inlet gas velocity of >25m/s (>82ft/s) (outlet ferrules) at 100% load was the design goal to minimize fouling on tubes by particles from the reformer refractory or catalyst.

The analysis of the natural circulation ratio by commercial software led to a redesign of the downcomer and riser system. Especially the first ("hot") section of the PGC had to be feed with feedwater sufficiently to prevent low circulation rates which may led to steam blanketing and overheating. Therefore, riser and downcomer nozzles were installed as close as possible to the tubesheet to remove the steam as quick as possible from this area. For some spools the inner diameter was increased to reduce the pressure drop. The nozzle positions on the PGC were adapted to the new length. The pipe support system and installation sequence had to be changed accordingly. Since the feedwater preheater is supported with constant pipe hangers along the PGC this system also had to be recalculated and modified.

Figure 3. New PGC in installed condition

Chronology of Failures and Hot Spots

The new PGC as shown in Figure 3 was installed and commissioned in September 2013. Six months after initial startup an increase in metal surface temperature was detected by thermography at the interface between the conical inlet section and the tubesheet flange. The temperature stayed below the acceptable limit until the next planned plant SD in 2014. The inlet was then inspected from the inside and a small visible gap in the refractory was patched. After start-up the temperature on the outside shell was again in the design range. In 2015 the shell temperature at the welding lip seal gasket on the tubesheet flange started to increase again and the same repair procedure with patching the gap was performed in 2016, this time without success. The decision was taken to renew the refractory at a small section in the area of the gap. The repair included the removal of the refractory at around 150mm of the conical section, re-welding of anchors and reinstallation of the two layer refractory system. The gap itself was filled with ceramic fiber of high Al_2O_3 content. The lower area was casted while the upper area was hand packed. Dry out with electric heating pads was performed. After start-up a hot spot developed in the upper area of the cone which lead to an immediate plant shut down. The cone was again inspected from the inside and the fibers in the upper area turned out to be completely blown out. The refractory in that area was demolished and reinstalled by gunning. The plant was again taken into operation, this time without a dry out of the refractory. At around 85% plant load, two smaller hot spots developed at the interface between the gunned and the casted area, but the surface temperature remained within an acceptable range. As an additional safety measure the surface was cooled with an air blower and permanent surface temperature measurement was installed.

During inside inspection in 2017 a large deformation of the central bypass tube inliner was detected together with indications of metal dusting on the inliner surface. It was therefore decided to purchase a new conical inlet section with new refractory lining and to replace the existing ferrules and bypass inliner of the PGC with new, more metal dusting resistant material Alloy 602CA (formerly Alloy 800H). To perform those works, the refractory of the tubesheet had to be renewed. During the regular plant shut down in 2018 the conical piece was removed and the refractory of the tubesheet demolished for ferrule and bypass inliner replacement. During those works it was observed that a black layer was spalling off from the tubesheet surface. The layer was discovered to be disintegrated, coked metal from the tubesheet. In some areas the coked layer was up to 10mm in thickness in an area where the tubesheet thickness is 50mm (2") in total. A picture of the damaged tubesheet during demolishing can be found in Figure 4.

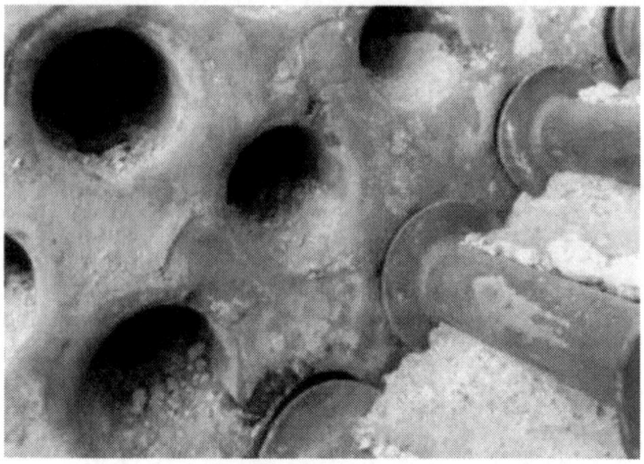

Figure 4. Spalled of layer from tubesheet during refractory demolishing

Immediately a task force including experts from the manufacturer Bertsch was formed to analyze the cause for the failure and assess the possibilities to put the boiler back into operation.

General Findings

The damage was more severe in areas close to the central bypass. The further the distance from the center, the better was the condition of the tubesheet. Close to the center the initially sharp edged holes in the tubesheet now looked like having fillets. The central bypass inliner showed heavy deformation along the length and at the section where it was casted with refractory. Further marks of metal dusting over the length of the inliner were observed. Figure 5 shows the inlet tubesheet holes with the visible abrasion at the edges.

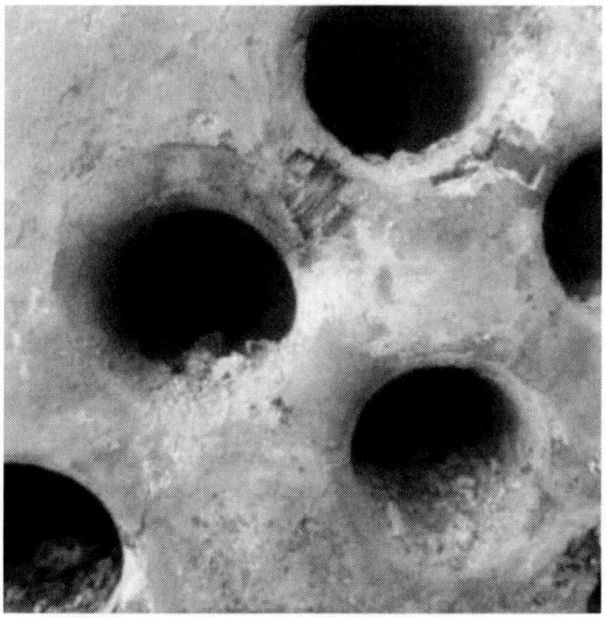

Figure 5. Tubesheet close to the center – The edges are rounded and spalled off material is visible

Process Abnormalities

No process abnormalities indicating tubesheet damage were detected. The temperatures at the inlet remained stable over the operational period.

Metallurgical Investigations

The team together with an external expert from VGB Power Tech (Germany) and the notified body decided to perform the following tests:

- Micro-hardness tests at different depths of the tubesheet
- Replicas to analyze the microstructure
- Crystallographic analysis of spalled off oxides by means of X-ray diffraction
- Liquid penetrant tests (PT) at locations where the oxide layer was grinded until no more coking was observed
- Tubesheet thickness measurements by means of ultrasonic measurements

Samples of the spalled material were sent to an external laboratory (VGB) to analyze the crystal structure as shown in Figure 6 and Figure 7. The result (as shown in Figure 8) showed that the material was subject to temperatures of around 580°C (1076°F) indicated by the presence of around 46% magnetite and around 50% wustite. It was not possible to determine the exact time it had taken to form that layer, but it was estimated to have been more than 12 months. This indicated that the metal must have been subject to the increased temperatures for a longer period.

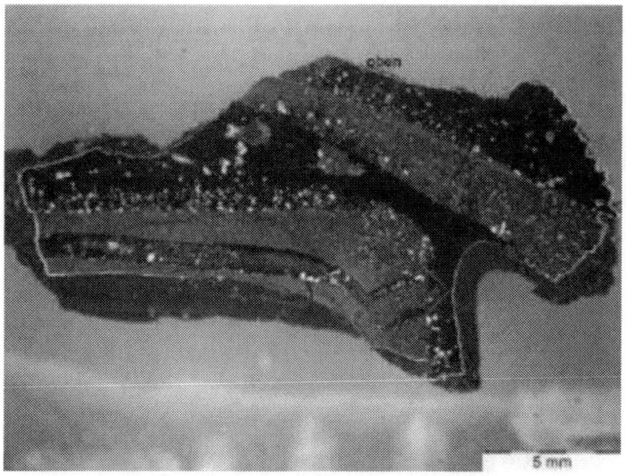

Figure 6. Sample of spalled off oxide

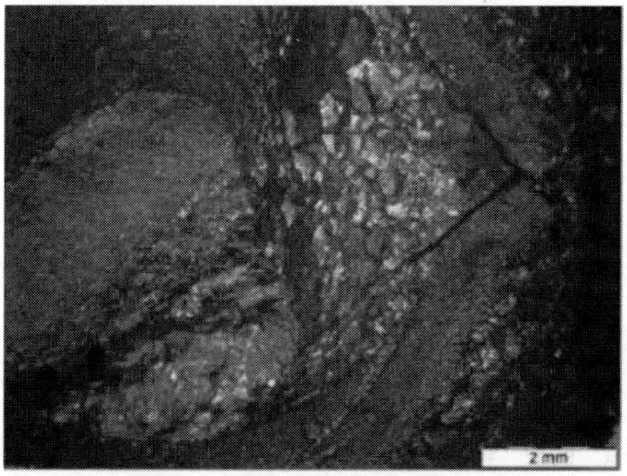

Figure 7. Sample of spalled off oxide

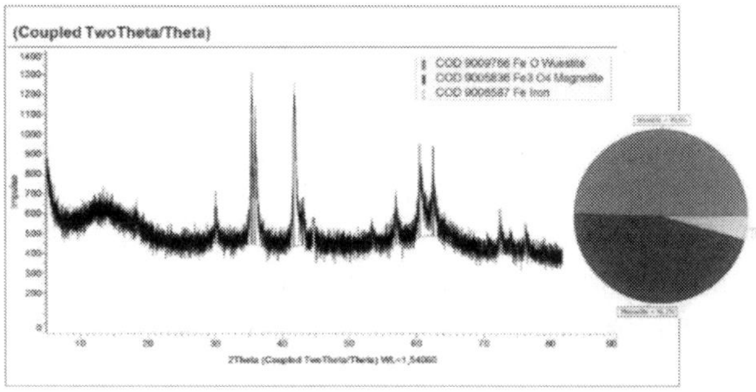

Figure 8. Result of XRF analysis

Replicas were taken at grinded sections on the tubesheet at three (3) different locations right below the removed coked layer. Micro-hardness was measured at the same spots. The results indicated a ferritic microstructure quite different from what one would expect for a 1Cr-0,5Mo steel. The metallographic structure was heavily decarbonized. This was in line with the hardness values measured. They were reduced from an originally 175HV10 to around 120HV10. It was therefore concluded that he yield strength of the material was reduced from its original value in the layer right beneath the coked material. The condition of the microstructure was suspected to improve over thickness since the shell side surface is permanently cooled with boiling water and will not have experienced those high temperatures.

Crack initiation or embrittlement was not observed which was proved by penetration testing (PT) and the hardness measurements together with the microstructure. The damaged microstructure as shown in Figure 9 was similar to a ferritic steel (e.g. X2CrNi12) which has lower creep resistance but better ductility than 1Cr 0,5Mo steel.

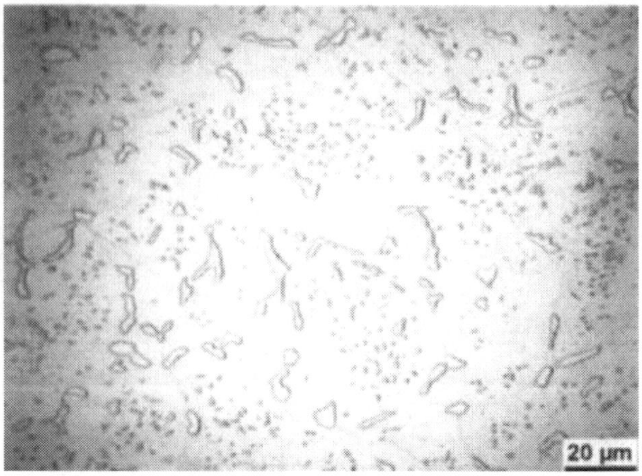

Figure 9. Microstructure of TS – 8mm below original surface showing ferritic structure (Hardness: 120HV10)

Wall thickness measurements on several locations (Figure 10) revealed slight deviations from the dimensions indicated on the manufacturer drawing. This was considered to result from manufacturing tolerances on the one hand and measurement uncertainties on the other. Nevertheless, the lower thickness values were used for the finite element analysis model.

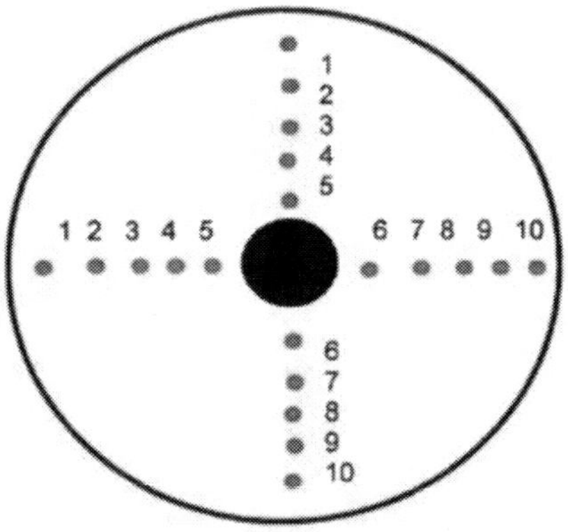

Figure 10. Position of wall thickness measurements

Root Cause Analysis

Summing up the results from the different analyses, the following was concluded:

- The tubesheet surface must have been subject to a temperature above 580°C (1076°F) for a long period (>12 months)

- The overheating was more severe in the central region of the tubesheet
- The gas flow must have been able to redirect itself from its original path around the refractory along the tubesheet through the tubesheet holes – highest flow must have been in the central region

First the used refractory materials were analyzed, and it became obvious that in the original design fibres with a high SiO_2 content had been used. These fibres are well known to be good insulators but dissolve in the reducing gas atmosphere at temperatures >850°C (>1562°F). This explains the slow growth of the hot spot at the lip seal gasket area. In fact, when the first repair in the gap area was performed, no fibre was found in the gap. Nevertheless, it didn't explain why the gas had found its way behind the tubesheet refractory.

Pictures from an inspection performed in 2017 revealed that there was a mechanical problem with the inliner of the central bypass tube. Huge deformations were found at the two neck sections (Figure 11 and Figure 12) which gave the indication that the inliner may have been restricted in axial movement. This generated a force in that direction pushing the refractory of the tubesheet against the direction of flow. In this area, the inliner was originally also wrapped with paper that was not resistant to the atmosphere. The third observation was an open gap between the ferrules and the tubesheet holes.

This behaviour could not be explained at first, but when reviewing the start-up procedure with the manufacturer, the investigation team found out that the procedure was insufficiently communicated by Borealis leading to additional stress and wrong assumptions during design.

Due to welding requirements, the hole must have an inner diameter bigger than the outer diameter of the tube, which leaves a relatively big gap between the ferrule and the hole.

The conclusion drawn from above findings closed the loop on why the tubesheet was overheated.

- The fibre materials used (high SiO_2 content) were not resistant to the reducing gas atmosphere and started to dissolve over time, opening gaps
- The inliner of the central bypass was restricted in axial movement and pushed the refractory away from the tubesheet
- The gap between the ferrules and the tubesheet holes was not closed with resistant material and allowed the gas to flow through (Figure 13)

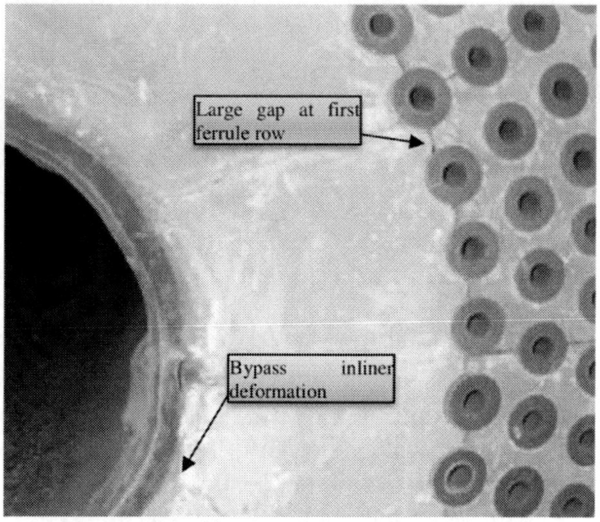

Figure 11. Observed damages before demolition of refractory lining

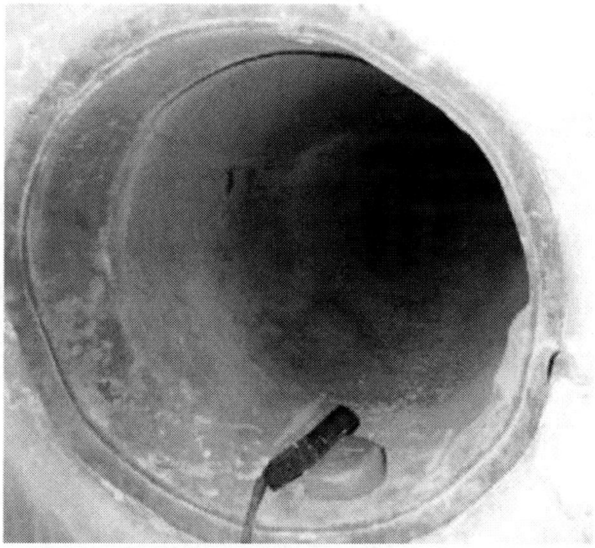

Figure 12. Inliner deformation before removal

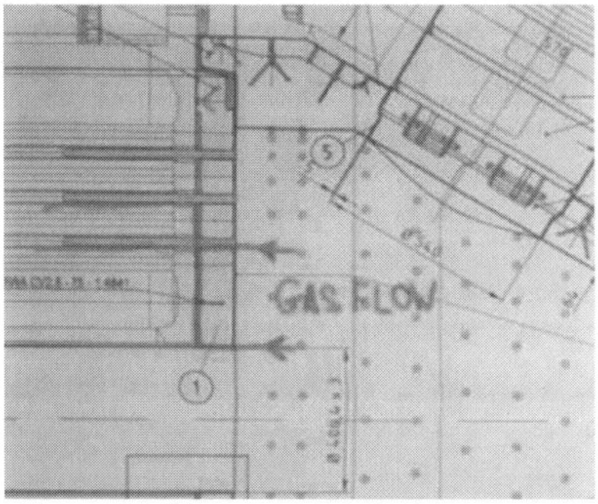

Figure 13. Unwanted gas flow due to damaged refractory

Fitness for Service Analysis

Since the root cause and the failure mechanism were clear it had to be evaluated whether the PGC could still be used until a replacement was available.

A repair was not an option since this would have led to cutting the tubesheet and re-welding all tubes as well as the shell. It was therefore decided to analyze whether a safe operation with the damaged tubesheet was still possible. The first step was executed by the vessel manufacturer, who recalculated the stresses in the existing tubesheet. The design dimensions were substituted in the finite element model by the actually measured values. Since the finite element analysis (FEA) model existed from the design stage this work was performed simultaneously with the metallurgical investigations. An overview of the FEA model with the different wall thicknesses (different colors) is shown in Figure 14.

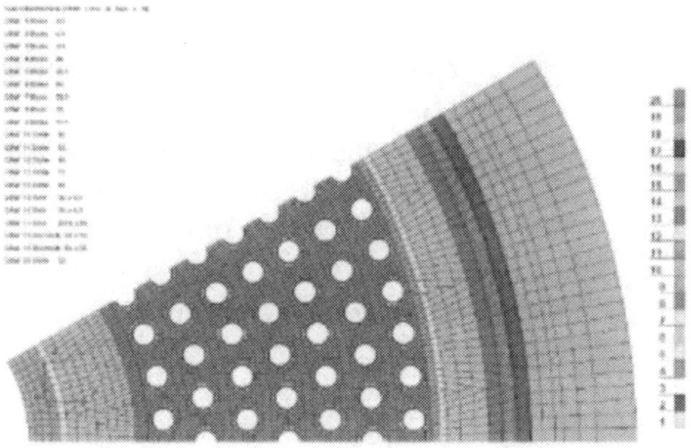

Figure 14. Overview of FEA dimensional model with actual wall thicknesses of tubesheet

The calculated stresses were then evaluated against the theoretically (as per EN 10222-2) allowable values. After discussion with VGB experts and the notified body it was decided to proceed with the conservative assumption that the hot tensile strength of the material is 20% lower than the value used for design. The tubesheet was assumed to have lower yield strength over the whole thickness. This is also a conservative assumption since in reality simulations showed that at the shell side the temperature was close to the boiling water temperature and the material strength will therefore improve over thickness.
Calculations were performed for different load cases, the most severe being the (theoretical) case of the shell and the tube side experiencing safety valve setting pressure (138barg/38,5barg | 2002psig/558psig). The allowable secondary stress for the tubesheet is 316MPa (45542psi) (already reduced by 20%) for this load case. The calculated max. value is slightly below that value which fulfills the design criterion. An example for the stress distribution along the tubesheet is shown in Figure 15.

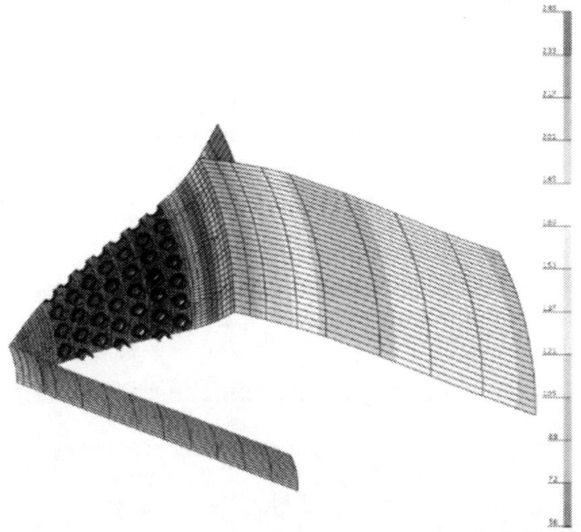

Figure 15. Stresses for most severe LC [MPa]

Figure 16. Tubesheet with refractory paper

Figure 17. Gap between ferrule and tubesheet hole

Embrittlement of the tubesheet as well as crack initiation by the coked layer was ruled out by metallurgical evaluation and tests.

To fulfill both legal and internal safety standards, it was therefore agreed with the notified body (authorised inspector) that the vessel could be operated until replacement equipment was available and measures had been implemented to ensure that no further overheating of the tubesheet occurs in that operational period.

Improvements on Refractory and Bypass Tube Inliner on Existing TS

The materials for the new tubesheet refractory lining and the gap to the conical section have the requirement to contain >96% Al_2O_3 and less than 1%SiO_2. In addition, to compensate the cavities created by the spalled off layer, refractory paper with high Al_2O_3 content was glued onto the surface (Figure 16). Since there was no other alternative, the gap between the ferrule and the tubesheet hole was filled with patch mass (Figure 17). High focus was put on the tolerances of the refractory between the cone and the tubesheet connection (±2mm | ±0,079inch). This was achieved by using the mold from the cone for the production of the tubesheet mold.

After casting the tubesheet in two sections and installation of the conical section the gap was additionally patched from the inside.

The inliner and ferrules were completely renewed. They were manufactured of Alloy 602CA, which is more metal dusting resistant than the former used material Alloy 800H which showed signs of metal dusting. To overcome the issue with the restriction in movement the outer diameter was selected lower than for the original design and the straightness of the tube was analyzed during installation. In addition, the inliner was wrapped with two layers of refractory paper as shown in Figure 18.

Figure 18. Inliner during installation wrapped with refractory paper and tape

Operational Permit

By implementation of the improvements on the refractory design and the result of the FFS analysis together with a positive shell side pressure test it was agreed that the boiler can stay in operation until end of 2019, which is the timeframe for delivery of a new PGC.

Improvements on the New PGC

Due to the good cooperation and excellent knowledge of the equipment manufacturer, Borealis was again awarding Bertsch with the contract for the PGC for installation during the shutdown in 2019. At this point it has to be stated that the engineering and performance of the tube sheet protection was subcontracted, and design and execution was reviewed by both companies. Main improvements on the new design were worked out by engineers from Bertsch and Borealis for the refractory, the ferrules and the bypass tube inliner. Further, the design with flanges on the inlet and outlet chamber was put into question. The advantage of having the possibility to dismantle the chamber comes with the disadvantage of a joint that may fail during operation and is difficult to manufacture. It was therefore decided to remove the flanges and directly weld the chamber. To ensure that future tubesheet repairs and ferrule exchanges are possible, a refractory filling nozzle was added at the top of the chamber close to the tubesheet.

Refractory materials were specified with following restrictions:

- Ceramic fibre and paper: SiO_2 content < 3%
- Refractory layer in contact with process gas: SiO_2<0,2% || Al_2O_3>94,5%

Detail design of bypass tube inliner was improved to reduce the possibility of restricted axial movement. Therefore, measurements during manufacturing have to be performed to ensure a perfect fit between the carbon steel tube inner diameter and the inliner outside diameter. In addition, all welding seams on the inside central bypass tube were grinded and nubs were added on the outside of the inliner to ensure a proper gap. The inliner neck is fixed on the front tubesheet.
The ferrules are expanded at the rear end with the purpose of reducing the gap between the tube and the ferrule in hot condition. In case the gas finds a way through the tubesheet hole, the resulting flow will be limited by the small gap. (Figure 19)

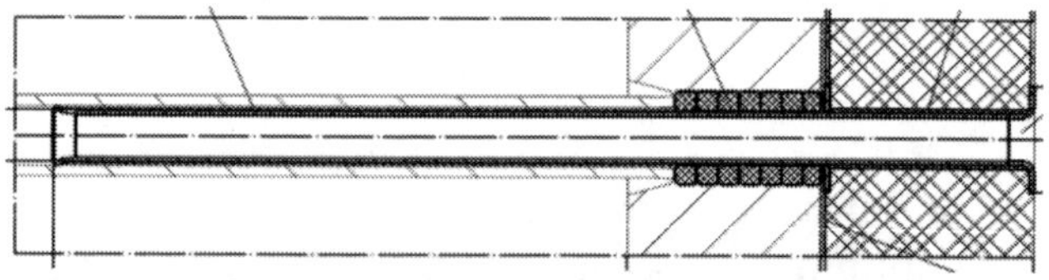

Figure 19. New ferrule design with expanded end. TS hole was closed with ceramic rings.

To monitor the temperature of the tubesheet, holes for the insertion of thermocouples were drilled by deep hole drilling technique (Figure 20). In case the temperature increases above the calculated value, this gives an indication that the refractory may have failed.

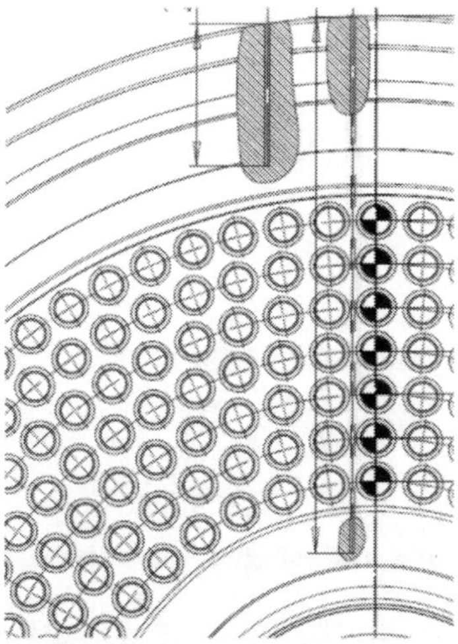

Figure 20. Deep hole drilling for T-measurement

Conclusion

The aim of this paper is to demonstrate the complexity of designing and operating a process gas cooler. The following learnings were taken consequentially by Borealis:

- Damages along the tubesheet (gas side) cannot be detected during regular inspections without having the refractory removed.
- If the gas bypasses the refractory it may cause overheating of the metal shell, leading to coking and material loss. Also, the risk of high temperature hydrogen attack arises.
- Fitness for service analysis can determine whether a PGC with a damaged tubesheet is further usable.
- During equipment design an exchange of all operating parameters including start up and shut down conditions is extremely important as design parameters are easily violated during those operational stages.
- During manufacturing and repair a high focus must be put on refractory design and execution as well as details regarding ferrule tolerances and the bypass inliner.

Due to the complexity of this topic a good and open cooperation with competent partners like Bertsch is crucial to perform a successful FFS analysis as well as an adequate redesign taking all details into account.

A Smart Combustion™ Approach to Improve Operation and Efficiency of Reformers

A Smart Combustion approach to reformer operation and optimization identifies the need to understand, in real time, more detailed information about the combustion process than is typically available. Increasing visibility of key parameters within the firebox, combined with real-time insights about the health and operation of each individual burner, can enable operations to make simultaneous improvements of emissions, safety, reliability, and profitability.

Chad Carroll, Rayford Forest, Eric Gebhard
John Zink Hamworthy Combustion

Introduction

Steam methane and hydrogen reformers have many critical parameters. Optimizing these systems beyond a base-case can be especially challenging. Optimization efforts are often focused on a single area of interest, and the steps taken can result in subsequent issues that negatively impact safety and reliability. Many combustion-related safety and reliability issues can be difficult to observe directly, and they often increase in severity over time.

In this paper, a case study by John Zink Hamworthy Combustion (JZHC) of a reformer optimization will be presented using JZHC's Smart Combustion solutions. Although reformers consist of multiple combustion zones throughout the process, the approach described in this paper will only focus on the primary reformer radiant (arch burners) section.

Smart Combustion Overview

John Zink Hamworthy Combustion's Smart Combustion solution approach to heater optimization combines combustion expertise, traditional and new measurement techniques, and advanced data analytics to provide the operator with information for each individual burner and the heater that was previously unavailable. One Smart Combustion tool used in this paper is the ZoloSCAN in-furnace laser measurement system for combustion balancing. This system is permanently installed in strategic locations throughout the reformer radiant and convection sections and provides real-time measurements of O_2, temperature and CO which can be used to make targeted improvements. Another tool used in this paper for combustion optimization is Ember™, JZHC's cloud-based analytics tool. Ember utilizes existing plant measurements to complete a set of proprietary calculations about the burners to provide operators with insights into the health and operation of each burner.

Combining the insights from ZoloSCAN and Ember together with advanced CFD modeling as a unified Smart Combustion solution provides visibility to combustion health and enables simultaneous improvements resulting in increased efficiency, safety, reliability, and profitability.

Typical Reformers

Steam methane and hydrogen reformers historically are large heaters with high quantities of burners, many with over 100 burners, installed in the roof of the heater and down-fired. It is common for these systems to be operated with both a forced and induced draft fan (balanced draft system) and preheated combustion air. The burners are installed in multiple lanes, or rows, with process tubes located between each burner lane. This is illustrated in Figure 1 with the rows of process tubes identified as the green lines. The burners in the outer lanes, left and right side of Figure 1, are typically designed to fire between 50-65% of the heat release of the inner lane burners.

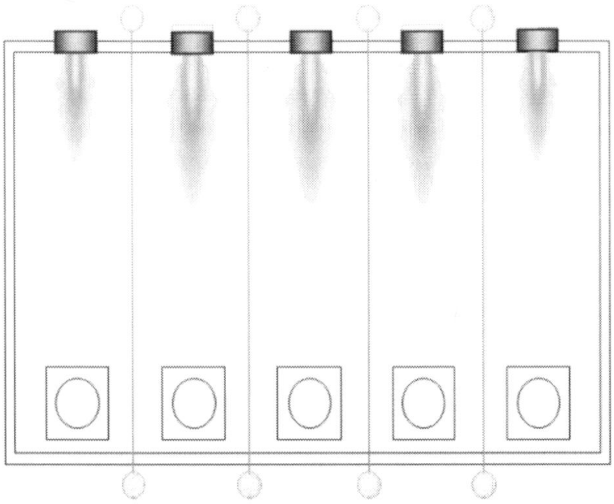

Figure 1. Typical Down-fired Reformer

As burner technology has changed over the last several decades from pre-mix or conventional raw-gas burners to ultra-low NOx burners, the basic design of reformers has stayed the same. To meet more stringent emissions requirements, the changes in burner design has resulted in longer flame lengths that are more susceptible to adverse internal flue gas recirculation patterns.

Common Issues

There are several common issues observed in reformers. Many of these issues are related and cause compounded issues such as decreased efficiency, flame impingement and localized hotspots on tubes that cause decreased tube life and limited production rates.

Perhaps one of the most common issues found in combustion systems is the maldistribution of the combustion air delivered to the burners. Maldistribution of the combustion air results in variation of the air-to-fuel ratio for each burner. In some forced draft systems, these air-to-fuel ratio imbalances can be observed from one row of burners to the next, as well as down the length of each burner row. The impact of varying air-to-fuel ratios per burner can result in some burners operating below the stoichiometric limit of air required causing large flames and increased CO emissions levels. Other burners operating with too much excess air can result in increased NO_x emissions.

Another common issue found in combustion systems are adverse flue gas patterns. Unproportioned burner momentum per available firebox area often result in flue gas patterns which cause the flames from the outer lanes to be pulled through or across the process tubes to adjacent inner lanes. In Figure 2, the flue gas is represented using particle streamlines in CFD. Momentum is illustrated by color with the scale transitioning from high momentum (red streamlines) to lower momentum (blue streamlines). The higher momentum of the inner lane flue gas (right side of figure) draws the flue gas from the outer lane through the process tubes near the roof of the reformer. In many cases, adverse flue gas patterns create undesirable hotspots on the process tubes.

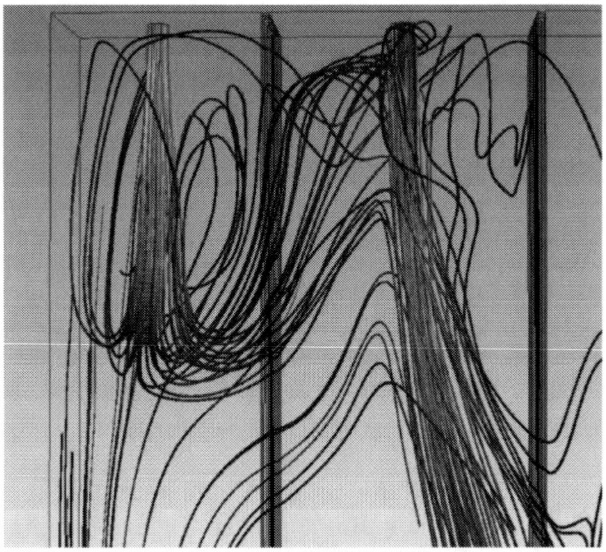

Figure 2. Unproportioned Flue Gas Flow Path

The combination of varying air-to-fuel ratios per burner and adverse flue gas patterns create additional challenges to optimization. This is compounded by the high quantity of burners installed in the arch. A single adjustment to one lane of burners can influence the rest of the reformer and can result in many iterations and trials to achieve the desired outcome.

Optimization and Outcomes

One typical operational target for a combustion system is to reduce the global excess oxygen (O_2) to increase efficiency. When optimizing for a low global excess O_2 level without knowledge of the air flow through each individual burner, an operator could unknowingly create a situation where burners have a sub-stoichiometric amount of air. This can result in an increase in burner flames lengths and potential impingement on process tubes, unburned fuel being drawn into a different zone of the reformer before combusting, and ultimately reduced efficiency of the reformer due to improper heat distribution on the process tubes.

Case Study

Reformer Description

The primary ammonia reformer consists of 120 arch burners arranged in 6 lanes and is split into two halves between the north and south, creating 12 half lanes of 10 burners each. Unlike many reformers, the fuel gas flow to each of these half lanes is controlled individually while the total fuel gas flow is controlled to achieve the desired process gas outlet temperature. The combustion air delivery system is designed to deliver air to both halves of the reformer. The layout of the reformer and the combustion air ductwork can be seen in Figure 3.

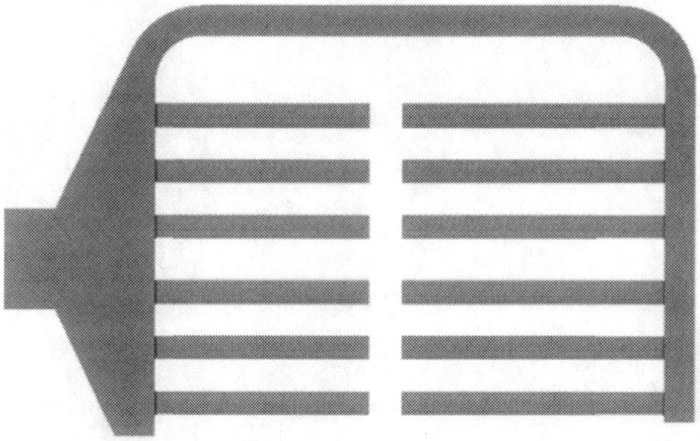

Figure 3. Ductwork for Reformer

Challenges and Modifications

During a major upgrade project, the arch burners were replaced with ultra-low NO_x burners to reduce NO_x emissions. At the same time, the forced draft ductwork was evaluated using CFD to ensure the burners would receive equal air distribution. As a result, perforated plates were installed at each half row inlet to better distribute the amount of combustion air fed to each burner half lane. Even taking this and other precautionary steps, the impact of the flue gas patterns in this reformer, coupled with the close burner to burner spacing, resulted in localized poor flame quality in the outermost rows of burners. It was visually observed that long merged flames were being pulled through multiple adjacent lanes. CFD analysis of the radiant section also confirmed the visual observation. This is shown in Figure 4 where the outer lane of the reformer is on the left and the center is on the right.

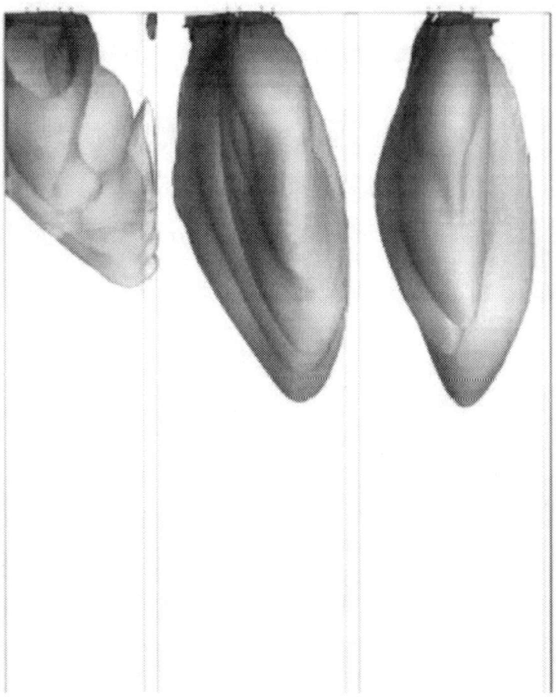

Figure 4. CFD Results of Ultra Low NOx Burners

Prior to the burner retrofit, a ZoloSCAN combustion monitoring system was installed on the reformer as shown in Figure 5. The ZoloSCAN system simultaneously measures the flue gas temperature, volume O_2 percent, CO level on a ppmv basis, and the H_2O concentration in the flue gas directly, in real time. By providing an accurate and reliable representation of the combustion gases within the firebox, it enabled operators to more evenly balance the oxygen and process gas outlet temperatures in individual burner rows. While balancing burners prior to the retrofit was successful, the issue on the end rows from the new burners with feedback from the system was unable to be corrected.

Figure 5. ZoloSCAN System Installation

As a next step, the burners were retrofitted with new fuel gas tips to tighten their flame pattern. This eliminated the burner-to-burner interaction and resulted in decreased flame lengths, as shown in Figure 6. However, the underlying issue of uneven process outlet temperatures remained.

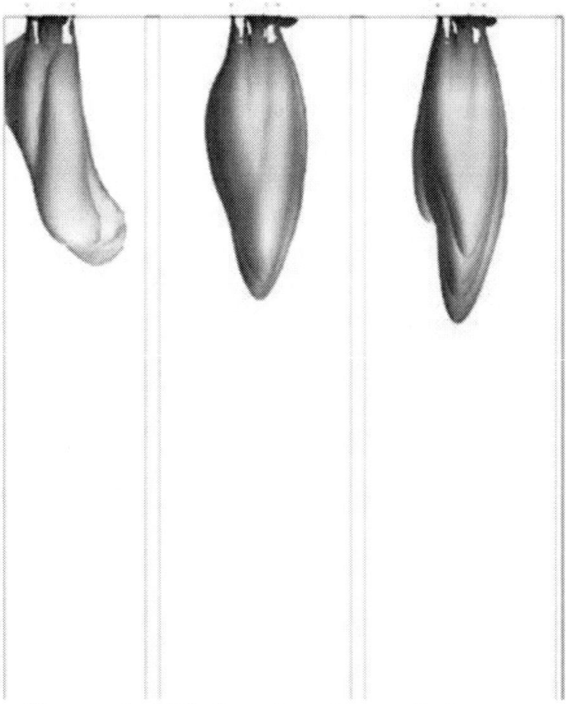

Figure 6. CFD Results of Retrofit Burners

After several months of operation in this state, it was evident that a more holistic approach was required to truly optimize the performance of the reformer. First, CFD modeling was done to understand the current issues being experienced. Those results are shown in Figures 7, 8 and 9. In Figure 7, several planes are shown displaying temperature imbalances and particle streamlines crossing through the process tubes. This confirmed what was visually observed in the reformer. These results were also verified with physical measurements taken by the ZoloSCAN system, Figure 8 and 9. Even with correcting all burners to have equal fuel and air flow this modeling confirmed there was still maldistributions and adverse flue gas recirculation patterns in the furnace due to a fundamental momentum imbalance.

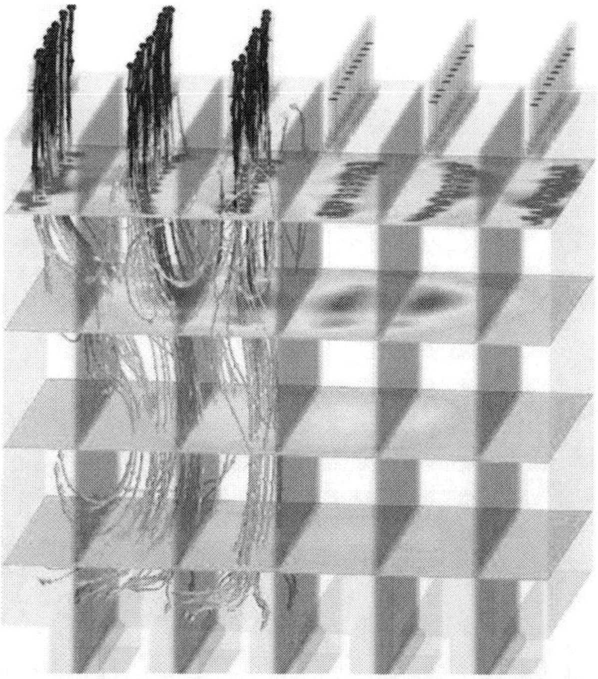

Figure 7. Multi-plane CFD Results of Imbalanced Temperature with Particle Streamlines

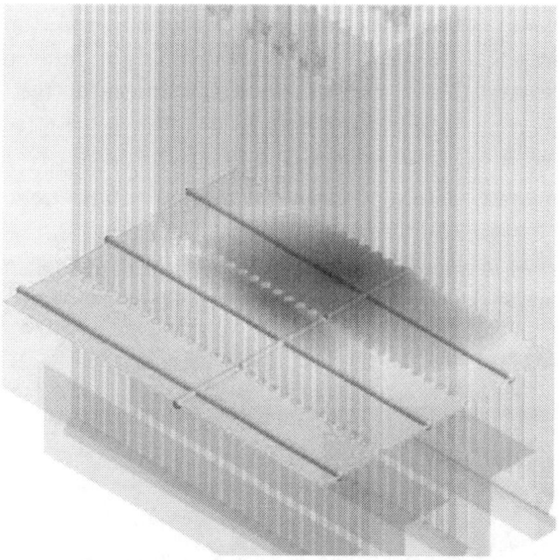

Figure 8. CFD Results of Imbalanced Temperature at ZoloSCAN Installation Plane

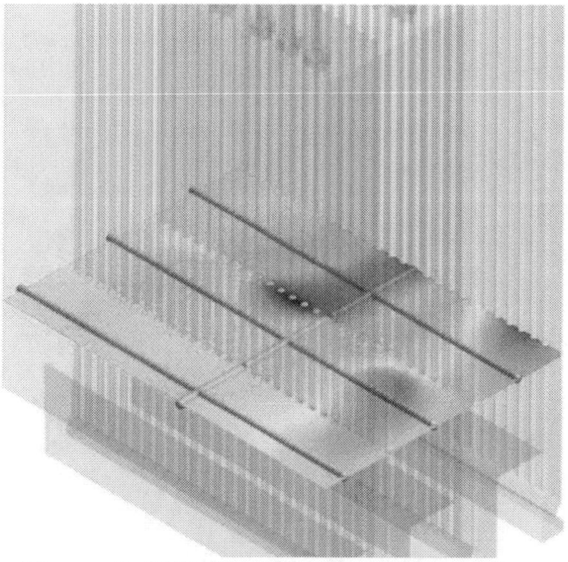

Figure 9. CFD Results of Imbalanced O_2 at ZoloSCAN Installation Plane

Smart Combustion Solution

A combined approach was selected using CFD modeling and Ember with real-time feedback from ZoloSCAN. More in-depth evaluations of the existing burner firing rates, combustion air distribution through the ductwork, and flue gas recirculation patterns were completed using CFD and results such as temperatures and O2 percentage in burner lanes were verified against measured ZoloSCAN results. It was verified that to resolve the effects of adverse flue gas recirculation patterns, the firing distribution in the reformer on a burner-half-row by burner-half-row basis would need to be adjusted and that the proper amount of combustion air would need to be supplied to each individual burner. Figure 10 shows one of the half rows of outer lane burners before the reformer was balanced.

Figure 10. Outer Lane Imbalanced Operation Before Ember

Ember was applied to the reformer utilizing existing measurements of fuel flow, burner fuel pressure, and global O_2 percentage; individual burner insights include individual burner excess O_2 percent, the recommended burner air register settings, and calculated burner gas tip health by comparing half lane firing rates from fuel flow rates and calculated firing rates from burner gas pressure (indicates burner tip plugging or damaging), all in a single location.

After Ember was implemented on site, the process engineer responsible for the operation and reliability of the reformer adjusted the fuel flows to the predetermined firing rate split per the CFD analysis, and the air registers were set based on outputs from Ember while monitoring the excess air levels of each burner in the reformer. Figure 11 shows the same half row of outer lane burners as does Figure 10 but was taken after the reformer was balanced. This approach enabled the balancing of the reformer to occur without going through multiple iterations or relying only on visual observations to determine if burners had the appropriate amount of excess air. Once Ember indicated that all burners had sufficient levels of excess air, the process outlet temperature of each tube row was uniform and within plus or minus 10 °F of the average, hotspots were no longer present, the O_2 percent in each burner lane was uniform, and the fuel gas flowrate for the same process throughput was decreased by several thousand standard cubic foot per hour.

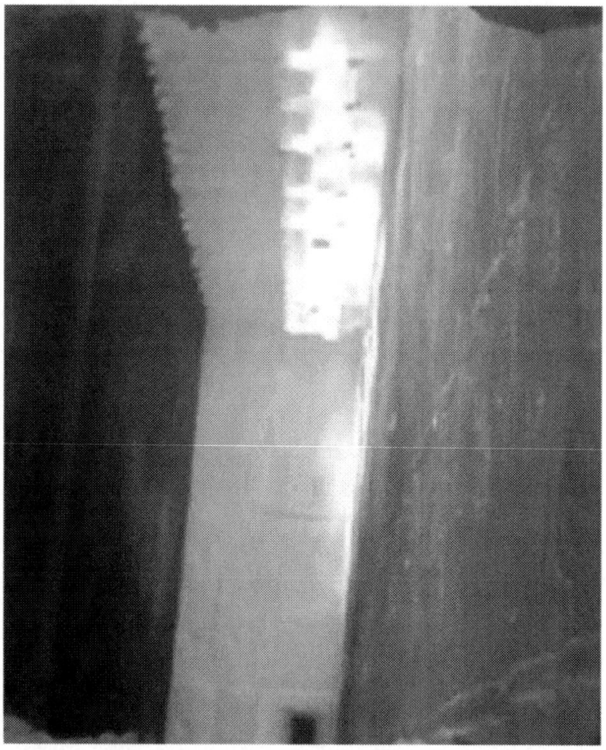

Figure 11. Outer Lane Balanced Operation After Ember

Optimized Operation

The initial installation of ZoloSCAN netted operational improvements of fuel savings, increased productivity, safer and faster start-ups and increased asset component life totaling approximately $1,000,000 per year.

Implementing Ember has enabled the operators to maintain a better optimized state of the reformer with increased reliability, safety, efficiency, and production.

Estimated additional fuel gas savings from decreased required flow is about $350,000 per year. Additionally, with the heat distribution improvements made to the reformer from implementing the full Smart Combustion solution, the site was able to increase production by 150 tons per day above what had been the previous limited capacity, nearly a 20% increase in production.

When CFD and ZoloSCAN are utilized together, ZoloSCAN readings can be mapped to indicate both positive and negative heater performance. In addition, when evaluated through CFD prior to ZoloSCAN installation, the location of the system can be optimized to provide the most insight and accurate representation of what is occurring inside the heater.

To reach an optimized state of operation, the root cause of existing operational issues must first be understood, the method to resolve the problem must be determined and planned, the correct level of visibility of key parameters must be put in to place, and the right system must be implemented to empower individuals with the tools necessary to keep the heater in the optimized state. John Zink Hamworthy Combustion's Smart Combustion solution approach provides a streamlined method to execute this task.

Successful Repair of a Severely Damaged HP Scrubber Sphere

February 2018, a leak was visible in the spherical dome part of a HP Scrubber in a urea plant. The plant was stopped immediately to attend the leak. Upon opening of the HP Scrubber sphere, it became clear that severe corrosion occurred on the liners. The liner was covered with huge amount of corrosion products and many cracks were visible. Unfortunately, these cracks resulted in leaks which were not noticed. Therefore, corrosion of the carbon-steel pressure part commenced undetected. The paper describes the root cause of the corrosion damages as well as the successful repair. This case demonstrates the importance of a reliable and robust leak detection system to operate critical high-pressure equipment safely.

Alex Scheerder, Jan Geerlings
Stamicarbon BV, The Netherlands

Introduction

The urea plant is a 1750 tpd urea plant based on Stamicarbon CO_2 stripping technology. The plant was commissioned in 1998 and the high-pressure synthesis section comprises four pieces of process equipment; namely the Urea Reactor (1), HP Stripper (2), HP Carbamate Condenser (3) and HP Scrubber (4), see Figure 1.

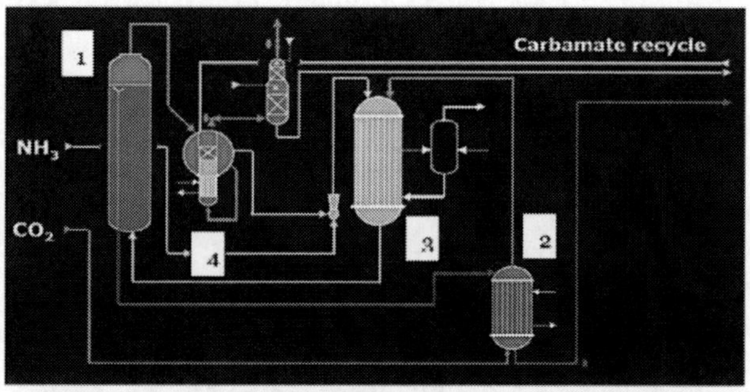

Figure 1. High pressure synthesis loop of a Stamicarbon CO_2 stripping process: [1] Urea Reactor, [2] HP Stripper, [3] HPCC, [4] HP Scrubber

The process conditions in the high-pressure synthesis loop are extremely corrosive due to the presence of ammonium-carbamate as an intermediate process step in the synthesis of urea from ammonia and carbon dioxide.

High pressure equipment utilizes a corrosion resistant barrier as a liner on the process side to protect the carbon steel pressure shell from corrosion. In the said urea plant, for the liner 316L UG (UG = Urea Grade) and X2CrNiMo25-22-2 austenitic stainless steels are used. It is important to passivate these stainless steels by adding passivation air to avoid onset of active corrosion (high corrosion rates).

Furthermore, these critical high pressure equipment must be monitored continuously for leakages of the stainless steel barrier (lining) by use of a leak detection system. The plant is equipped with a leak detection system based on the detection of ammonia vapor. The leak detection pipes are dipped in a bottle containing an ammonia sensitive reagent (Phenolphthalein solution), see Figure 2.

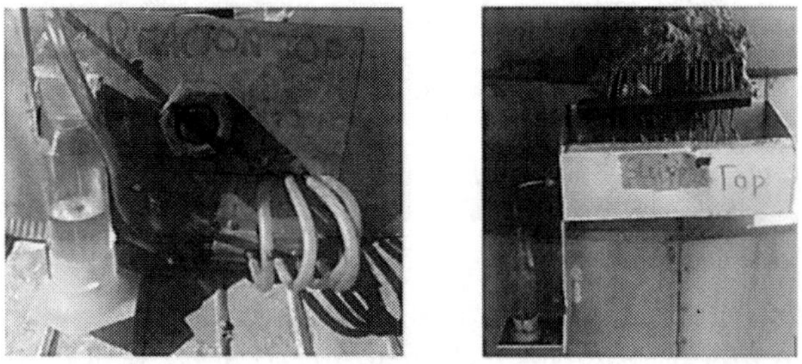

Figure 2. Leak detection system; bottle is filled with Phenolphthalein solution. Colorless at low pH, turns pink at pH>8

At the beginning of February 2018, a leak to the atmosphere was observed in the spherical dome part of the HP Scrubber, which resulted in the plant being stopped to attend to the leak, see Figure 3.

Upon opening the high-pressure HP Scrubber sphere, the corrosion damage was shown to be quite severe, which required a quite complex and comprehensive repair.

Figure 3. leak spot at the outside of spherical dome

It was also concluded that the leak in the corrosion resistant barrier was not detected, which resulted in late detection and consequently severe corrosion of the carbon-steel part. The incident can be classified as a near miss. Fortunately, in this case the corrosion of the pressure retaining part resulted in a so-called "leak-before-break" scenario (LBB), allowing the plant owner to timely shut down the plant to avoid catastrophic failure.

This paper describes the root cause of the corrosion damages as well as the successful repair. The plant was started up successfully April 6th, 2018.

Case history

The HP Scrubber (see Figure 4) consists of a heat exchanger and a spherical scrubbing part. The sphere is constructed of 20MnMoNi45 DIN 17201 carbon-steel plate; mono-wall, actual thickness is 53 mm (2.087") protected at the inside with a 9 mm (0.354") 316L UG liner.

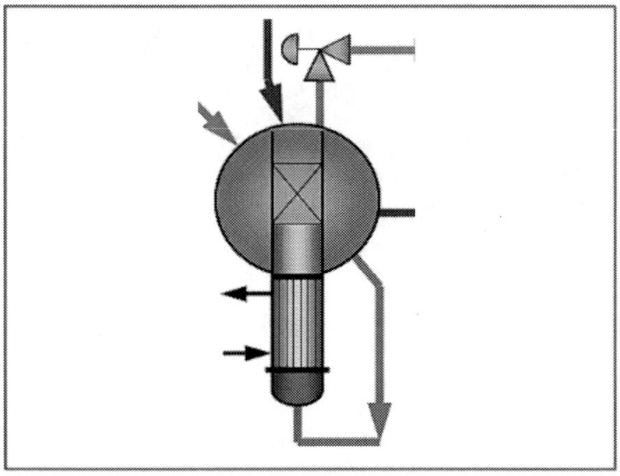

Figure 4. HP Scrubber with dome

Upon opening the spherical dome (February 7th), severe corrosion and large amounts of corrosion products was observed on the 316L UG liner, as can be seen in Figure 5.

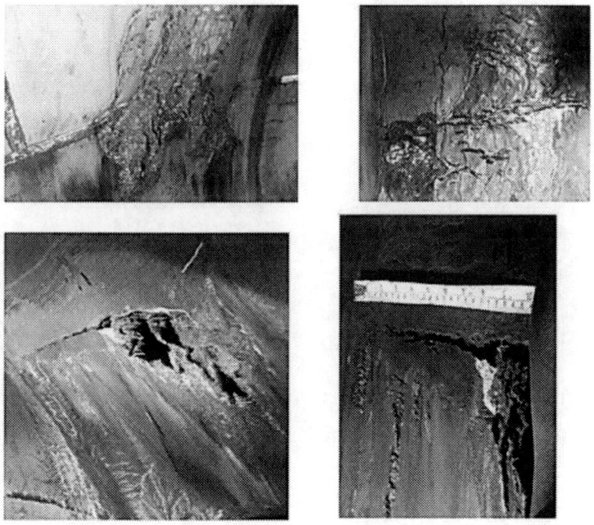

Figure 5. 316L UG Liner plates covered with large amount of corrosion products

All corrosion products were removed by grinding. Dye Penetrant Testing (DPT) revealed large number of cracks in the liner, see Figure 6.

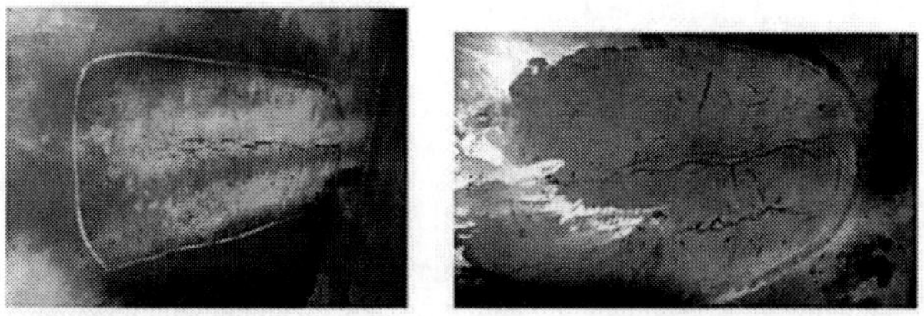

Figure 6. DPT of the liner revealed many cracks in the 316L UG liner

Based on these observations it was decided to remove and replace all liner segments. After removal of all liner segments the Wet Fluorescent Magnetic Particle Test (WF-MPT) revealed many cracks in the carbon-steel pressure shell, as can be seen in Figure 7. Also, it became clear that the leak detection system (grooves and holes) was partly blocked by corrosion products, making the system unreliable.

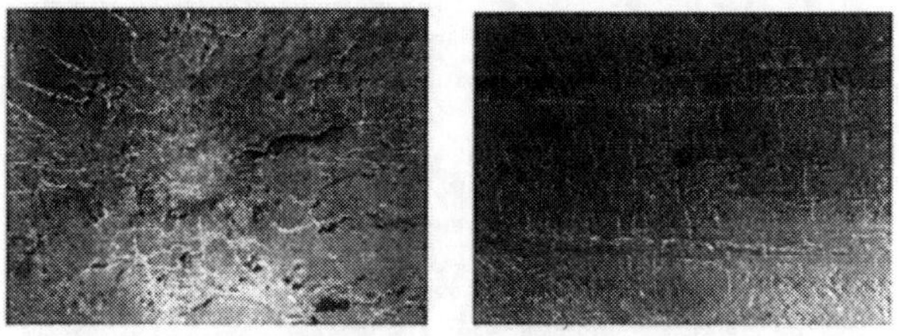

Figure 7. WF-MPT revealed many cracks in the carbon-steel pressure shell

The repair strategy was developed based on a fit-for-service (FFS) assessment. An important aspect of the repair was also to restore an effective leak detection system.

Root Cause Analyses

Metallurgical examinations of following samples were carried out, see Table 1:

SAMPLE ID NUMBER	DESCRIPTION
S09176	Liner plate sample: 316L UG including liner weld
S09177	Boat samples c-steel shell part (20MnMoNi45)
S09178	Corrosion products
S09179	Boat samples c-steel shell part (20MnMoNi45)

Table 1. Samples

Investigation results

Liner plate sample (S09176)

On the process medium side, the liner plate was partly covered with a blue oxide layer. Local absence of an oxide later indicates condensation of gasses on the liner surface (see Figure 8).

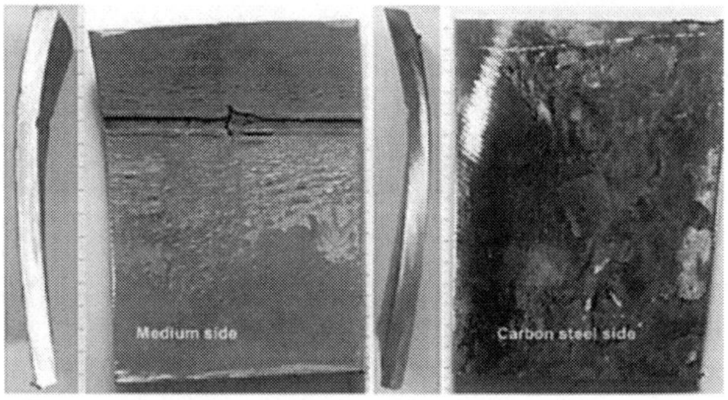

Figure 8. Liner sample (S09176); medium side covered partly with blue oxides. Backside covered with black oxides

The back side of the liner plate was covered with black corrosion products. The cracks started from the process medium side and progressed almost perpendicular to the back side, see Figure 9.

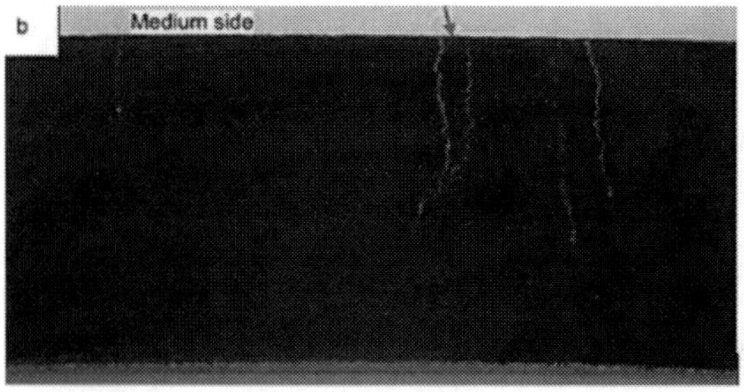

Figure 9. Cracks starts at medium side

Microscopic examination of the cracks shows that the cracks are intergranular stress corrosion cracks, which are filled with corrosion product (oxides), see Figure 10. This indicates that the cracks propagated slowly.

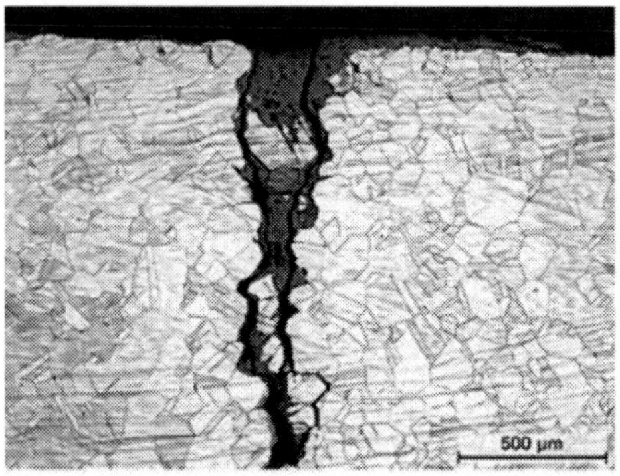

Figure 10. Intergranular stress corrosion cracks with oxides inside the cracks

At the liner plate surface (process medium side) typical deformation patterns in the austenitic grains are visible, see Figure 11. This indicates strain in the material.

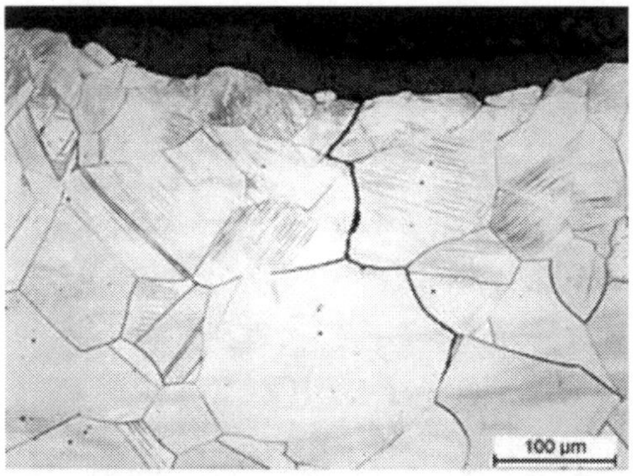

Figure 11. Typical deformation pattern in the austenitic grains indicating cracks starts at surface plastically deformed

This plastic deformation of the liner is mainly due to the bending of the liner plates during manufacturing. However, also temperature and pressure cycles (start-stops) could play a role as well.

Examination of the corrosion products found at both the process medium and the backside of the liner shows they consist mainly of iron oxides (corrosion product). However, at the backside of the liner also Sulphur, Copper and Nitrogen are found, see Figure 12.

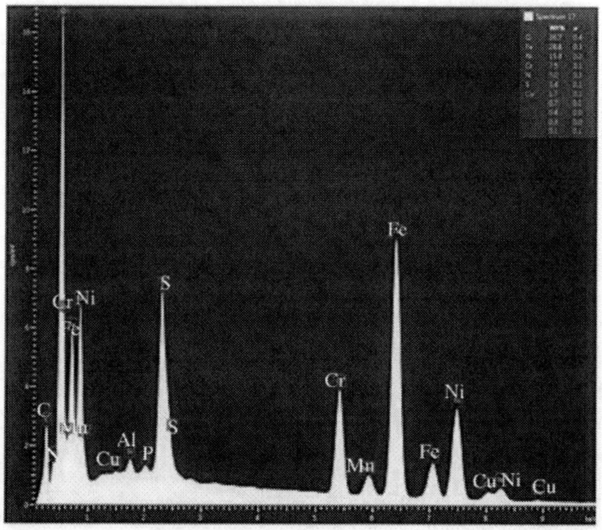

Figure 12. EDX spectrum of the elements found in the corrosion deposits at the liner backside. Apart from the normal elements found in 316L UG material, foreign elements such as Sulphur, Copper and Nitrogen are found

The origin of the Sulphur and Copper is unclear. The presence of ammonium-carbamate trapped behind the liner also attacked the liner at the backside; i.e. intergranular corrosion attack is observed.

Boat samples carbon steel shell (S09177 and S09179)

Two boat samples were removed from the carbon-steel sphere for metallurgical investigations, see Figure 13.

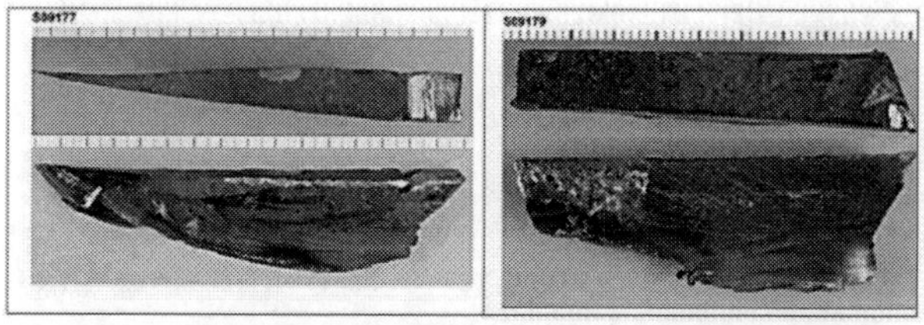

Figure 13. Two boat samples from carbon-steel sphere; left hand side S09177, right hand side: S09179

Metallographic examination of both boat samples indicated branched trans granular stress corrosion cracks, see Figure 14. The corrosion products were mainly containing iron oxides, but also traces of Sulphur were found. In the crack itself oxides are found, which also indicates that the crack propagated slowly.

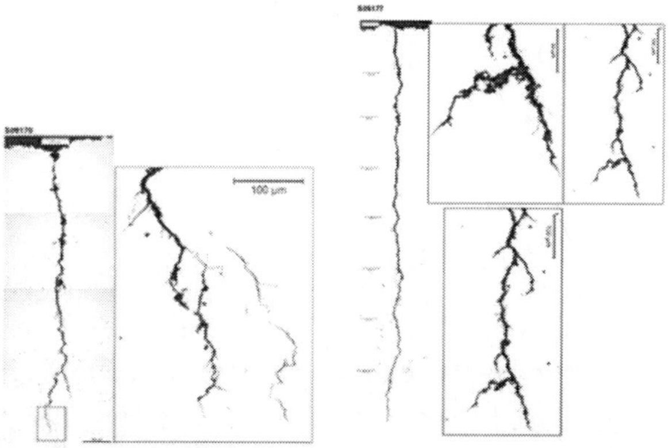

Figure 14. In both boat samples trans granular stress corrosion cracks are visible.

Root Cause of the Damages

The leakage in the 316L UG liner was due to intergranular corrosion cracking, initiated at the process medium side of the liner. Since the inner surface was plastically deformed (strain) and the presence of oxides in the cracks indicated that the failure mode was so-called Strain Induced Intergranular Cracking (SIIC). This failure mode has been observed in other urea equipment lined with 316L UG liners in the gas phase area.

SIIC is an electrochemical corrosion mechanism supported by mechanical deformation. Condensation of ammonium carbamate gas will create an electrolyte on the liner surface. Plastic deformation of the liner material was evident and also the presence of Sulphur was proven. Sulphur is believed to enhance this failure mode.

Ammonium-carbamate entering the space behind the liner started to corrode the carbon-steel. This resulted in localized wall thinning of the carbon-steel pressure shell as was observed in the sphere. The observed cracks in the carbon-steel are believed to be related to hydrogen. The hydrogen is originating from the corrosion of the carbon-steel due to the presence of ammonium-carbamate. Following reactions play a role:

$NH_2COONH_4 \longleftrightarrow NH^{4+} + NH_2COO^-$ (Acidic behaviour ammonia carbamate)

$2Fe + 2\,NH^{4+} \longrightarrow 2Fe^+ + 2H_{atm}\,(H_2) + 2NH_3$ (dissolving Iron and producing hydrogen)

Atomic hydrogen (H) is capable of diffusing through steel and the cracking failure mode is called hydrogen induced stress corrosion cracking (HISCC) or hydrogen induced cracking (HIC). The presence of Sulphur found on the back side of the liner sample promotes the entry of atomic hydrogen into the steel. The atomic hydrogen diffuses in the lattice to areas with increased tensile stresses brought about by the presence of notches, the microstructure (martensitic), geometric discontinuities and flaws in the lattice. The presence of hydrogen in the metal lattice reduces the toughness and resistance to fracture. Especially the high-tensile steel used for the Sphere is susceptible for hydrogen induced cracking. Hydrogen induced cracks are also known to propagate slowly.

REPAIR

The repair of the Scrubber sphere can be divided in two phases:
Phase 1: Removing the damaged liner and repair the carbon-steel pressure shell
Phase 2: Installation of the new liner segments

Phase 1: Repair of the C-steel pressure shell

First the existing liner segments were removed from the sphere. The inner surface was cleaned from rust and debris, see Figure 15.

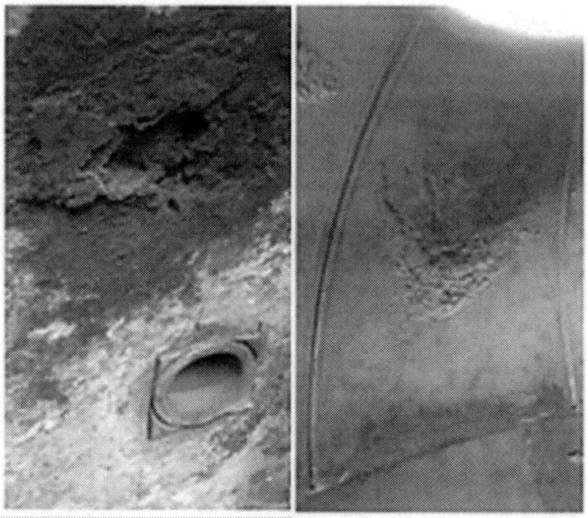

Figure 15. Left side before cleaning, right side after cleaning

All cracks were removed by grinding under guidance of WF-MPT testing, see Figure 16. Since the carbon-steel was loaded with hydrogen, grinding was done carefully in order not to propagate the cracks or the formation of new cracks. The time needed to remove all cracks was approx. 11 days.

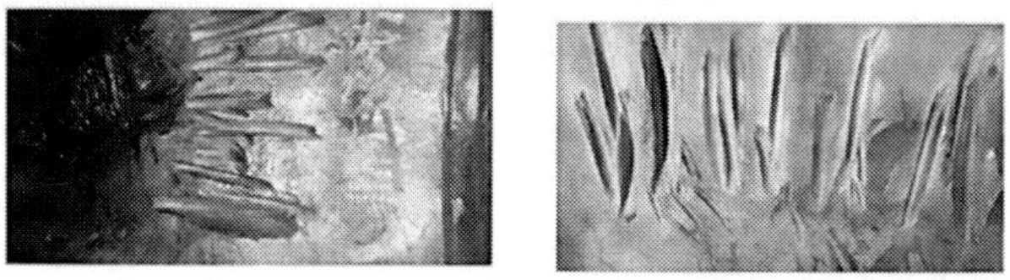

Figure 16. Grinded areas; at this location minimum remaining wall thickness is 10 mm

Stamicarbon performed strength calculations (FFS) to determine the minimum required wall thickness. This varied between 37 mm (1.457") up to 45 mm (1.772") depending on the location of the grooves. The buttering layers to attach the new liner plates were restored by grinding and /or welding. The leak detection grooves and holes were cleaned and repaired when necessary.

To avoid crack propagation during repair welding, the pressure shell was heated for approx. 2 hours to outgas any hydrogen still present in the carbon-steel.

Repair welding was done at a pre-heat temperature of approx. 120 °C (248 °F). A heating system was installed at the outside of the sphere, see Figure 17.

Figure 17. Insulation wrapped on all ceramic fiber heating elements

Covered electrodes (SMAW: AWS E7018 type) were used with low hydrogen content to minimize the uptake of hydrogen during welding. An impression of the repair activities is presented in Figure 18. The time needed to repair all grinded areas was approx. 6 days.

Figure 18. Repair welding in progress

After completing the repair welding, the temperature was increased again to soaking temperature (approx. 300 °C, 572 °F) for some time to outgas any uptake of hydrogen during welding. Finally, the temperature was lowered to ambient and after waiting for 48 hours the repaired zones were checked for cracks. No cracks were found, and the hardness was well within the maximum allowable values.

Subsequently a post weld heat treatment (PWHT) was performed in accordance to the design code for stress relieving and tempering. After the PWHT the repaired areas were checked again, and no cracks were found. Finally, the wall thickness in the welded grooves was restored to original wall thickness (53 mm; 2.087") by applying a two-component metal-epoxy compound, in order that the liner was supported in these regions to avoid plastic deformation.

Phase 2: Installation of new liners

The liner in the HP scrubber sphere is divided in 16 pre-shaped segments, as can be seen schematically in Figure 19.

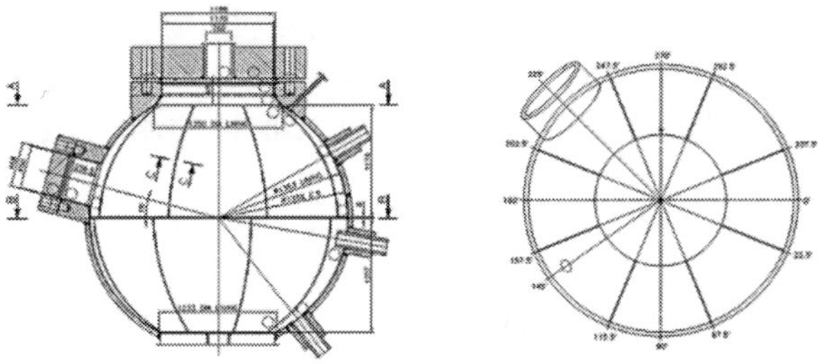

Figure 19. Schematic overview of the 16 liner segments to be installed. Right hand sketch shows 8 segments in top half

The new liner plates are 6 mm thick X2CrNiMoN 25.22.2. The liner plates were delivered with an over length. The plates were cut to fit the existing buttering layers. Welding was done by Gas Tungsten Arc Welding (GTAW) method using matching filler wire. The installation and welding of all liner segments took about 11 days. Some impressions of the liner installation are presented in Figure 20.

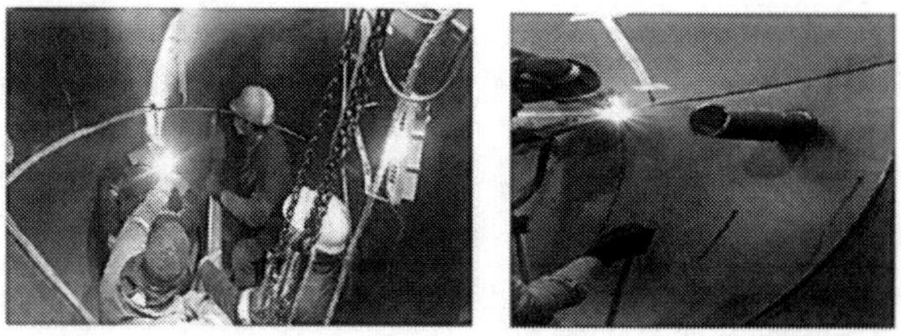

Figure 20. Installation and fit-up of liner segments in top part of the sphere.

After the pressure test, the HP Scrubber sphere was opened again to perform an ammonia leak test, no leaks were found. Before boxing up, the internal pall ring cylinder was welded in.

The plant was started up successfully on April 6th, exactly 2 months after the leak was detected and the plant was shut down (February 5th).

Discussion and Recommendations

The severe corrosion of the pressure bearing part of the sphere is due to the fact the liner leakage preceding this corrosion could not be noticed probably due to chocking of leak detection holes. This demonstrates the importance of a reliable and robust leak detection system.

Nowadays Stamicarbon recommends using state of the art leak detection system, which is based on continuous forced air flow fed to an ammonia analyzer (so-called pressurize system), which can be connected to the DCS system (see also Reference [1]). The principle of the pressurized system is presented in Figure 21 and 22.

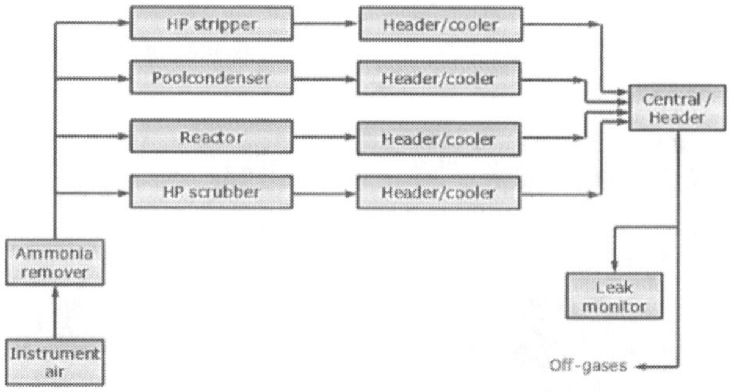

Figure 21. Principle of the pressurized leak detection system (schematic overview for a Stamicarbon pool condenser plant)

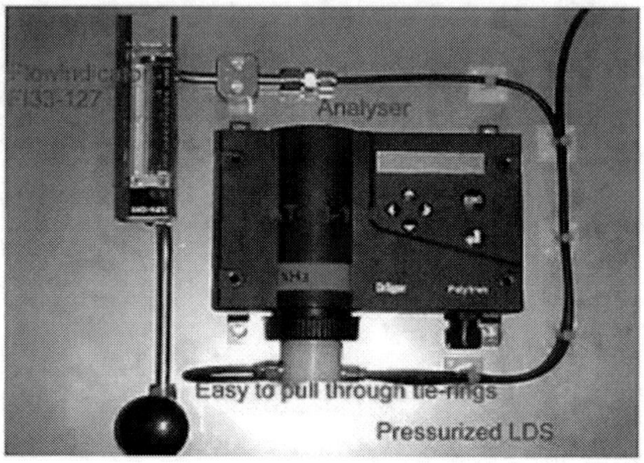

Figure 22. Ammonia analyzer

The leakage in the liner occurred due to so-called strain induced intergranular corrosion (SIIC). This is the result of condensation of reactor off gasses in combination with plastic deformation of the 316L UG liner. The plastic deformation is mainly the result of bending the liner plates during manufacturing. However frequent temperature and pressure cycles may play a role as well. Condensation of the hot reactor off gases cannot be avoided completely in the HP Scrubber sphere, despite the presence of a steam tracing. Plastic deformation of the 316L UG liner material cannot be completely avoided either. Due to the liner leak, ammonia-carbamate started to corrode the carbon-steel pressure shell. Besides wall thinning, also hydrogen induced cracks (HIC) were developed in the carbon-steel. Unfortunately, the selected steel (20MnMoNi45 DIN 17201) for the HP Scrubber sphere is prone to this failure mode (HIC) due to the high strength properties.

Corrosion of the carbon-steel pressure retaining part should be avoided always. This is managed by having a reliable leak detection system in place. Furthermore, the risk of hydrogen induced cracking can be reduced by applying so-called low strength steels (typically ASME P1 types; such as for instance SA-516 Gr70 or SA662 Gr. C) instead of high strength steels.

References

[1] Safeguarding Stamicarbon high pressure urea vessels. AIChE Ammonia and Related Facilities Safety Symposium, San Francisco, California Sept. 12-16, 2010. Alex Scheerder, Stamicarbon bv, The Netherlands.

Damage to the Combustion Air Ducting of an Ammonia Primary Reformer

Severe damage to the parent material and structural integrity of the Combustion Air Ducting (CAD) arrangement was experienced on an Ammonia Plant Primary Reformer. The direct cause of the damage was postulated to be reverse flow of excessively hot combustion air/flue gas from within the radiant section of the Primary Reformer through the CAD. In addition to the cause of the damage sustained, this paper presents the sequence of events and recommendations to prevent recurrence.

Shawn Sankarsingh, Teressa Martinez
Industrial Plant Services Limited (IPSL), Trinidad and Tobago W.I.

Dan Barnett
BD Energy Systems Inc. Houston USA

Introduction

The combustion air ducting located within the penthouse of an ammonia primary reformer, sustained major heat damage following a plant outage in May 2016. This event occurred on a 1,850 MTPD (2,040 STPD) ammonia plant operated by IPSL which is a subsidiary of PROMAN, a leading global methanol and fertilizer producer. IPSL manages, operates and maintains fourteen (14) petrochemical plants in Trinidad and Tobago. The plant on which this event occurred, had been in service for 14 years without damage to the reformer. This paper describes the damages incurred, sequence of events that led to the damage and the lessons learnt.

Equipment Layout

The Primary Reformer is a gas fired processing furnace containing radiant and convection sections. The radiant section of the reformer consists of two hundred and eighty (280) catalyst tubes arranged in five (5) rows, each row containing fifty-six (56) tubes. Heat transfer to the catalyst tubes is provided by ninety-six (96) down fired radiant section/arch burners, mounted equally within six (6) rows of CADs in the penthouse of the Reformer. Each burner is made up of (i) a primary burner which burns natural gas, and (ii) a secondary burner which burns reject gas recovered from other areas within the process plant. Below each burner row, six (6) flue gas collection tunnels at the floor of the radiant box each have a tunnel burner firing horizontally into the flue gas stream to provide supplemental heat to the downstream convection section. A negative pressure/draft within the radiant section of the Primary Reformer is maintained during normal operations via a steam turbine driven Induced Draft (ID) fan. The combustion air to both the arch and tunnel burners is supplied by a motor driven Forced Draft (FD) fan.

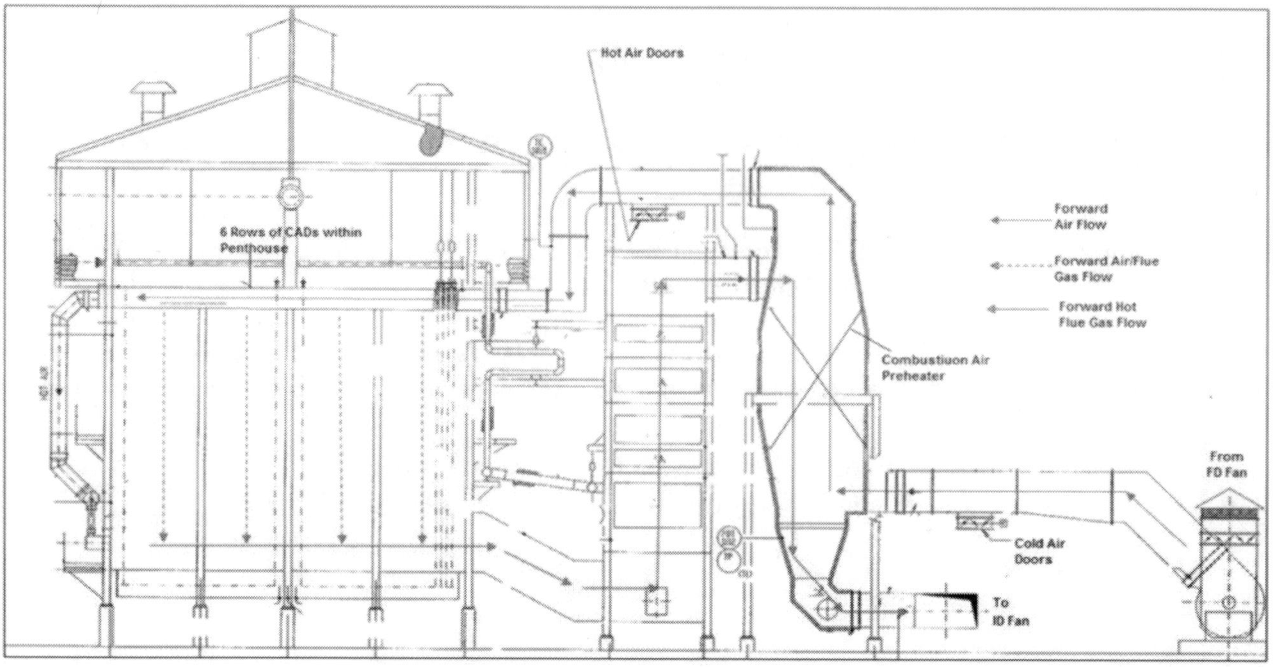

Figure 1. Simplified layout of Primary Reformer showing air/flue gas flow through the combustion air ducting from FD Fan to ID Fan

The main combustion air supply ducting includes two sets of air doors (cold and hot air doors). These doors are situated in the combustion air ducting upstream (cold air doors) and downstream (hot air doors) of a combustion air preheater and serve to supply combustion air using only the ID Fan during start up and during instances of FD Fan unavailability. Figure 1 shows the general equipment layout and combustion air/flue gas flow from the FD Fan to the ID Fan under normal operating conditions. Figure 2 shows the make-up of the arch burner, its position in relation to the CAD and the secondary burner tip projection required per design.

Sequence of Events (Incident to Repair)

On the 4th May 2016, a severe gas curtailment was imposed on the plant resulting in a plant shutdown. Prior to the shutdown, there were no reported issues on the front end of the plant and

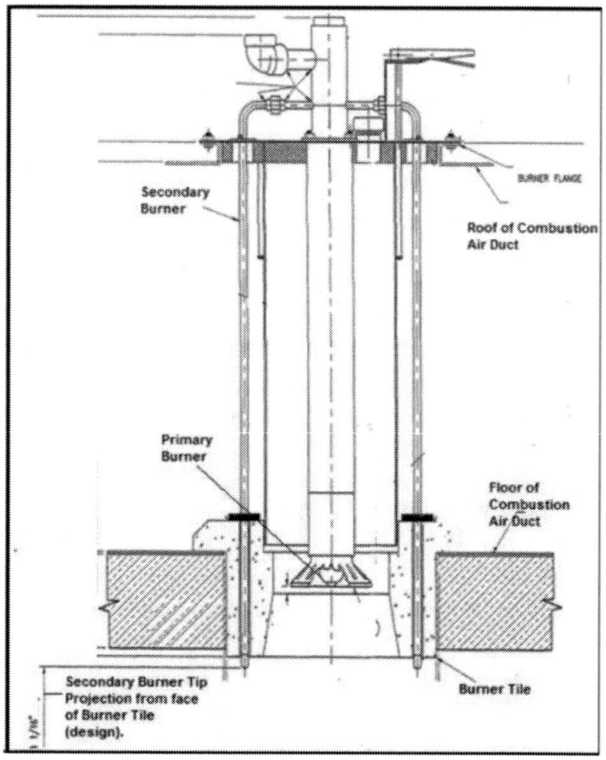

Figure 2. Arch Burner installation within the CAD.

the Primary Reformer was performing as designed. The opportunity was taken to conduct maintenance works on plant equipment during the outage, however no major maintenance works was scheduled as necessary to be performed on the Primary Reformer. Upon plant restart and shortly after achieving production, routine visual inspection of the firing within the Primary Reformer radiant section revealed that the majority of the arch burners were exhibiting highly erratic firing. The erratic firing was characterized by bright yellow opaque flames the length of which could be seen as far down the harp as the tunnel roof in some instances. A general hazy appearance was observed on all burner rows indicative of incomplete combustion. Also noticeable was that each burner's secondary burner tips were seen somewhat recessed within each burner tile, which indicated that the floor of the CAD may have been sagging. This however could not have been confirmed while the plant was online. Several other tests, comprising of fuel to air burner adjustments and surveys, CAD main damper adjustments and burner replacements were performed from May 2016 to a scheduled plant turnaround (TAR) in January 2017, with no significant improvement in burner firing observed.

TAR Inspection Findings/Repairs

During the January 2017 TAR, major maintenance works were carried out on the Primary Reformer which included a total reharp of all five (5) rows of catalyst tubes in addition to the internal inspection and repairs of the six (6) CADs labelled A to F within the penthouse of the reformer. The repairs were successfully conducted by Proman AG (Trinidad) Limited over a period of twenty two (22) days without any safety incidents.

Internal inspection of the CADs during the 2017 TAR revealed widespread scaling on all walls, floor and roof consistent with an overheating event. The sides, floor and roof of each CAD were noted to be significantly warped. A large amount of sagging was observed towards the center of each duct with a maximum deflection of two and a half inches (2.5"(63mm)) noted on the worst affected CAD. Figure 3 shows typical CAD condition.

Figure 3. Typical scaling on all metal surfaces as well as warpage of side walls, roof and floor of CAD

It is relevant to note that based on the evidence that the floor of the CADs was sagging as indicated by the recessed secondary burner tips noted just after the May 2016 plant restart, the decision was made to procure and prefabricate the floor of the CADs using new ASTM A 516 Gr 70 carbon steel plate material, in the event complete change out of the flooring of all CADs became necessary.

During the 2017 TAR, in order to address the damages observed given materials on hand, the time constraints and practicality, the floor of each CAD arrangement was removed leaving approximately five inches, 5" (125mm) on either side. Prior to the floor removal, the walls were vertically supported with chain blocks and laterally braced to prevent inward buckling of the ducting. Due to the extent of warpage experienced on the floor of the CADs, aligning the new prefabricated floor plates with the existing floor edges required that various sections of the side walls be slotted, and the existing floor edges heated and then mechanically straightened to achieve a level surface. The top of the floor plate was seal welded to the existing floor edges whilst the bottom was stitch welded with six-inch (6 in. (150mm)) welds at eight-inch (8 in. (200mm)) centers. Figure 4 shows the typical alignment of the new floor plates to the existing duct.

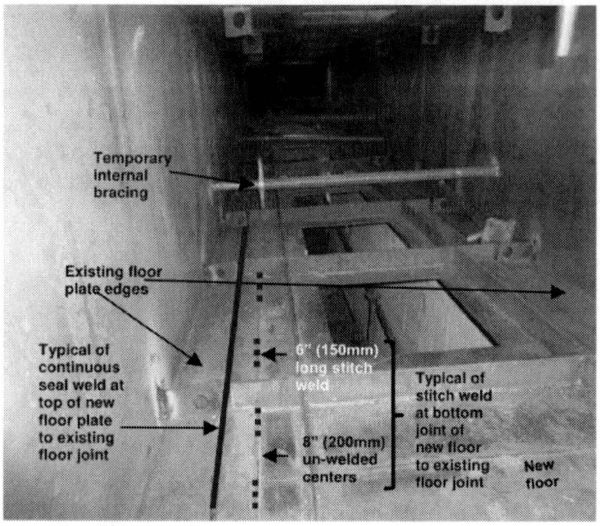

Figure 4. New floor plate installation with burner box cut-outs and internal bracing

Although the floor of the CADs were now aligned and fairly straight, the distortion of the roof plate posed a challenge for proper burner/burner box alignment. Each of the ninety-six (96) burner box assemblies required individual fit up and adjustments. The method of installation first involved removal of the flange bracket. Figure 5 highlights the flange which needed to be removed.

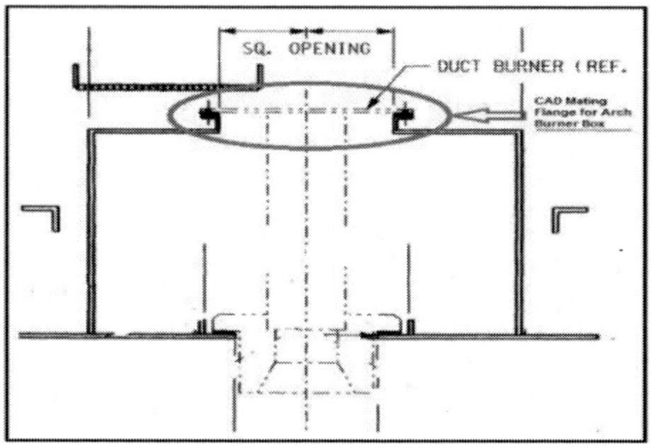

Figure 5. Location of flange on CAD roof which was removed

New flange brackets were fabricated to fit into the remaining slot. With the burner tiles mounted onto the new floor plates, the entire burner assembly was installed. A temporary plate was installed at the base of the tile in order to maintain the required 1 1/16" secondary burner tip projection. With the secondary burner tip projections fixed, the new flange bracket position was marked, and the entire burner assembly was removed. The new CAD flange bracket was then reinstalled, and seal welded from the inside. Figure 6 shows a sketch of the new flange bracket installation after confirming required burner position within the CAD.

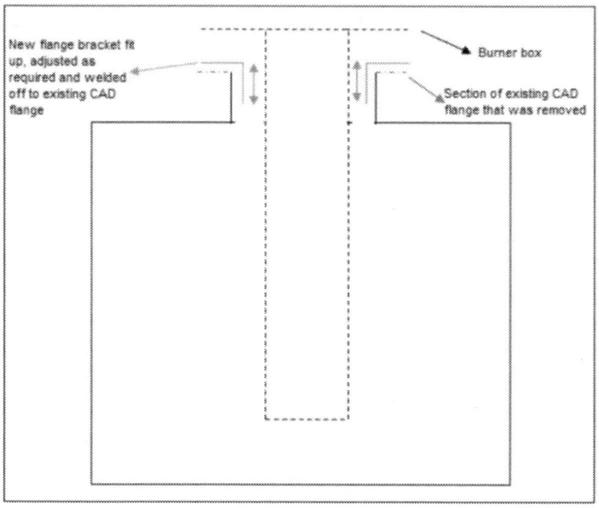

Figure 6. Typical of adjustments required for new flange bracket installation

Arch/Tube seal plates which were originally affixed to the floor of each CAD were observed to be warped on all rows. These were cut out and replaced with ASTM SA-516 Gr.70 plates of equivalent size. Figure 7 depicts the damage sustained on the arch/tube seal plates.

Figure 7. Typical warped Arch/Tube seal plates which were replaced with new material

On the outer CAD rows, A and F where the duct is supported by the northern and southern radiant box walls, several large gaps were observed. These gaps were reduced in size by welding two inch (2 in. x 2 in. (50 x 50mm)) angle iron along the entire length of the affected wall (measuring approximately (30 ft. (9000mm)). Figure 8 shows the gaps noted on the outer CAD to wall joints.

Figure 8. Typical gaps noted between outer CADs and radiant wall joints.

All six (6) expansion joints on each duct within the reformer penthouse, were damaged and warped. In all cases, the expansion joints have exceeded their design expansion both axially and laterally. Both inward and outward deformation was noted on the expansion joints. For joints which exhibited inward deformation, the internal sliding plates were mechanically straightened maintaining a half inch (0.5 in. (12.5 mm)) gap between both sliding plates. For joints exhibiting outward deformation the top sliding plates were replaced with new plate allowing for a half inch (0.5 in. (12.5 mm)) gap. The fiberglass wool was repacked once the required gaps were achieved. Figure 9 details the typical damage noted on the CAD expansion joints within the penthouse.

Figure 9. Excessive movement and deformation of CAD expansion joints noted.

Discussion

Investigation into the cause of the severe damage noted on the CAD was performed in conjunction with a furnace designer. The initiator of the event was determined to most likely be during the plant shutdown in May 2016.

During a controlled plant shutdown of the Primary Reformer, the FD Fan is taken offline. Control logic causes both the Cold Air Door and Hot Air Doors to open. The ID Fan would then remain online to assist in cool down of the entire reformer train.

It was noted in May 2016 that the FD Fan was taken offline followed by an ID Fan trip within a few minutes. With the Cold Air Door and the Hot Air Doors opened and no forced or induced draft, hot flue gases would have travelled from the convection section to the radiant section, up through the CADs to the Hot Air Doors, and towards the atmosphere. All six (6) CADs in the primary reformer penthouse would have experienced flue gas temperatures which are typically in excess of 1800°F (980C). Due to lack of temperature indicators within the ducting, it was not possible to determine the actual temperature that the CADs were exposed to. Figure 10 shows a general arrangement drawing of the normal flue gas flow (black) vs reverse flow (red) within the Primary Reformer.

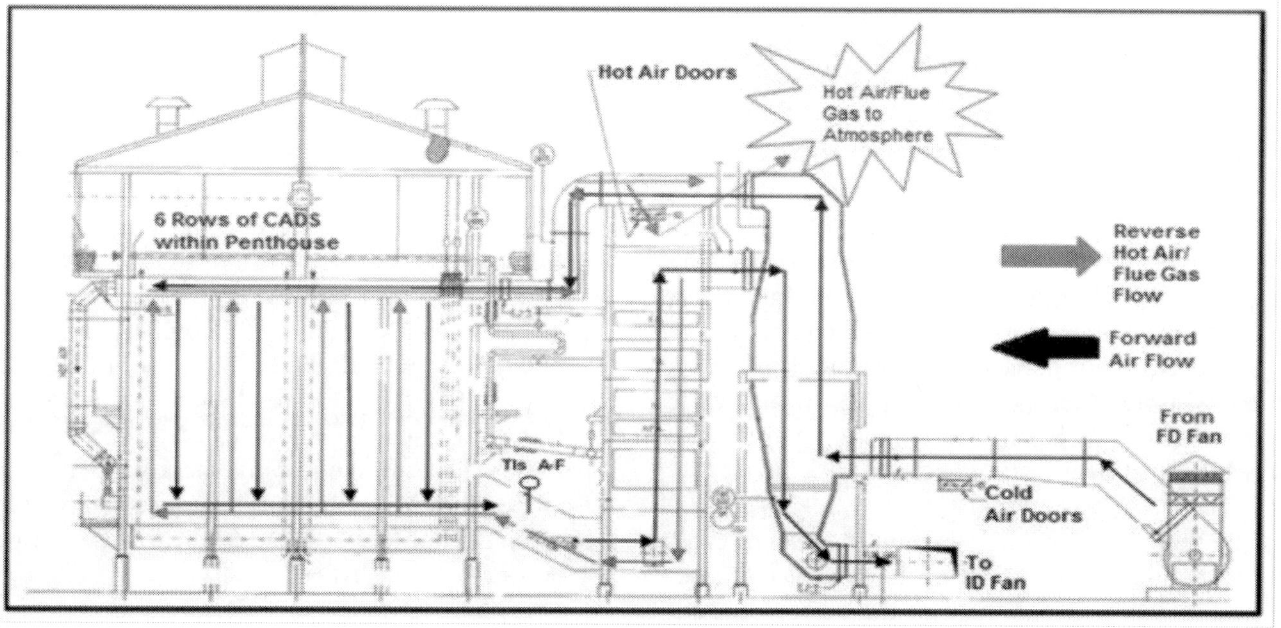

Figure 10. Reverse flow of hot air/flue gas.

However with both fans offline, based on the temperatures observed on convection/transition section temperature indicators TI - A/B/C/D/E/F, flue gas temperatures experienced an initial increase from approximately 1400°F (760°C) to 1500°F (815°C) over a thirty minute (30 min.) period after which a temperature decrease from approximately 1500°F (815°C) to 750°F (400°C) was experienced over nine hours (9 hrs.). Figure 11 shows the process trends of the flue gas temperatures after both the ID and FD Fans were shut down.

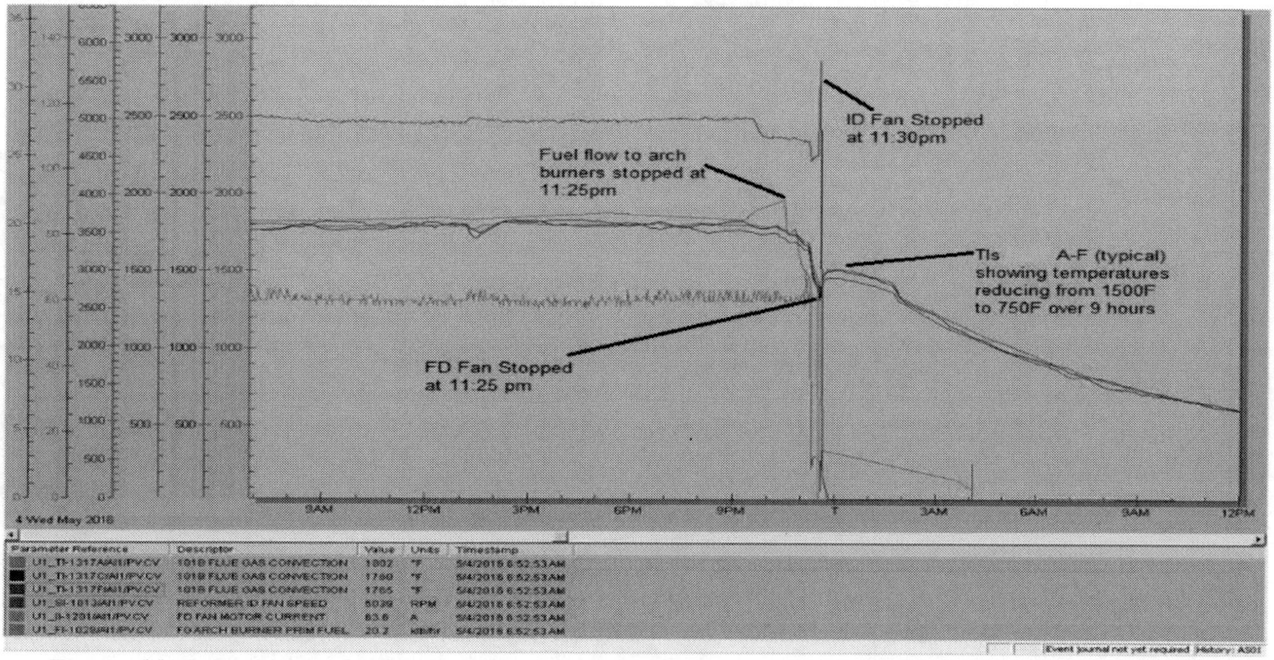

Figure 11. Process trends showing flue gas temperatures within convection section being between 1500^0F (815°C) to 750^0F (400°C) for approximately 9 hours after ID fan was stopped.

The primary reformer CADs were fabricated using ASTM SA-516 Gr. 70 plates for which allowable stresses (factor of yield stress) decrease significantly when exposed to temperatures above 750°F (400°C). As temperatures increase above the CAD design temperature of 650°F (345°C) to 1000°F (540°C), the allowable stresses decrease from 18.8 ksi (130 MPa) to 2.5 ksi (17 MPa) respectively. Therefore, susceptibility to warping or sagging increases with temperature and time. It is this sagging of the CADs which resulted in the recession of the secondary burner tips within each burner tile. The main burner would have now also been positioned higher within the tile resulting in overheating of the tiles.

The secondary burners function to assist in complete combustion as part of the low NOx burner design. The reinstatement of the design secondary burner tip projections were therefore critical during remedial works in ensuring serviceability of the reformer. As per burner requirement, a projection of approximately 1.0625 (26.9875 mm) is required. With the secondary burner tips completely or partially recessed, incomplete combustion can occur leading to fuel (haze) residing within the radiant box. Flame impingement on burner tiles and tubes would have also occurred and was observed. In addition to this, the burner flame would have resided higher within the tiles thus exposing the ducting to a higher than design heat flux. Figure 12 shows photos of the burners in May 2016 (firing issues began) and in January 2016 (prior to/no firing issues). It can be seen that in May 2016, flame impingement was occurring on the burner tiles and large gaps were noted between the burner tile halves indicating sagging of the floor of the CADs.

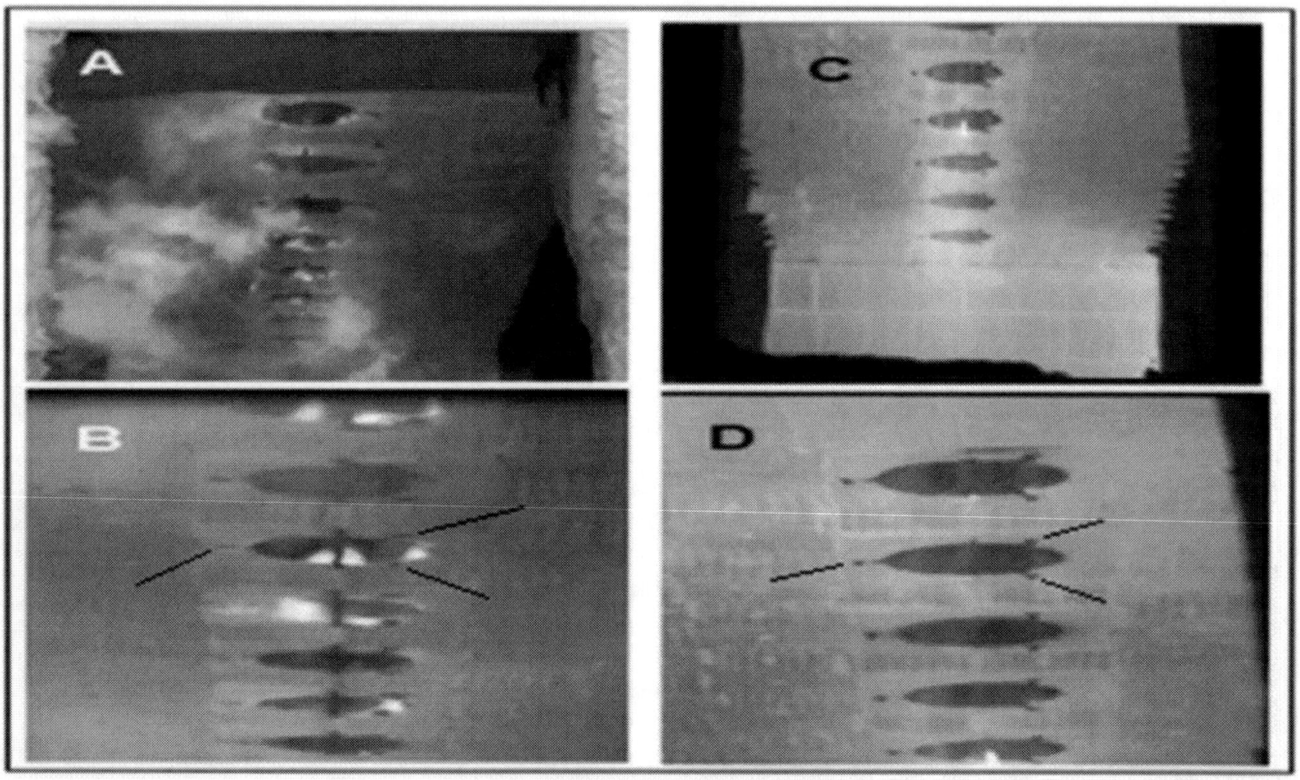

Figure 12. A-May 2016 showing erratic burner firing, B-May 2016 showing secondary burner tips (3 ea. typical) recessed in tile, C-Jan. 2016 showing normal burner firing, D-Jan. 2016 showing normal projection of secondary burner tips (3 ea. typical) through tile

Metallographic Replica Testing (MRT) done on the CADs during the January 2017 TAR revealed moderate spheroidization within the perlite grain structure of the duct which is consistent with heat degradation. As per API 571(Damage Mechanisms Affecting Fixed Equipment in the Refining Industry), spheroidization is generally observed to occur in steels that have been exposed to temperatures of 850°F to 1400°F (455°C to 760°C). Based on the high flue gas temperatures trended within the convection/transition section by temperature indicators TI-A/B/C/D/E/F after the ID Fan was shut down, it was surmised that the CADs was exposed to this temperature range of 850°F to 1400°F (455°C to 760°C).

Recommendations

To prevent a reoccurrence, the following recommendations were made:

1) Adjust the Cold and Hot Air Doors Logic: During a controlled plant shutdown of the Primary Reformer, the FD Fan is taken offline. Control logic causes both the Cold Air Door and Hot Air Doors to open. The induced draft fan would then remain online to assist in cool down of the entire reformer train. In the event the ID Fan trips during a plant shutdown, control logic should automatically close the Hot and Cold Air Doors to mitigate against the potential for reverse flow through the CADs.

2) Amend Standard Operating Procedures (SOP's):

   Restart of the ID fan is now included in the SOP if the ID fan trips during a plant shutdown. This would allow for controlled cooling of the primary reformer train from the radiant section through the convection section. In the event the ID fan cannot be restarted, ensure that the Hot and Cold Air Doors are closed until restart of the ID fan.

3) Install Temperature Indicators (TIs) on all six (6) CADs within the primary reformer penthouse:

Due to the absence of TIs within the CADs, the plant operators were unaware of the fact that the CADs were experiencing excursion temperatures during the plant shutdown. TIs were subsequently installed on the CADs with high temperature alarms of 750°F set to assist with operator intervention in the event of temperature excursions.

Conclusion

The consequence associated with temperature excursions within the primary reformer CADs can be significant as outlined in the aforementioned experience. The importance of continuous operator monitoring during plant shutdowns and start-ups can be crucial in identifying and addressing exigencies in a timely manner. Evaluation of installed instrumentation on the primary reformer train as well as a review of SOP's for primary reformer shutdown should be pursued in light of the experience related.

References

API 571 Damage Mechanisms Affecting Fixed Equipment in the Refining Industry.

ASME Section II: Materials - Part A – Ferrous Material Specifications

ASME Section II: Materials - Part D – Properties

Secondary Reformer Loss of Containment Due to Air Pipe Failure

A loss of containment scenario occurred on the Secondary Reformer in an Ammonia plant resulting in an emergency shutdown and a fire emitting from the top head of the vessel. The direct cause of failure was a cracked weld joint on the internal air distribution pipe. This caused a flammable mixture to leak internally in the secondary reformer before the burner tip and perpendicular to normal flow. Flame impingement on the internal components eventually deteriorated the liner, refractory, and outer wall. Further analysis indicated that cracking on the air distribution pipe was caused by thermal fatigue. This paper will present the incident and investigation findings as well as the actions that resulted.

Jenni Dickson
Koch Fertilizer Enid, LLC

Daniel Shelley, PE
Koch Fertilizer Enid, LLC

Introduction

During normal operations of Ammonia 1 plant, the Secondary Reformer vessel shell experienced a loss of containment which resulted in a total plant shutdown as well as damage to the surrounding area. This paper will discuss the failures that occurred, investigation into the cause and actions that resulted.

Background

The ammonia unit is an original Kellogg 1,000 short TPD Ammonia plant built in 1972.

The process flow enters the chamber at the top of the Secondary Reformer and is directed downward through the diffuser ring towards the combustion zone of the air mixer. Preheated air is introduced into the process through a nozzle located just below the diffuser ring. Auto ignition takes place in this zone and this provides the heat input needed for the endothermic reaction to take place across the secondary reformer catalyst.

A waste stream (approximately 5 GPM/ 1 m^3/hr) of the process condensate removal section is injected into the airline as well to be burned. See Figure 1 for a schematic of the process flows.

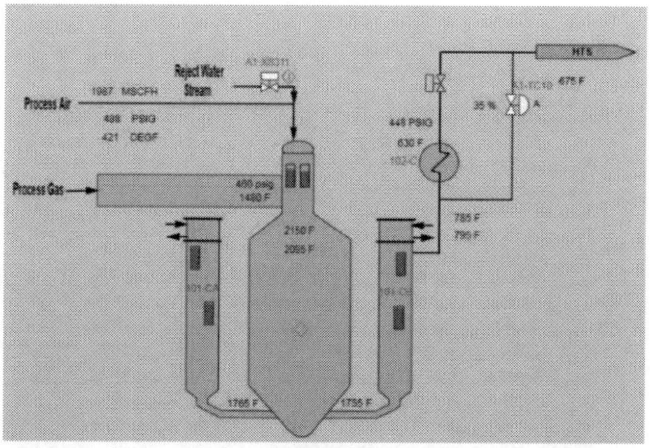

Figure 1. Secondary Reformer Equipment Diagram

This vessel shell is SA-516-70 carbon steel (1-3/8-inch wall thickness with 0.125-inch corrosion allowance). The shell has internal bubbled alumina refractory. The refractory is internally lined with Incoloy 601 material. The design conditions for the top head are 561 psi and 950 °F (39 bar and 510 °C). The vessel below the flange bottom is designed for 486 psig and 400 °F (33.5 bar and 205 °C). The lower part of the secondary reformer vessel is also water jacketed to provide additional heat protection.

Plant Health Monitoring

Between October 23-28, 2017, there was an observed step change increase in the differential pressure across the secondary reformer of approximately 10 psi (0.7 bar). The following events showed up on the monitoring system at about the same time:

- middle catalyst bed temperature dropped approximately 100 °F (55 °C).
- increase in differential pressure across the secondary reformer
- upper bed temperature dropped approximately 20 °F (11 °C)

The trends for the differential pressure are shown in Figure 2. The differential pressure across the vessel had historically followed the production rate. However, during this period there was a deviation from the normal trend.

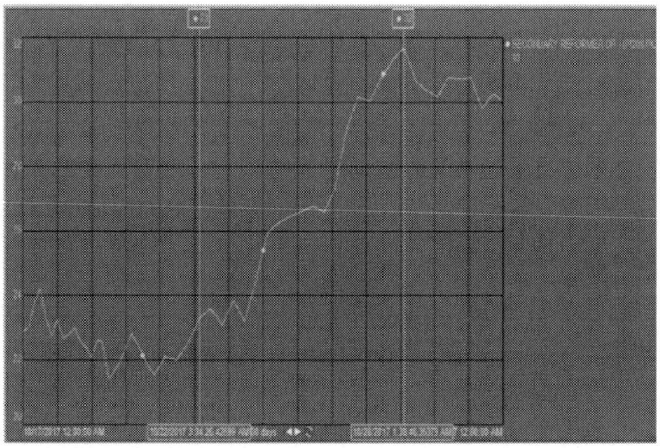

Figure 2. Graph of Secondary Reformer dP

The site process team, operations team and subject matter experts were consulted at the time of the increase in differential pressure. Historically, there had been multiple observed anomalies on the tip of the air mixer. In these cases, similar process conditions and process changes were observed. Based on the historical conditions, current performance and perceived risk, the decision was made to continue production.

Incident Summary

On November 10, 2017 at approximately 11:48 am, a rocket-like noise could suddenly be heard throughout the facility. Operations witnessed fire emitting from the side of the secondary reformer in the #1 Ammonia unit. An emergency shutdown and plant evacuation were initiated. The unit was successfully depressurized and brought to a stable state for further investigation.

Equipment damage occurred near the top N/NE side of the secondary reformer shell just below flange bottom including damage to the penthouse walls / insulation, instrument conduit, and charred valve handles. Inspections were performed (including hardness testing) in conjunction with equipment repairs.
Cost impact estimates was $1.05MM (lost production/schedule, property/equipment, labor/time). No other occupational safety or environmental incidents were reported.

Investigation

A hole was found in the side of the secondary reformer pressure shell.

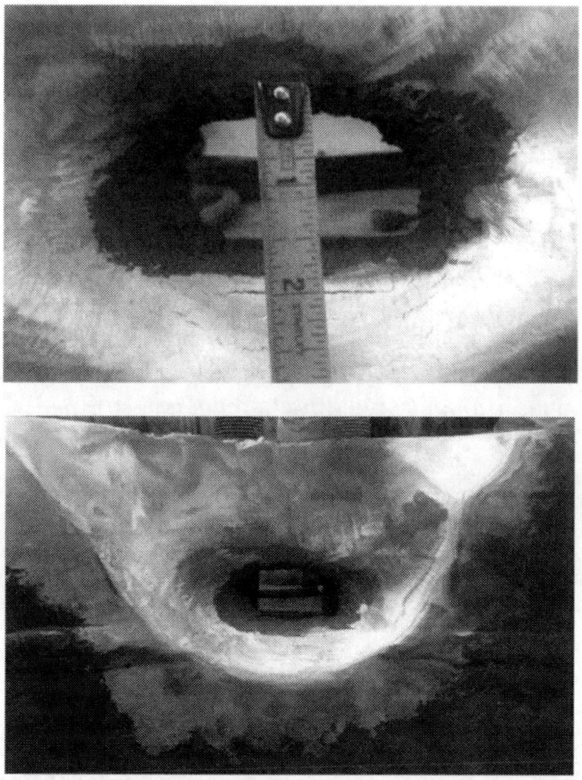

Figure 3. Hole in the side of the Pressure Shell

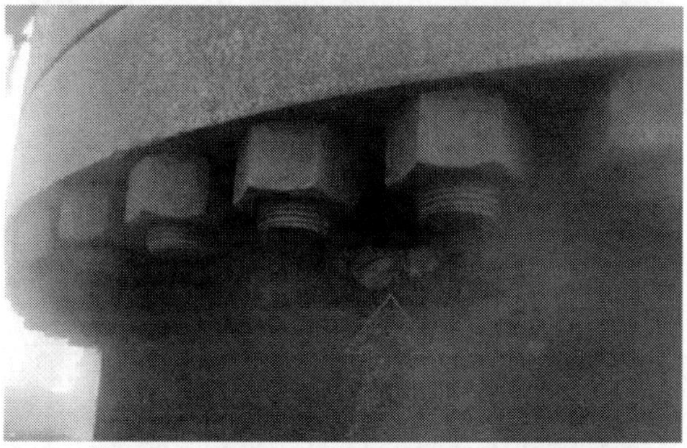

Figure 4. Lower Flange of Top Head

As the top head was removed, a circumferential crack in the air pipe was immediately apparent.

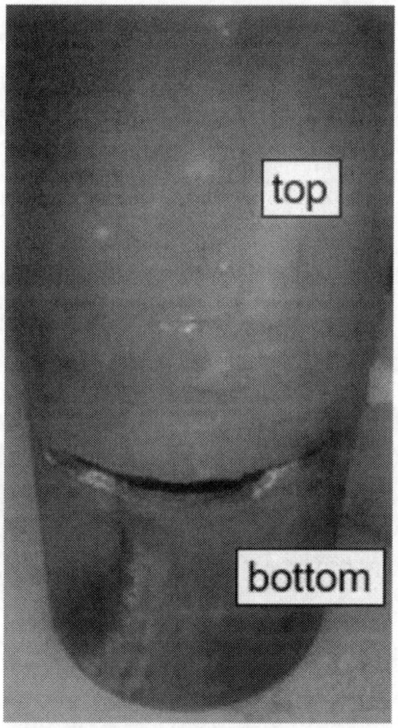

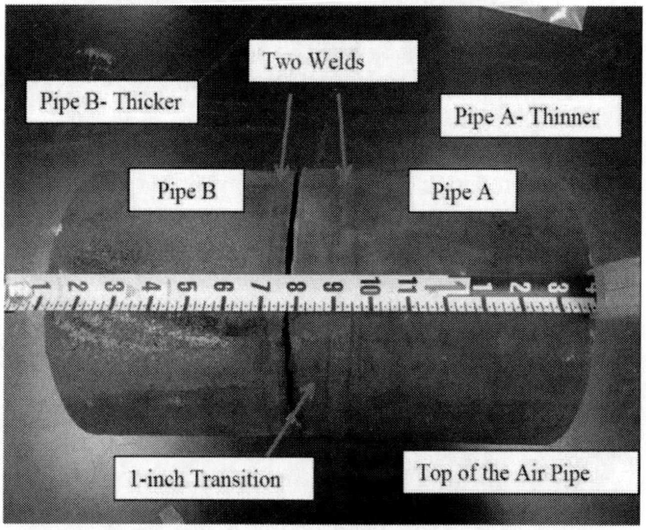

Figure 5. Cracked Air Pipe

The bottom end of the pipe was still attached by a tiny final area. It broke the rest of the way upon removal while setting the head assembly on the ground.

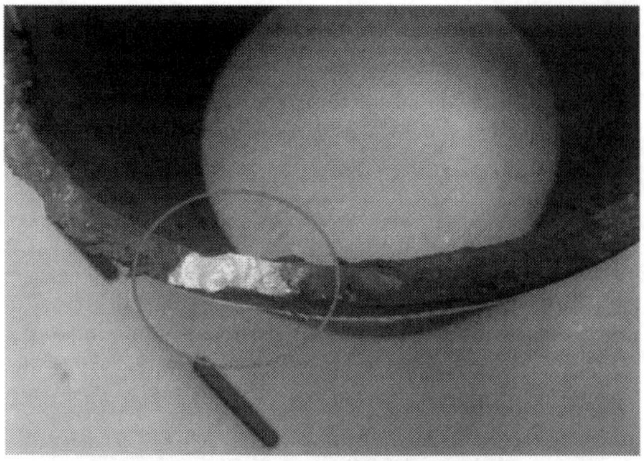

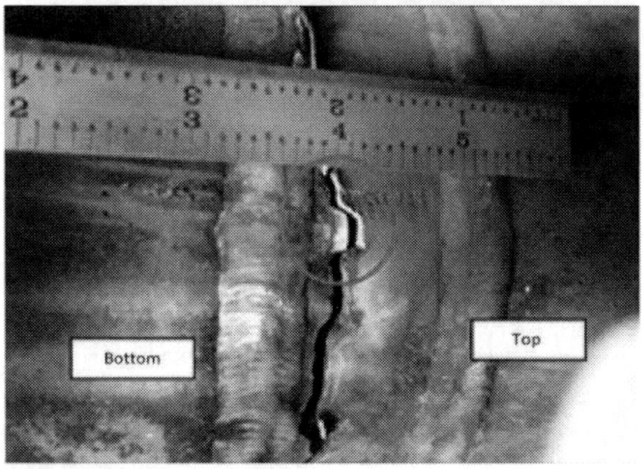

Figure 6. Final Fracture Location – Cracked Air Pipe

The presence of this circumferential crack caused premature ignition perpendicular to normal flow. Flame impingement eventually destroyed the liner, refractory and breached the pressure shell.

Root Cause

Minimal QA/QC was performed for this repair (prior to failure) because it was improperly categorized as a non-pressure pressure containing internal component.

The air pipe developed thermal fatigue cracks in the thinnest section of the transition piece between the welds. The cracks were caused by differential thermal expansion restrained by the dissimilar pipe sizes and non-smooth weld zones.

Thermal cycling normally occurs during start-up and shut-down of the system but also potentially by the introduction of reject water stream through the air pipe into the burner tip. The injection point is about 6 feet above the crack site. The Secondary Reformer was cycled (brought in and out of production) approximately 24 times in the life of this air pipe from 7/2015 to 11/20/2017.

There was a lack of complete fusion in the weld which accelerated cracking, however, some thermal fatigue cracks initiated and propagated entirely outside of the weld zone suggesting the same failure would have eventually occurred even if quality welding had been performed.

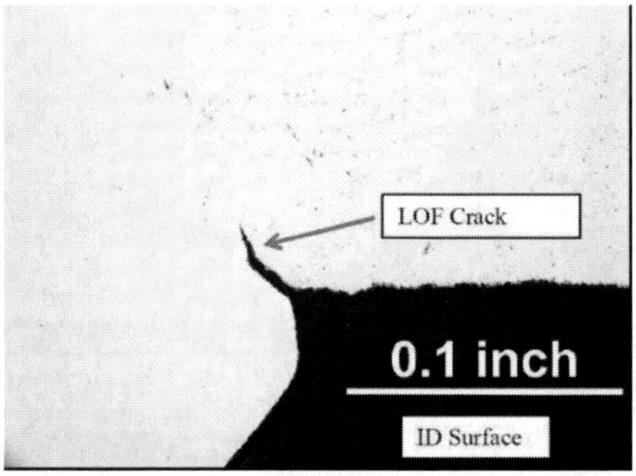

Figure 7. Cracks (Lack of Fusion)

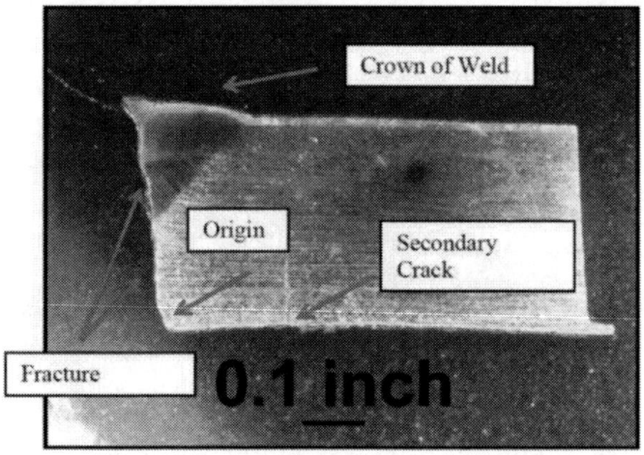

Figure 8. Cracks (Thermal Fatigue)

Though intergranular cracking was found on the inner diameter (ID) of the air pipe, it was local to the surface of the pipe and did not seem to contribute to the failure.

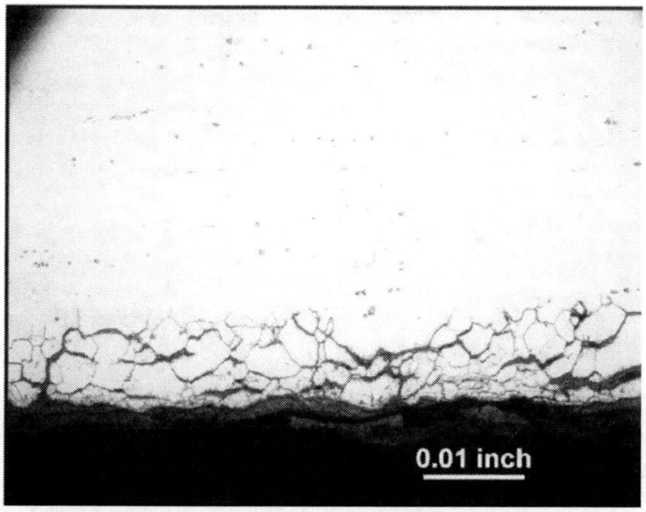

Figure 9. Intergranular Surface Cracking

Conclusions

Strict quality control (QA/QC) is required for repairs on a system like this. We recommend following the original designers' notes for this joint.

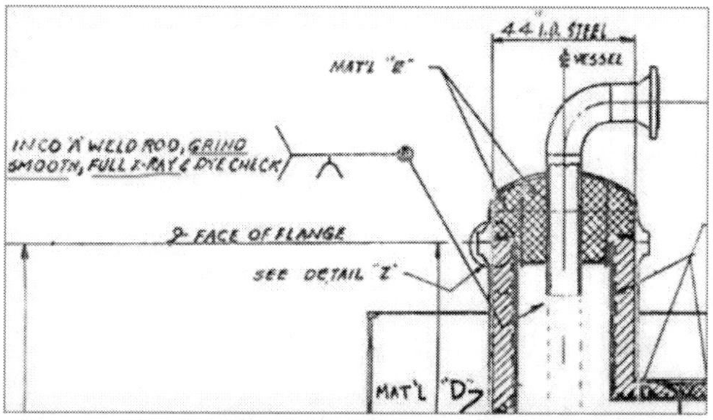

Figure 10. QA/QC Requirements from Original Design

A quality repair has since been successfully completed. Key design considerations were as follows:

- match pipe size per design
- fully x-ray welds
- ensure proper spacing of welds (outside of heat effected zone)
- grind welds flat and smooth
- This air pipe is part of a nozzle assembly in a burner system and is not classified as a simple sparger or distribution pipe. Appropriate non-destructive evaluation (NDE) is required due to high risk associated with premature intermixing and combustion of the process fluids.

Every practical effort should be made to reduce the number of start-ups / shut-downs to limit thermal cycling.

Surface intergranular cracking seems to regularly occur with Incoloy 800, but this mechanism has not historically resulted in unacceptable service life of the liner or air pipe. Incoloy 600 may resist this type of surface cracking but costs more. Incoloy 800 appears to be the preferred design material for this application.

Ensure a formal risk analysis or RCA is performed on significant process changes to avoid overlooking hidden failure modes. While it is good to rely on past experience, be careful to not rule out alternate outcomes without sufficient evidence or investigation. This proactive approach to abnormalities will assist in identifying and dealing with issues before they arise.

References

Benac, Daniel. *Failure Analysis Report - Koch Air Pipe_05789-012 -17*. Baker Engineering and Risk Consultants, Inc. December 8, 2017. 3330 Oakwell Court, Suite 100, San Antonio, TX 78218-3024.

Inconel weld damage and metal dusting in NH_3 front end equipment

This paper will describe the findings from an internal inspection of the transfer line connecting to a front end superheater, both prior to and after superheater bundle removal.

This paper will also discuss the failure analysis results, as well as the metal dusting found during the inspection and subsequent repairs. It will further discuss warning signs leading up to the turnaround and being prepared for potential repairs. Recognizing the risks of hot spots found during routine monitoring or operations and the importance of infrared monitoring routines and their significance in ensuring continued integrity.

Chris Butlin
Nutrien
Fort Saskatchewan, Alberta, Canada

Samuel Okulaja
Nutrien
Calgary, Alberta, Canada

Introduction

In 2018 Agrium completed a merger with PCS (Potash Corp. of Saskatchewan) to form a new company called Nutrien. The Nutrien Fort Saskatchewan Nitrogen Operations facility was commissioned in 1983 and produces ammonia and urea fertilizer for the Western Canadian and export markets. The site includes a nameplate 1000 metric tonne per day Kellogg ammonia plant and a nameplate 907 tonne per day Stamicarbon urea plant.

The subject transfer line and front end superheater are an integral part of the ammonia plant and were part of the original plant commissioned in 1983.

Background

The first main component of discussion is the transfer line from the primary front-end waste heat boiler (101C) to the superheater (102C). This transfer line has a plant equipment number of 112D and will be referenced as such from this point forward. 112D has been in operation since original commissioning in 1983 (35 years in 2018).

The 112D transfer line is constructed of a carbon steel shell (SA-516-70), that is water jacketed. The internal surface is protected by a 3.75" refractory layer which also has a 20" ID protective shroud or liner. Both the original and repair refractory materials are Alumina based. See Table 2 for the product break down. This liner is an Inco 800H material with a thickness of 0.250" (6 mm).

The second main piece of equipment referenced in this paper, is the superheater. This Sulzer superheater unit has a plant equipment number of 102C and will be referenced as such from this point forward. The 102C shell has been in operation since original commissioning in 1983 (35 years in 2018), and the bundle had been in operation since 2006 (12 years in 2018).

The 102C superheater is a shell and tube exchanger in which the process gas moves from the 112D transfer line, through a refractory lined conical section and into the bottom of the tube bundle and rises vertically amongst the tubes to its exit. The heat from the process gas then transfers to the boiler feed water on the tube side as part of the high-pressure steam system. The shell side of 102C is designed to ASME Section VIII Div. 1, and the tube side is designed to ASME Section 1.

The primary shell section of 102C is protected by an internal annulus that channels the hot process gas through the tube bundle before it can exit onto the high temperature shift converter. The flow of this gas can be controlled or throttled by an instrument bypass valve at the top of the bundle. The bundle is comprised of two tube materials, one being SA213 T22 (2 ¼ Cr-1 Mo) and the second is SA213 TP347 (Type 347). These two tube materials are welded together using ERNiCr-3 (Alloy 82) filler material.

At the bottom of this shell, is a conical section and a vertical section which is connected to the 112D transfer line. Like the 112D transfer line, the vertical and cone sections are refractory lined with an Inco 800H protective shroud, though not water jacketed. The 102C shell material of construction is SA-387 Grade 12 Class 2 (1Cr-0.5Mo).

Internal access to the horizontal run of the 112D transfer line, and bottom riser and cone section of 102C is achieved through removing a blind flange that has a refractory filled "plug". This manway plug is designed to seat inside the 20"ID transfer line. The remaining void between the plug and transfer line liner is filled with a manually applied "KAOWOOL" ceramic fiber blanket insulation during every reinstall. The plugs metallic can material is also Inco 800H.

Without removing the 102C bundle, the only internal inspection access to the bottom cone and bundle is through the 112D manway and transfer line. The 20" ID access makes repairs to the vertical section of the liner or shroud and accessing the cone section extremely difficult.

Online Inspection Activities

After an incident at the Nutrien Redwater (formerly Agrium) facility in 2007 involving refractory lined equipment, Infrared Thermography (IR) routes were introduced at legacy Agrium facilities. The routes initially were conducted on a weekly interval, but eventually settled on an interval of one month. The primary purpose of the IR route is to look at potential refractory degradation by way of locating hotspots in the equipment. We have also found that this is a good tool for indication of liner damage or loss as well.

The bottom portion of 102C which includes the conical and riser section to 112D is included in the IR route conducted at the Fort Saskatchewan plant. Unfortunately, due to the water jacket configuration on 112D, there is no way to accurately inspect the transfer line while the unit is online. The only way to assess 112D is internally during an outage.

Equipment	Design Temp	Operating Temp	Design Pressure	Operating Pres.
112D	1480 F (804 C) *	700 C (1292F)	550Psi (3792Kpa)	3405Kpa (494Psi)
102C Shell Side	800F (427 C)	700 C (1292F)	550Psi (3792Kpa)	3405Kpa (494Psi)
102C T/S (BFW)	910F (488 C)	432 C (810 F)	2000Psi(13790Kpa)	12200Kpa(1769Psi)

*Note: 112D Design temperature is the designated working temperature by design on the drawing, not design metal temperature. 112D Design metal temperature is 500F (260C), and this temperature is used as an "Alert" temperature on the 102C riser and cone.

Chemical composition of 112D process gas (Secondary Exit (mole%);

H_2	Ar	N_2	CO	CO_2	CH_4
55.54	0.28	23.45	12.09	8.26	0.38

Table 1. Design and operating conditions.

Original Refractory; Firelite – 95 Refractory Concrete

Bulk Density dried to 110C	Thermal Conductivity @ 800C	Component Elements
1.73 g/cc (0.0625 lb/in3)	0.570 W/m-K (ASTM C 417)	Al_2O_3 (94.5%), CaO (4.3%)

Repair Refractory; KAST-O-LITE 97L Precast Brick

Bulk Density dried to 105C	Thermal Conductivity @ 871C	Component Elements
1.46 g/cc (91 lb/ft3)	0.87 W/m-C	Al2O3 (94.4%), CaO (5.1%)

Table 2. Refractory Information

2014 Inspection Results/Repairs

112D - As part of a major turnaround (T/A) in 2014, the 112D manways were opened to allow for an internal inspection. The internal Inco 800H liner was found to be "worn" through in two locations exposing the refractory behind the liner directly to process. This is identified in Figure 1.

Figure 1. 112D – 2014 As found condition

During this T/A, the horizontal run of 112D was repaired by removing the liner and installing precast brick to replace the refractory. A new section of Inco 800H liner was then installed over the entire horizontal run. The completed transfer line repair in 2014 is identified in Figure 2.

102C – In 2014, the bundle was not removed to allow internal access to the shell and bottom cone section. The bottom of the bundle was inspected visually from the riser off 112D. The protective liner was found to distorted and beginning to "buckle" at the cone. The connecting welds were also found to be in poor condition. The as found condition in 2014 is identified in Figure 3.

The decision was made by site leadership that no repairs of the 102C cone and riser were to be conducted in 2014 due to timing and scope. The bundle itself was hydrotested as part of the front end high-pressure steam system due to other work being conducted.

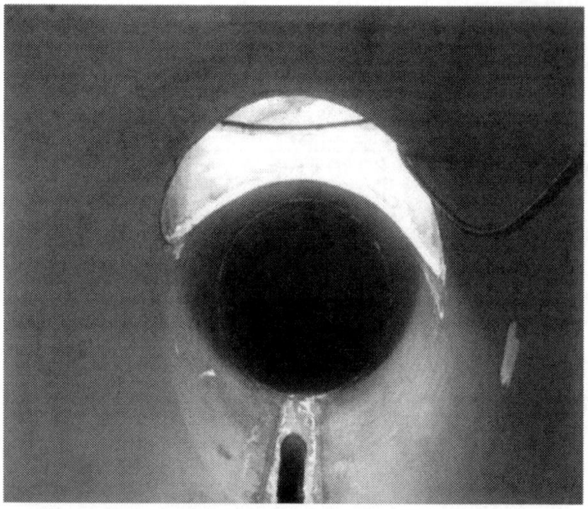

Figure 2. 112D – 2014 Repaired condition

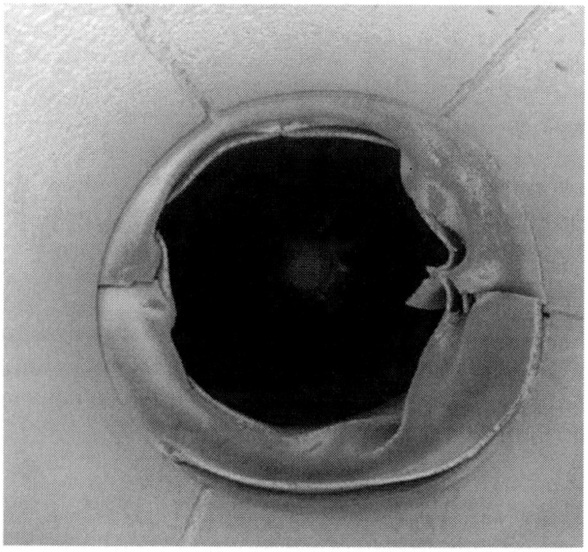

Figure 3. 102C – 2014 Riser/Bottom cone

2017 Early Indication of Problems

In November of 2017 during the monthly IR route, the unit inspector discovered a hot spot on the riser between 102C and 112D. The highest skin temperature recorded in November 2017 was 777°F (413°C). The frequency of IR was increased immediately, and temperature readings progressively got hotter. For example, on February 27, 2018 the recorded highest temperature was 934°F (501°C). The initial finding in November 2017 was roughly seven months from the planned 2018 T/A start date. This hot spot is shown in Figure 4.

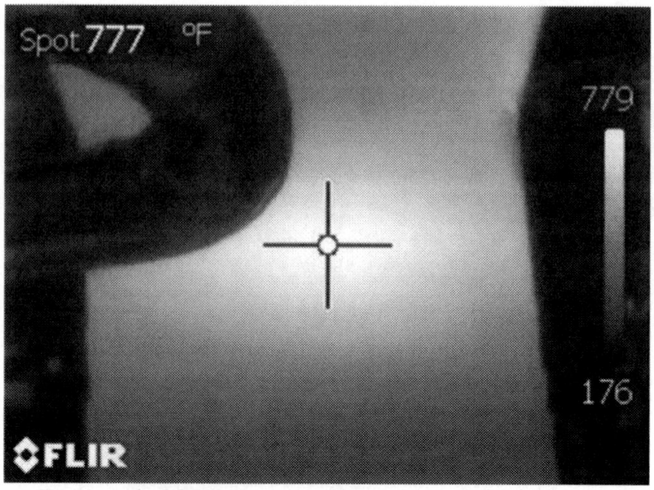

Figure 4. 102C Riser – November 2017 Hot Spot

Once the hotspot was initially discovered, steam was applied to the area to attempt to cool the location. A Fitness For Service (FFS) was completed and the maximum allowable temperature was determined to be 950°F (510°C) compared to an original design maximum temperature of 800°F (427°C). Thermocouples were also applied directly on the surface as the use of steam was interfering with the IR results and monitoring. This was considered a safer option, as the temperature was then recorded continuously. The area was monitored, and the steam was left on until the unit was brought down for T/A.

2018 Inspection Results/Repairs

112D – During the scheduled 2018 T/A, the manway plugs were removed from 112D to allow internal inspection access. The initial inspection found that the horizontal liner section was in very poor condition with multiple locations deteriorated and exposing the refractory. At the transfer line section connecting to the riser to 102C, there appeared to be pieces of liner physically missing. The as found condition of 112D is shown in Figure 5.

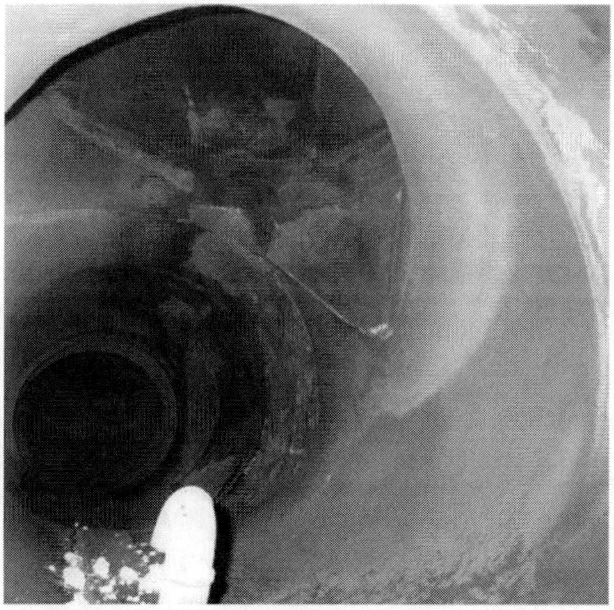

Figure 5. 112D – 2018 As found condition

Figure 6. 102C 2018 Riser, Internal view of hot spot

As in 2014, this horizontal run of the 112D transfer line had its entire liner and refractory replaced. The liner was replaced with Inco 601 instead of 800H to address the continuous damage found each T/A. Also, the manway plug was repaired, as some of the welds on the plug were broken. The Inco 800H material was also beginning to distort. It should be noted that this liner section had been repaired or replaced in previous T/A's before 2014 as well.

102C – As in 2014, the inspection began with the riser between the 112D transfer line under the 102C cone. At the location of the hot spot being monitored, a section of liner had broken off and there were significant amounts of refractory missing. This is identified in Figure 6.

It was also noted that liner at the top of the riser as it transitioned into the cone was severely damaged and had begun to constrict the opening into 102C. The cone section liner pieces were also broken apart inside the cone. In some cases, the sections of liner had broken free and were only secured by their welds at the top by the bundle itself. This is shown in Figure 7.

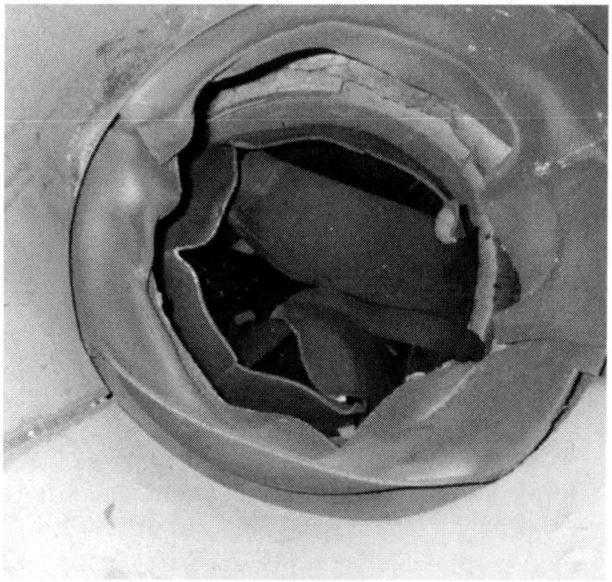

Figure 7. 102C – 2018 Riser/Bottom cone

A mechanical contractor began removing the liner and refractory to allow access into the cone section. This work was completed in a very confined environment. Once the sections of liner were removed, the inspection group could enter the cone section and assess the condition of the 102C bundle. The bundle was found to have refractory debris, and multiple pieces of the 800H liner lodged amongst the tubes. An example is shown in Figure 8.

The return bend sections of the tube bundle were found to have sustained mechanical damage from the debris striking the tubes while in service as shown in Figure 9.

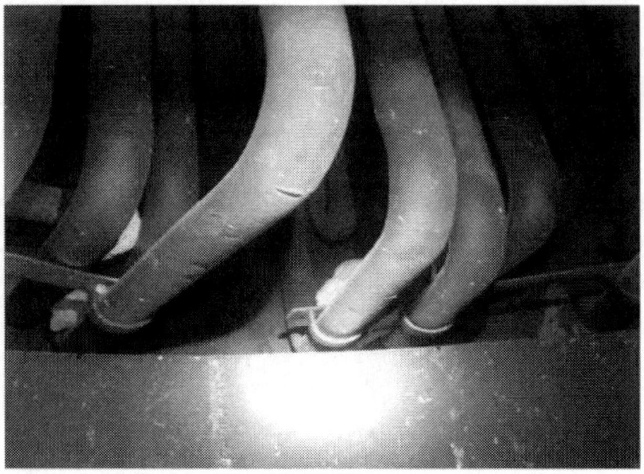

Figure 8. 102C – 2018 Damage from debris

Due to the mechanical damage on the tubes, the decision was made to remove the bundle from the shell and install a spare in kind bundle with the same metallurgy and welds, that was being kept in storage at the plant site. This bundle removal involved cutting the BFW inlet/outlet piping on the bundle (this is considered ASME Section 1 Boiler Proper).

The bundle removal also allowed for a more thorough and detailed inspection of the cone section due to access. Because of the high temperatures while in service, in situ metallography was performed on the base material of components once refractory was removed. This was done to ensure the material had not sustained any permanent damage. The riser section between the cone and 112D was replaced entirely due to the high temperatures on the original hot spot with the same material.

Figure 9. 102C – 2018 Liner plate lodged in bundle

Once the bundle was removed, the inspection of the bundle found that the bimetallic weld previously described was in poor condition on multiple tubes and close to failure. This bimetallic weld had failed in 2006 in the previously installed bundle. A failure analysis in 2006 concluded that the weld had suffered metal dusting. The condition of the welds in 2018 had very similar damage as the same welds in 2006. An example of the dissimilar weld is shown in Figure 10.

The total repair scope for 102C completed in 2018 was as follows;

- Bundle replacement
- 100% Refractory removal/replacement in bottom section.
- 100% replacement of all refractory pins.
- New riser between cone and 112D, which included a hydrogen bake out of existing material, PWHT and 100% MT/RT.
- New Inco 800H liner in cone and Inco 601 in the riser connecting down to 112D. (100% replacement)
- Weld repair of internal annulus due to deterioration at joints.
- 100% replacement of all refractory pins.

- Tube side hydrotest once new bundle was welded in place (ASME Section 1 Boiler Proper).

Figure 10. 102C – 2018 Removed weld

102C Bundle Dissimilar Weld Failure Analysis

As in 2006, multiple samples of the dissimilar welds were removed from the bundle for analysis. The results of the failure analysis were consistent with the 2006 results and it concluded that the bimetallic weld had suffered preferential attack due to metal dusting. The effect of metal dusting on the ERNiCr-3 weldment is visible in Figure 11.

Metal Dusting and Cause

Metal dusting, the determined damage mechanism, is a form of carburization resulting in accelerated metal loss in the form of localized pits. Metal dusting occurs in process streams containing Carbon and Hydrogen. Surface pits in metal dusting may contain soot or graphite dust, as was the case with the affected 112D material and bimetallic weld in the 102C bundle.

Initially it was thought that process changes caused such excess metal dusting between 2014 and 2018. This was due to the condition of the horizontal liner in 2014 vs. 2018. Also, in 2014 a PRISM hydrogen recovery system was installed in the back end of the Ammonia plant. We initially thought maybe this had changed the operating conditions enough to cause the extent of the metal dusting that was found in 2018. However, there was no significant process change between pre- & post PRISM installation. The operating temperature and pressure were lower than before the 2014 PRISM installation. Also, the process gas composition is comparable in both operating scenarios, although the Ammonia plant did produce ~25 metric tonnes per day more after the 2014 PRISM installation.

Though metal dusting is known to attack Nickel based alloys, as well as low alloy steels and 300 series SS, the Inconel weldment in the bimetallic weld seemed to be prone to preferential attack. The tube parent materials of SA 213 T22 and SA 213 TP347 appeared to be spared from metal dusting but the filler material or weldment of ERNiCr-3 (Alloy 82) was affected in this service environment.

Future Mitigation Strategies

112D – There had been discussion surrounding a possible material change in the transfer line section that continues to deteriorate due to metal dusting. The issue of high process temperatures of ~700°C (1292°F) at that location is the primary concern in terms of material options. Metal dusting is known to occur in the range of 482°C(900°F) – 816°C(1500°F). During the 2018 T/A, portions of 800H liner were replaced with Inco 601, specifically, the horizontal run of 112D and the riser up to cone section. This approach of using Inco 601 material with higher nickel and chromium content was taken to see if the liner would be able to sustain the process conditions for longer periods than the 800H.

102C – With the dissimilar weld, a spare bundle is currently being designed and the location of the weld is being reviewed. Also, the discussion of removing the weld all together has been put on the table. Removing the weld would then require an increased cost in tube material for the course of the bundle. Ideally removing the weld all together would be the best course of action as the SA 213 TP347 and SA 213 T22 materials have both performed well in this service environment. In the case of leaving a bimetallic weld in the bundle, the filler material is also being reviewed.

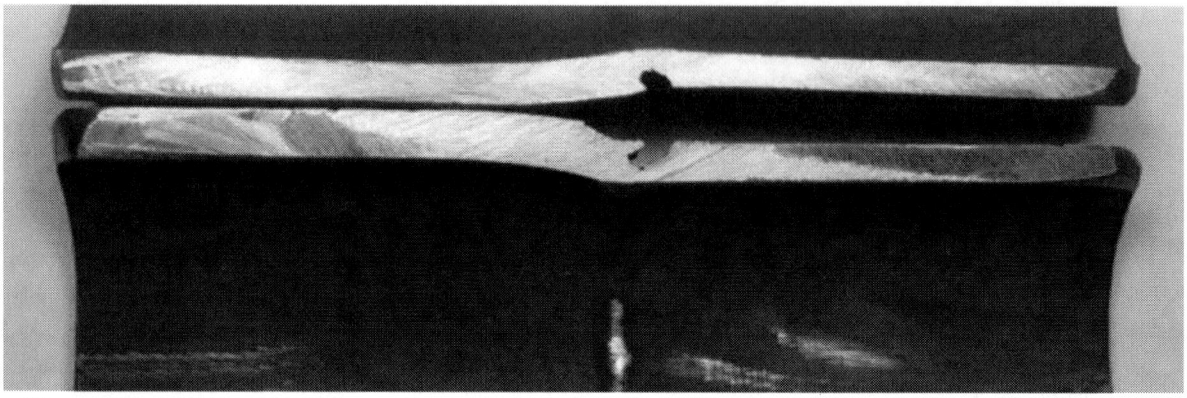

Figure 11. 102C – 2018 Cross section of dissimilar weld showing Metal Dusting of ERNiCr-3

Conclusion

Through multiple meetings and discussions with process engineers and operations leadership, there has not been a "smoking gun" in terms of what may have caused this issue operationally. Historical records indicate that in 2012, 2010, and previous outages before that, the horizontal section of 112D and the change of flow into the riser to 102C have been problem areas in terms of maintenance and repair. This could be due to a possible process gas turbulence or "eddy effect" while in service.

Infrared or IR routes have proven to be very valuable when monitoring our in-service refractory lined equipment. It has served as an early warning system and enabled our inspections group to initiate plans and repair scenarios for equipment. In the case of refractory lined equipment with a metallic liner such as 112D and its 800H, if a hot spot is found, it may be an indication of lost or missing liner material. This should prompt you to plan to inspect downstream equipment for the same mechanical damage that was found in the 102C bundle. This may mean having to expand inspection scope in the planning phase of your T/A.

When planning for T/A's and outage repair work, it is favorable to have any required materials on hand and formed to expedite any required field work. Inco 601 and 800H plate materials may be long lead items depending on location and forming them to suit liner applications is also required once you've received them. Materials Engineers should be consulted when choosing materials to combat metal dusting, such as possibly using Alloy 693. Planning for the worst could save you days in the end.

Safe Startup for World Scale H2 Plants with Topsoe Furnace Manager

Large, world-scale H2 plant steam methane reformer (SMR) fireboxes represent safety challenges due to size and complexity, such as personnel safety issues while interfacing with the firebox during the startup and light off of hundreds of burners

Ken Wohlgeschaffen
Chevron

Scott Sexton
Haldor Topsoe Inc.

Introduction

In 2018, Chevron finished construction and started up two, world-scale hydrogen plants in one of its refineries.

Hydrogen (H2) supply is vital to the refinery operation, and steam production from the steam-methane-reformers (SMR's) is a very important utility to the refinery. Simply, it is critical that the new hydrogen plants were constructed and operated reliably and safely.

The heart of these plants were the reforming furnaces. Full understanding of the design and operation of the furnaces was necessary for seamless integration of the new hydrogen plants. Furnace protection is especially important from a business asset perspective since this equipment represents about 35% of the total hydrogen plant asset value.

Plant Configuration and Startup

The SMR's in the new hydrogen plants are top-fired, or down-fired, with burners in the arch only. The burners are relatively large to minimize the number of them and required a standard staged lighting across the entire SMR during initial startup to provide even heat distribution.

Startup of the SMR's utilized natural gas fuel only, with nitrogen (N2) circulation to remove process heat during startup. Steam for the process was self-generated by the SMR's, and available for circulation in the process once SMR heat up had achieved temperatures where adequate steam supply was available.

The new H2 plants were designed with the typical integration of pressure-swing-adsorbers (PSA) for purification of the syngas generated by the SMR's and front end of the plant. The PSA off-gas is rejected during PSA regeneration and recycled back to the SMR fuel system, providing a significant amount of fuel during normal operation. The balance of fuel is natural gas supplied by the local utility company.

Several hundred process control loop response checks, alarm checkout, and process loop tuning tasks were completed by experts assigned 24 hours every day leading up to initial SMR burner light off. During the initial burner lighting, and during the subsequent heat up and dry out of the SMR's, preparation was also made to have the PSA systems available as needed when syngas was produced. The coordination and integration of a large number of contractors and vendor suppliers' personnel with internal Chevron personnel required constant communication. Chevron recognized the numerous operation's tasks associated with large-scale SMR startup. (Figure 1)

H2 Plant Operator Physical Activities

Startup

- Satisfy interlocks & startup conditions
    - flame scanners, piping pressure integrity, firebox purge, process interlocks
    - Establish diluent (N2) flow
    - Establish boiler levels
- Light burners at heights above grade
- Adjust combustion air to burners
- Open/close valves at height & grade
- Start pumps & fans at grade
- Maintain boiler level at height
- View firebox for tube overheating
- Communicate with team
- Record process & equipment status
- Continue heat-up cycle for >6 hours
- Introduce steam after ~8 hours of heat up
- Back out diluent N2 from process & increase steam flow to process
- Introduce NG feed after ~12 hours & > 1300 F
- Line up process to flare
- Startup PSA: pressurize beds, sequence valves, & then start syngas feed flow & attain H2 purity
- Line up PSA off gas to SMR fuel & repeat some steps above without tripping plant or PSA offline!
- Increase plant rates to >50%

Figure 1. Operator Startup Duties on Large SMR's

To safely accomplish the SMR operational tasks for these new plants, Chevron utilized technology which would provide the right information to those who needed it at the time.

Firebox Technology Selection

Early in the construction process, Chevron H2 subject matter experts (SME's) were involved with operations to integrate technology for firebox management. Considering the physically large fireboxes, with burners located on the top of the firebox, the efficiency of operator surveillance and monitoring of the SMR's was evaluated. Several considerations were explored, including extent of coverage of the firebox by people looking into viewport windows, including firebox data collection, duration of firebox surveillance, and evaluation of results of burner or process adjustments.

Additionally, the frequency of firebox interactions by humans looking into viewport windows was considered. Minimization of this interaction was important from a personnel safety standpoint. Technology that could reduce this interaction frequency, and provide more persistent coverage of the firebox than short intervals (<1 minute in duration) of "eyeball viewing" through viewports were evaluated.

Haldor Topsoe worked closely with Chevron to integrate Topsoe Furnace Manager (TFM) into the SMR fireboxes to assist with firebox management. Chevron chose TFM in late 2017 for installation and startup in 2018.

Topsoe Furnace Manager (TFM)

Topsoe Furnace Manager (TFM) provides an alternative to direct human interaction with SMR fireboxes. (Figure 2) TFM is an array of permanently installed image collectors acquiring images every second, 24/7. TFM collects millions of data points annually, with deviation alarms, stored in an easy-to-use historian, and is remotely available to the organization outside of the plant. TFM monitors the firebox with minimal safety risks. The reduction of human-firebox interaction allows personnel to perform higher value work, such as image and data analysis, collaboration, and/or minor maintenance and turnaround planning.

Figure 2. TFM Installation—Image and Data Acquisition Units

In hydrogen plants, the SMR operation encompasses several hazards. Fire, loss of primary containment of process fluids, uncontrolled combustion, mechanical equipment failures, and process upsets are a few. Traditional hazard mitigation for SMR operation has required plant personnel to directly interface with the firebox to verify equipment and process status. This interaction has inherent risks. When personnel interact with a firebox to determine the status inside, they become human data collectors. It is recognized that this interaction is important, and tools have been developed to facilitate the best use of this interaction, including specialized personal protective equipment (PPE) and handheld data collectors. Even under the best of circumstances, the firebox interaction is typically a relative short duration, with only about 1000 hours or less of data collection annually. The amount of data collected in terms of temperatures and imagery is restricted by the duration of the personnel interacting with the firebox. To increase the duration of the interaction to improve data collection an increase in exposure to hazards occurs. Therefore, hydrogen producers are faced with a decision to sacrifice firebox data acquisition to reduce hazard exposure. Extrapolation and leveraging of a limited amount of firebox data is the result of the decision. This extrapolation of limited data is typically historical based, as is the operation and maintenance of the firebox. Notable firebox failures occur from gaps in historical knowledge.

TFM is a system-of-systems including electronics, optics, data handling, data analysis, placement design, visual display, self-protection, compliance, and human-machine-interface. Image acquisition is at the heart of the system. The image collectors send the images to the data collectors, which organize, transmit, store and analyze the data (Figure 2). Alarm safeguards are immediately available to the plant operators, and remotely to the organization outside of the plant supporting operations.

TFM Capabilities:

Image collection: Over 500,000 images per image collector per year

Data collection: Over 50,000,000 data points per collector per year

Alarms: Tube metal temperatures, with associated Integrity Operating Windows (IOW's), and burner flame quality, such as impingement, jetting, and poor mixing.

Benchmarked images of good examples and bad examples of burners and tubes

Historian: Time and date stamped images with data

Remote access to images and data

Wireless computer interface with firebox

24/7 persistence of the entire system

Self-protection for reliability

TFM Evaluation and Selection

Chevron entered into extensive discussions directly with Topsoe, and also worked through a large engineering firm with H2 plant expertise to fully evaluate TFM. Extensive evaluations of TFM were completed, including National Recognized Laboratory Testing (NRTL) certification for a Class 1, Division 2 rating. Commonly known as "UL or CSA" certification, NRTL certification requires extensive documentation and certification of all components in the system. Once registration with the NRTL is completed, documentation is submitted for initial design review. After NRTL design review, several rounds of testing and evaluation by a certified NRTL with qualified and certified safety inspection engineers is performed with the actual equipment. Following successful completion of the certification testing, then next step in the process involves on-site inspection and certification of the equipment manufacturing facilities.

Topsoe communicated with Chevron and the third party H2 plant engineering firm very frequently throughout the selection, testing and certification process of the TFM equipment and manufacturing facilities. Final certification by the NRTL was completed in Spring 2018.

Fastrack TFM Installation

When Chevron selected TFM in late 2017, the construction of the new hydrogen plants was moving on a fast-track basis to meet expected startup target dates in 2018. Simultaneous NRTL certification, TFM manufacturing, and TFM shipment for installation were ongoing. To manage this timetable, Topsoe, Chevron and the H2 engineering firm all worked from the same project schedule. Topsoe communications with Chevron operations was also important to ensure smooth implementation of TFM into the control room. Communications with all key stakeholders, including construction contractors, was obviously very important.

The first step in the installation process was selection of the position and location of the TFM image units on the SMR's. Topsoe reviewed placement options and visited the site working directly with SMR contractors so that the correct location of image units was chosen. Several considerations were taken into account including accessibility, OSHA compliance, proximity to utilities, and firebox view angles for final image unit placement on the SMR fireboxes.

The next step in installation involved mounting of the image units on the SMR fireboxes. This was completed in mid-2018 working with SMR contractors, Chevron, and third party H2 engineering personnel.

Finally, utilities were connected and continuity for image and data transmission was completed in the fall of 2018, in preparation for the first SMR light-off.

Topsoe fully integrated into the Chevron, contractor and third-party engineering team to make sure TFM was ready for the first burner lighting.

First Light off Flame Detection

The first hydrogen SMR burner lighting began in the fall of 2018. Startup of the first SMR also occurred while work finished on mechanical completion of the second plant.

In preparation for SMR burner light off, Topsoe installed TFM software and communications equipment working with Chevron and I&E contractors. An important consideration was proper placement and configuration of the keyboard-video-mouse (KVM) equipment so that TFM could be integrated into the operator's control room. Also, TFM remote viewing in the refinery's Hydroprocessing Division's main control room about a half-mile away was an important requirement involving extension of the TFM communication network.

During the initial light off of the first SMR, flame detection by installed flame scanners was intermittent. It was important to ensure that flame detection was reliable to prevent a furnace "bogging" (fuel rich) event. Installed flame scanners typically experience loss of signal, especially on the initial burner light off before they are properly adjusted and positioned. TFM provided a second source of flame detection, or second layer of protection, during the critical first burner lighting phase. (Figure 3).

Figure 3. First SMR burner light off

Burner Lighting and Heat Up

Burner lighting continued throughout the initial phases of heat up. TFM continued to capture images of all phases of this initial heat up period.

It was very important to have a quick view of all burners during the light-off period so that operator actions could be directed and organized.

TFM continued to monitor all burner rows during the initial heat up period ensuring that the burners were lit, had sufficient air (no after burning), and flames were not impinging upon catalyst tubes in the SMR. (Figure 4)

With continuous TFM burner monitoring, operators were freed to complete other important tasks in the plant, minimizing their ascent to the top of the SMR for burner and combustion air valve adjustments.

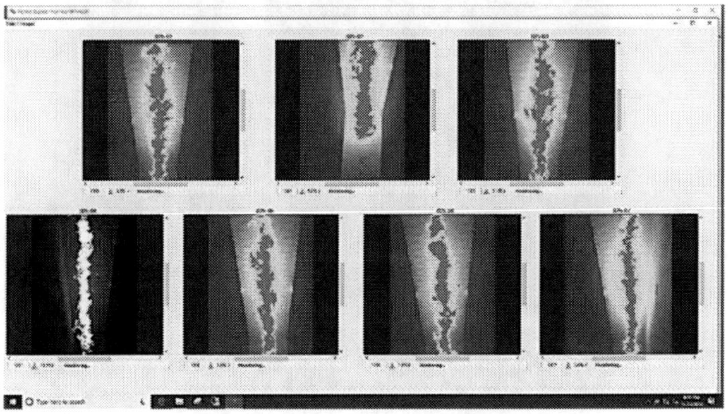

Figure 4. Burner Rows during Heat Up

Firebox Radiation After Plant Trip outs

It is normal, and to be expected, that during the very first startups of complex process equipment plant tryouts will occur.

For an SMR, it is critical to safety to understand the condition of the firebox during a plant tryout. With TFM, the firebox can still be "seen" with images even when burners have tripped offline. The radiation signature from the firebox continues to give off energy that TFM can detect.

This detection of the firebox radiation signature by TFM is important for safety to ensure that pockets of fuel are not present, and to ensure that flames are not present. TFM acts as an administrative safeguard confirming that the plant's burner management system (BMS) is active and has safely shut off fuel and flames. (Figures 5 and 6).

Figure 5. Firebox Radiation after Tryout Looking into Lower Viewport

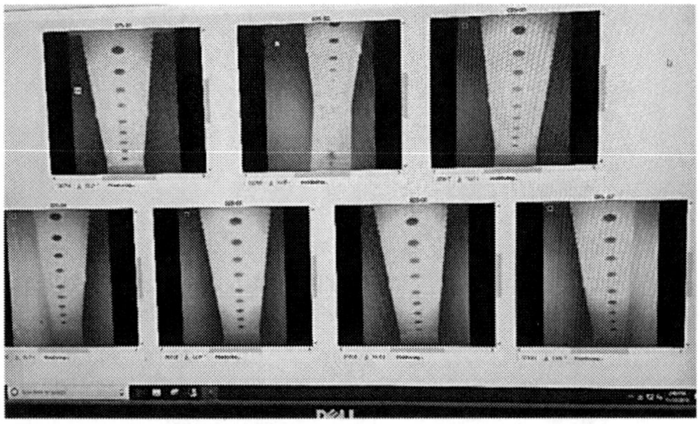

Figure 6. TFM Burner Row Views after Plant Trip out

Burner Heat Signatures

After the initial plant light off and heat up, the next phase of operations was a transition to normal operation with the PSA on line and the PSA off gas recycled back into the SMR firebox. During this transition period, N2 recycle was replaced with process steam. Then process natural gas was fed producing syngas which was then fed to the PSA after high temperature heat exchange, shift conversion of carbon monoxide (CO) to carbon dioxide (CO2) and H2, and cooling.

During this entire transition, TFM continued to monitor the firebox. Once the transition from natural gas fuel to PSA tail gas fuel occurred, the burner heat signatures appeared. (Figure 7)

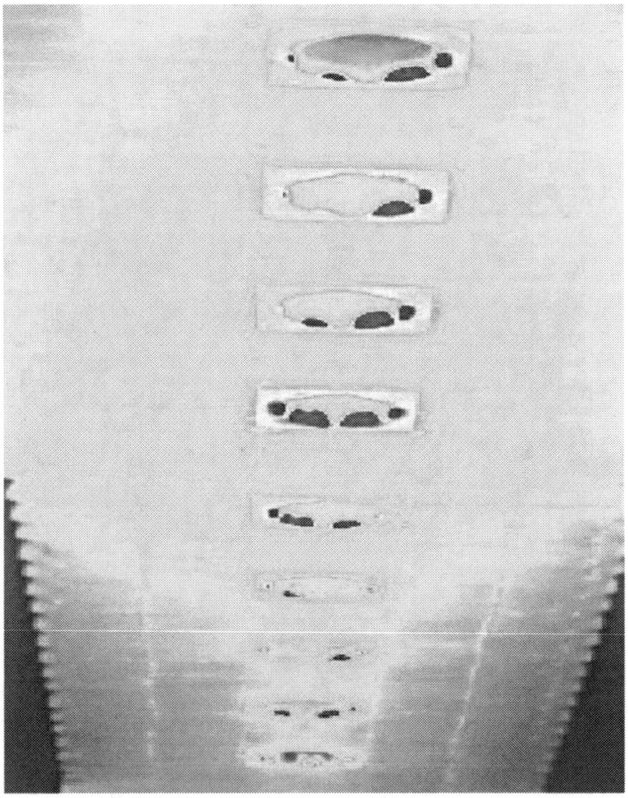

Figure 7. Burner Heat Signatures

TFM captures the subtleties of burner heat signatures, especially important at high temperatures. Without proper burner heat signatures, the mechanical reliability of the burners can degrade, as well as, the heat release and overall burner performance.

Multiple Inspections

After transitioning to normal operation with PSA off gas fed as fuel to the SMR firebox, TFM's movement functionality was utilized to physically view the lower part of the catalyst tubes operating at higher temperatures than the section near the burners. (Figures 8 and 9)

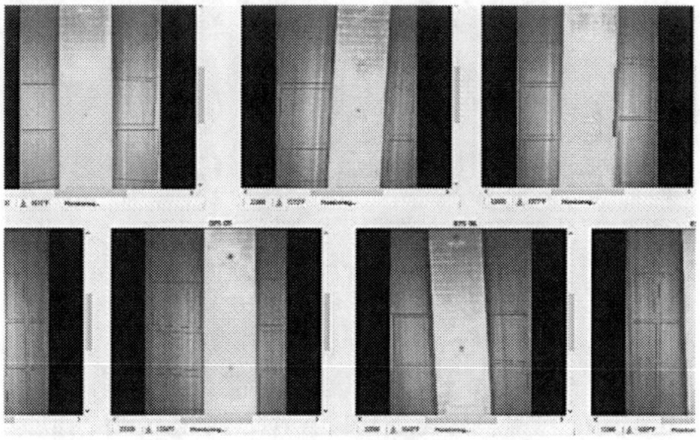

Figure 8. Catalyst Tube View

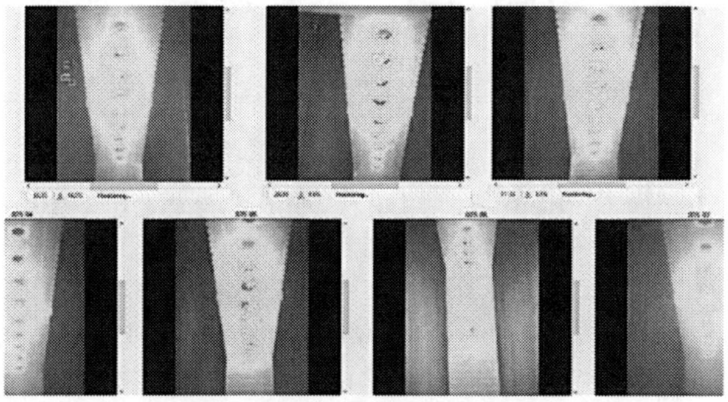

Figure 9. Burner View

Hot Tube Alarm

The ability to quickly see hot tubes is very important in SMR operation. Without persistent 24/7 coverage with imagery inside the firebox, it is not possible to quickly determine when and where hot tubes develop. TFM, with its persistent coverage of the firebox with multiple image collectors, captures the moment a hot tube develops and will also provide alarms to alert the operating team quickly (Figure 10).

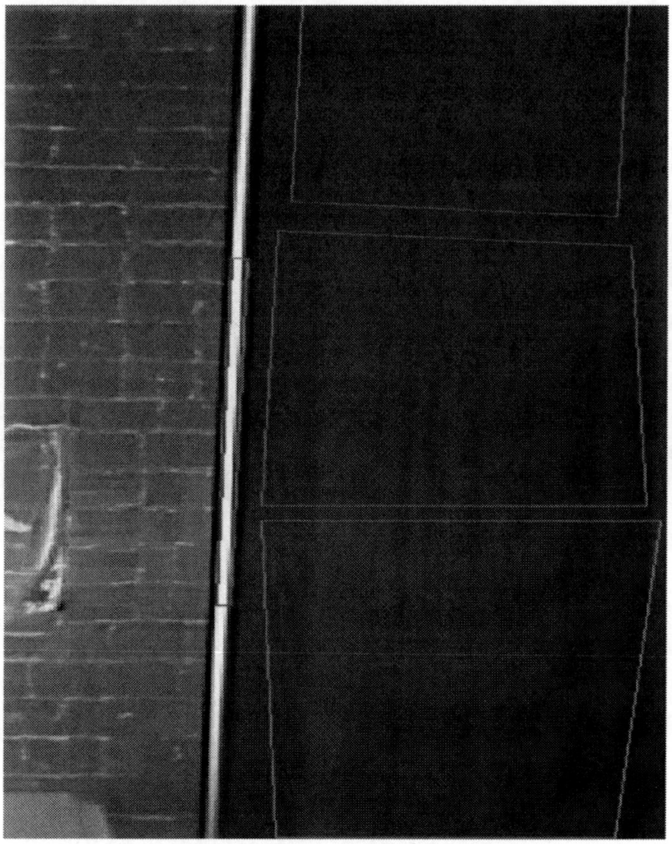

Figure 10. Hot Tube with Red Alarm Box

Tube Temperatures

After PSA off gas was fed as fuel, then plant rates were increased while adjusting various operating parameters such as steam-to-carbon (S/C) ratio and operating temperatures to begin the optimization process. It was very important during this time period to understand the status of the catalyst tube temperatures.

TFM's tube temperature correlation was utilized, based on historic images checked with pyrometer readings at specific locations at specific times.

The first step in the process of correlating tube temperatures is to fully understand the characteristics of each firebox. Topsoe's experience in firebox monitoring based on many years of collecting tube temperature data on all types of SMR's was leveraged to TFM's tube temperature correlation process.

For these SMR fireboxes, tube temperatures were very consistent from the beginning. Burner impingement was not observed, and burners did not require significant adjustments. Process blockage or other process-related problems were not significant. TFM regions-of-interest (ROI's) were established based on the observed temperatures in specific areas of the catalyst tube banks. Alarm settings were established in discussions with Chevron to provide an early warning notification about hot tubes. Where TFM identified specific areas of the SMR's that indicated a partial imbalance in heat distribution, Topsoe notified and discussed with Chevron regarding next steps.

A manufacturing defect was found in one of the furnace tubes in one of the furnaces during the construction phase. This tube was capped prior to startup. During operation it was used as a reference tube for temperature calibrations since it was always the hottest tube in the firebox.

Topsoe worked closely with Chevron to "fingerprint" the tube temperature profile of both SMR's at normal operating conditions. Individual temperature charts and graphs were created to assist Chevron in quickly evaluating tube temperature deviations whenever they developed. (Figure 11)

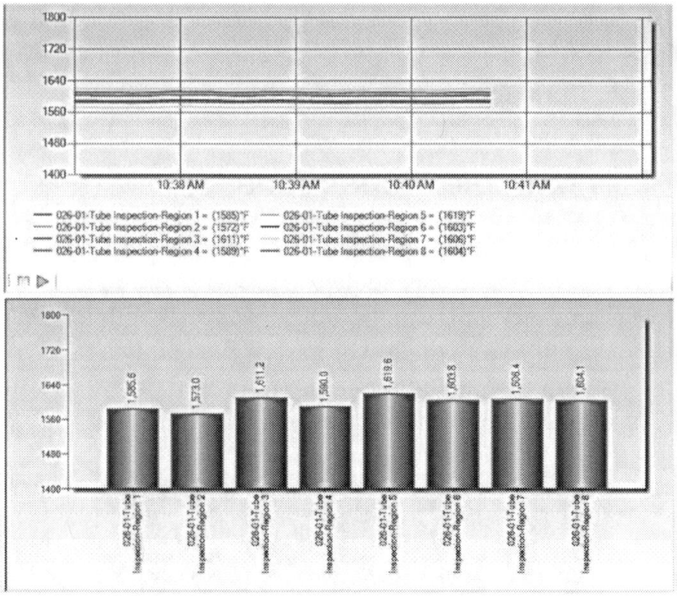

Figure 11. Tube Temperature Chart & Graph

Multiple Plant Firebox Monitoring

Another important requirement for Chevron was to have the ability to manage both H2 plants, including both large SMR's from a single console with one operator using technology for monitoring. TFM provides the imagery for continuous monitoring of both fireboxes.

TFM displays were set up so that both SMR's were monitored continuously in immediate proximity to the plant's DCS control system. (Figure 12) Any firebox upsets or changes required are observed and will be in alarm if outside of normal operating range. In this fashion, TFM provides great insight into firebox condition prior to dispatching an operator to the field to inspect and hunt for a problem by opening viewports.

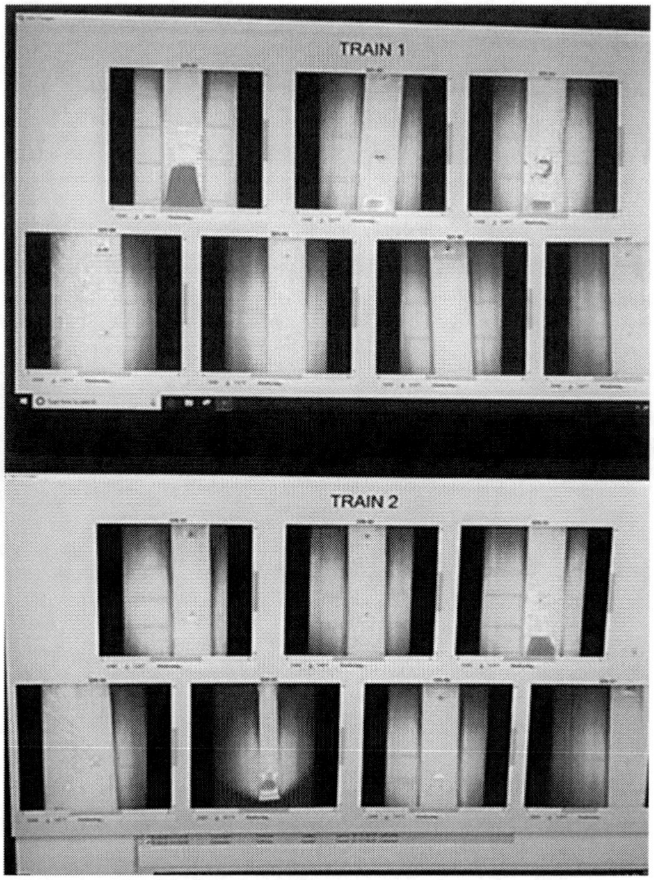

Figure 12. Multiple H2 Plant Firebox Displays

Historian and Event Capture

TFM not only provides instantaneous images of firebox condition, including burner and catalyst tube temperatures, it keeps images in its historian for future reference.

Conclusion and Summary

Chevron chose Topsoe TFM technology to ensure safe and reliable commissioning and startup of their two, new, world-scale hydrogen plants in 2018. TFM successfully monitored firebox conditions from the initial burner lighting of the first SMR up through normal operation of both hydrogen plants without incident or issues.

Rationalizing Safety Critical Systems

The first and foremost part of process safety management is defining which equipment is critical to the process and its safety. Of course, it is essential to take care of all equipment for safe operation but not all equipment is safety critical. If everything is safety critical, then nothing is safety critical. When non-critical assets are included in list of critical equipment, the term "safety critical" loses its true essence and sanctity. A, "consider all," approach will unnecessarily overwhelm maintenance manpower and can result in miss allocation of resources. Personnel started to treat safety critical equipment with less importance.

This paper explains the difficulties and disadvantages encountered to organizations of wrong interpretations, implementations of safety critical systems, and approaches used to revamp the whole system for its true implementation as per OSHA's PSM standard (29 CFR 1910.119) and best practices.

Usman Ali
Senior Process Safety Engineer
Fatima Fertilizers Company Limited (FFCL)

Ahsan Sarfaraz
Unit Manager HSE
Fatima Fertilizers Company Limited (FFCL)

Introduction

Fatima group, established in 1936, is one of the leading industrial groups of Pakistan which owns and operate three fertilizers complexes in Pakistan. FFCL is one of Pakistan's largest fertilizers manufacturing complexes that process natural resources like natural gas, rock phosphate, limestone and water into vital products for farmers and industrial customers. FFCL is primarily involved in manufacturing and marketing of fertilizers, capable of producing two intermediate products i.e. ammonia and nitric acid and three final products which are urea, calcium ammonium nitrate (CAN) and nitro phosphate (NP). The complex contains integrated utilities section for all 5 production units and 1 bagging unit for final product handling and dispatch. The fertilizer complex is a fully integrated production facility occupying 947 acres of land and a total of 1733 employee's including 250 management, 468 staff and 1015 contract employees.

The parent unit of the complex is a 1700 MTPD ammonia plant which is currently operational using natural gas as the basic raw material. This ammonia plant was relocated from Rozenburg, Netherlands to Pakistan in year 2007. The plant was originally designed by CF Braun in 1968 with name plate capacity of 1350 MTPD. Prior to relocation of ammonia plant from Kemira, Rozenburg to Pakistan; it had gone through several revamps and energy conversation projects resulting in an increase in production capacity from 1350 MTPD to 1650 MTPD. The plant was shut down in 2000 and purchased by Fatima group in year 2006 and relocated to Pakistan in year 2007.

The ammonia and its by-product carbon dioxide are used for the production of 1500 MTPD prilled urea. The urea plant is a Stamicarbon Urea 2000+ technology new plant and was commissioned in 2010. The nitric acid production facility has a capacity of 1500 MTPD (based on 100% conc.), which along with ammonia and limestone, is utilized for the production of 1200 MTPD prilled nitro phosphate and 1400 MTPD granulated calcium ammonium nitrate. The NP plant is designed by Chemical Fertilizer Industry and it was commissioned in June 2011. The ammonia, CAN and NA plants are relocated plants. In addition to the main process units, independent small units are installed to aid their smooth operation which include a nitrogen plant, instrument air and plant air plant, reverse osmosis (RO) plant, demineralized water treatment and steam generation unit. The complex is self- sufficient in power generation and also supplies a residential colony. Figure-1 below illustrates the main process units at the FFCL plant.

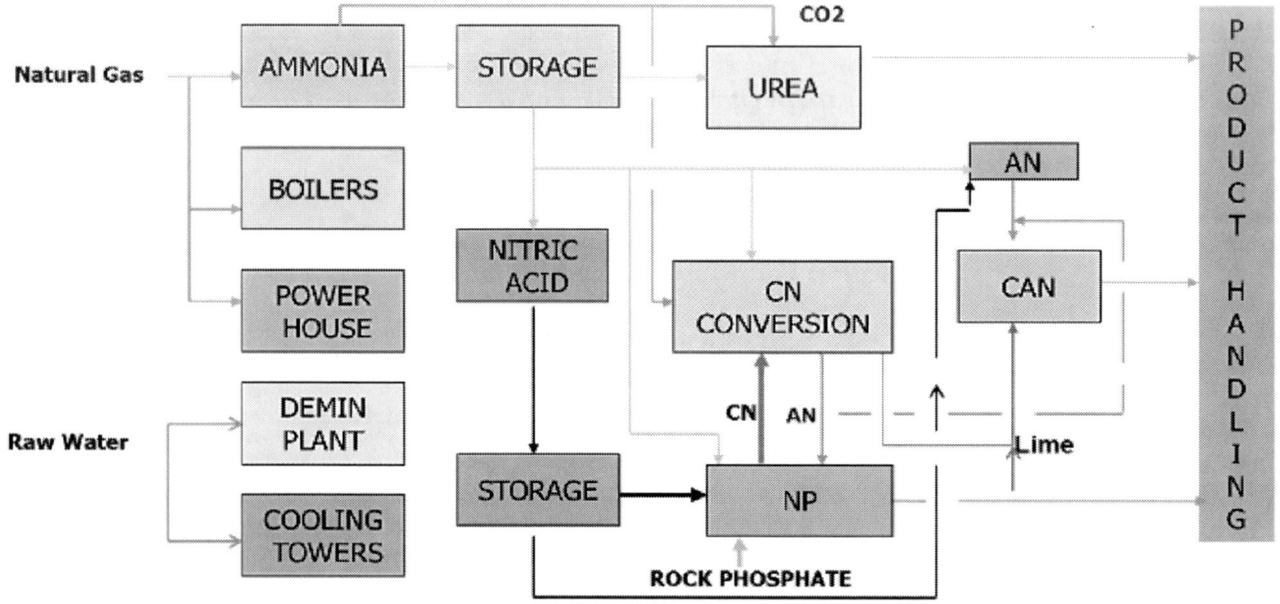

Figure 1. FFCL Plant Process block diagram

Paradigm Shifting

Conventional Safety Management systems were adopted initially which mainly worked around personnel safety performance and usually process safety is a low focus area, consequently overall site safety performance remained unsatisfactory with major setbacks in the form of severe process safety incidents (major fires & Tier Level-1 Process Safety incidents). Below are a few glimpses.

- 22 recordable injuries were reported in 2011.
- 4 Lost Time Injuries, 1 Restricted Work Case, 2 Medical Treatment Cases & 48 First Aid Cases reported in 2012.
- High number of fires with 29 process and non-process fires occurred in 2012
- Elevated Total Recordable Injury Rate of 0.51 (per 200,000 hours).

Major incidents like ammonia syngas compressor fire and catacarb incident occurred caused the plant to remain unavailable for more than 40 days.

Following the 2012 series of process safety upsets on the ammonia plant; the site management analyzed the situations and identified inadequate process safety management systems as the root cause. Subsequently it was agreed to align exiting systems and practices with Best in class Process safety model (DuPont PSM model) for sustainable business model and to enhance safety, reliability and productivity of site. Based on an initial gap analysis, a PSM task force was established comprising of in-house resources having sound experience with developing and implementing PSM systems in similar facilities in the past with a mandate to lead the implementation of a PSM system across the FFCL site as per a road map.

The whole PSM system was implemented at site in less than four years and the first 3rd party assessment was carried out by DuPont in May 2015 which rated FFCL at 3.5 from scale 1 to 5 which mean the PSM system is well implemented. The 2nd assessment was done by DuPont in November 2017 and rated FFCL site at Excellence level. Figure 2, 3 and 4 below depicts the site safety stats of FFCL after shifting focus on process safety management system.

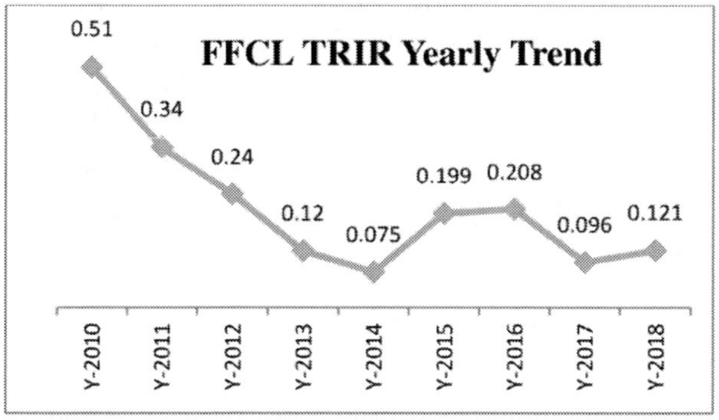

Figure 2. FFCL Total Recordable Injury Rate yearly trend

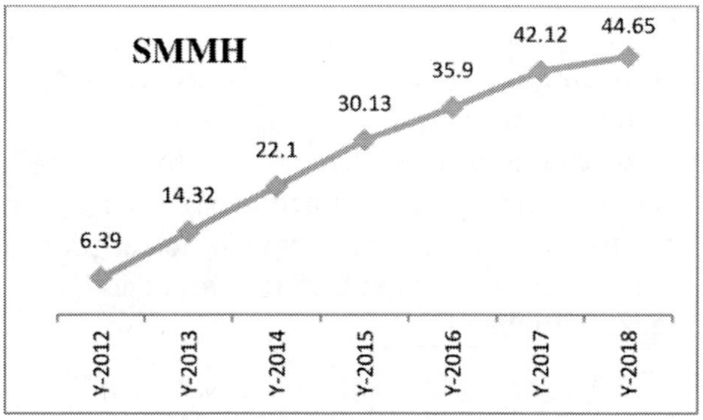

Figure 3. FFCL Safe Million Man-hours trend

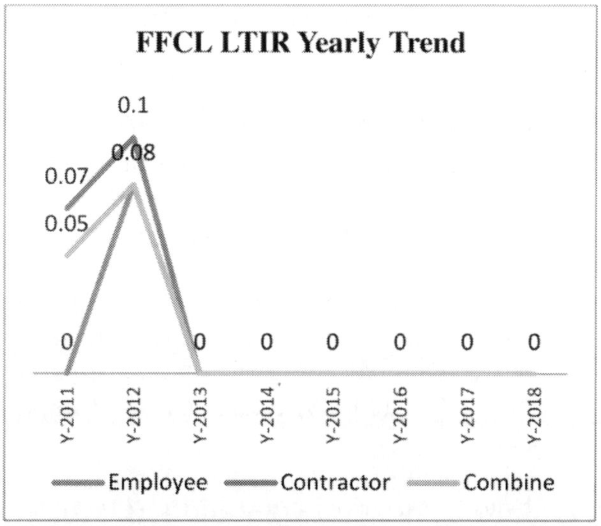

Figure 4. FFCL Lost Time Injury Rate Yearly trend

Safety Critical System Implementation

The safety critical system that was initially defined in the PSM 1st element Process Safety Information was devices, equipment or system whose failure could result in, allow or contribute to the release of or exposure to sufficient quantities of hazardous substances or their energy (i.e., fires, explosions, etc.) that could result in serious injuries or irreversible health effects, significant property damage or significant environmental impact. This includes:

- Any plant equipment (vessels, pipelines, valves etc.) which contains hazardous material in sufficient amounts.
- Systems, instrument and controls, interlocks etc. that prevent loss of containment of hazardous material
- Equipment to respond to or mitigate the effects of a loss of containment e.g., fire-fighting equipment, hydrants, hoses, safety showers, fire suppression systems, emergency response communication equipment etc.

As per Process Safety Information procedure guidelines, the whole system was distributed into two basic groups, these being HHP (Higher Hazard Process) and LHO (Lower Hazard Operation).

1. **Higher Hazard Process (HHP)** - A HHP is any activity (e.g. manufacturing, handling, storing or using) relating to substances that, when release or ignited, have significant potential for death or irreversible human health effects, significant property or environmental damage or off-site impacts due to acute toxicity, flammability, explosivity, corrosivity, thermal instability, latent heat or compression.

2. **Lower Hazard Operation (LHO)** – A LHO is any activity that exclusively manufacturers, handles, stores or uses substances without potential for death or irreversible human death effects, significant property or environmental damage or off-site impacts due to toxicity, mechanical hazards (including stored energy), or asphyxiation.

A very conservative and defensive approach was used initially during selection of safety critical devices. The term safety critical is very misleading and as safety critical is generally a sub-set of HHP area, so initially all HHP area components were declared as safety critical in our system. An "Include it all" approach was used because nobody wants to be responsible for overlooking a safety critical device. This results in misallocation and overloading of resources and cheapening of the term safety critical resulting in loss of true sanctity of safety critical system at our site.

To implement the Safety Critical (SC) system in its true essence and to improve Safety Critical Defeatable system compliance, resource optimization and Mechanical Integrity/Quality Assurance and Quality Control and Preventive Maintenance plans compliance at site, FFCL decided to carry out a detailed revamp of whole SC system. The revamp team did the gap analysis of whole SC system by consulting and referring

- FFCL Level-II SOP
- Industrial best practices
- OSHA & DuPont Guidelines

Safety Critical Equipment Identification

The team reviewed each category of safety critical components, equipment and systems and identified the major equipment and systems that will be Safety Critical.

Equipment Containing Hazardous Substances

Equipment which are part of the process circuit (enclosed by NODES that can be remotely and readily isolated) containing hazardous substances in threshold amounts (as determined in OSHA / EPA / NFPA 30 list of regulated chemicals) are included in the list of safety critical system. It includes all "Wetted Parts" e-g Vessels, Heat Exchangers, Piping System including NRVs and Manual Isolation Valves, Control Valves, Pumps, Turbines, Glass Level Gauges (GLGs) etc. of High Hazard Process (HHP) Loop.

Equipment containing hazardous substances is the last line of defense with respect to process containment and considered as Safety Critical due to its wetted part. Isolation of these equipment's cannot create any safety concern (with respect to the release of or exposure to sufficient quantities of hazardous substances that can result in serious injuries or irreversible health effects, significant property damage or significant environmental impact) due to which defeat protocol is not required for the isolation and handover of these equipment.

Examples include (if handling hazardous substance in threshold limit)

- Process and/or pressure vessels
- Process equipment - For example, heat exchangers (tube and shell sides) would be

Safety Critical if either the process material or heat transfer medium were a hazardous substance.

- Pumps and compressors
- Piping, including special piping (e.g., high temperature, high pressure), transfer hoses.
- Inline instrumentation i.e. Thermo wells of Local and remote indication instrumentation installed on Safety Critical piping
- Local Level gauges
- Instruments and impulse lines - Transmitters and associated small diameter impulse lines should be evaluated for incident potential before designating them as Safety Critical.

Containment Controls

Equipment ensuring containment control during normal plant operation or after shutdown will be placed under the safety critical system list for each corresponding unit. This includes for example:

- Relief Devices such as Process Safety Valves, Process Safety Devices etc. including their piping.
- Breather valves, conservation vents or pressure vacuum relief valves, emergency vents and overflow pots etc. if wetted by hazardous substances in threshold amounts.
- Check valves and excess flow valves of HHP loop.
- Cathodic Protection Systems of HHP

Note: All relief valves and rupture disks are treated as safety critical, regardless of the substance handled, because of the concern for equipment failure / rupture hazards.

Shutdown Controls

Equipment, instrument and systems of HHP loop acting as the last line of defense with respect to safe shutdown of the unit or some portion, failure of which could result in catastrophic event (with respect to human, property_or environment damage) will be considered as Safety Critical. This includes:

- Sensors, logic solvers, final control elements and their associated interconnecting hardware that makeup interlock loops that mitigate Class A and Class B hazardous events. Common industry terms for this collection of equipment are the Safety Interlock System or Safety Instrumented System (SIS).
- Inline instrumentation, not part of the SIS, which acts as the containment for hazardous substances, such as a thermo well or control valve and whose mechanical failure could result in a loss of containment with impact on personnel or the environment would be classified as Safety Critical due to its wetted part. Defeat protocol is not required for such instrumentation.
- Sensors, final control elements of HHP circuit and their associated interlocking hardware that make up interlock loops for the safe shutdown of plant / system to mitigate hazardous events.
- Manual emergency shutdown systems of HHP loop.
- Trip securities of machines e.g., high axial or radial vibration trip security of machine having hazardous substance as per OSHA / EPA / NFPA 30 guidelines.
- Mechanical/Electronic over speed trip securities of all steam and gas turbines up to final control elements such as the trip and throttle valves.

Notes:

- Instrumentation systems provided for basic control of the process is known as the Basic Process Control System (BPCS) in the industry should also be evaluated for control elements whose failure could result in a process safety incident.
- Equipment, instrument and system of HHP loop will not qualify for safety critical if it is just an indication of certain parameter on

DCS / in area that will not trigger certain sequence of events for safe shutdown of plant in case of process upset. Typical example is the temperature transmitter installed on ammonia line and line is Safety Critical. Temperature transmitters will not qualify for Safety Critical as per shutdown logic definition because it is not initiating the shutdown sequence. However, for containment reference thermo-well of temperature transmitter will be safety critical. Defeat protocol is not required for such instrumentations.
- Equipment, instrument and system of LHO ensuring shutdown control will not be considered as Safety Critical.
- Instrument systems that are declared as Safety Critical should be tested and maintained on a regular basis to ensure safe operations.

Examples:
- Typical example of Safety Critical device is boiler steam drum low level trip security. BFW flow to steam drum is very critical to maintain level in steam drum but low level is indicated by level transmitter on steam drum and actuates the shutdown logic in case of low level. So low level transmitter / switch will be safety critical as per definition of shutdown controls.
- Another typical example for safety critical shutdown control is the high-pressure trip security of synthesis section at urea plant. Ammonia to Carbon dioxide (N/C) ratio is very critical for plant operation but in case of disturbance in N/C ultimate effect will be in the form of synthesis section high pressure so the pressure transmitter will be safety critical as per definition of shutdown control. Ammonia to Carbon dioxide (N/C) ratio will not be safety critical.

Controlled Release Equipment / Systems

Operating equipment or systems associated with the controlled release of hazardous substances are Safety Critical. These include for example:
- Liquid separators from relief devices (PSVs, Rupture disks) of HHP loop
- Flare and flare header systems including the components essential to their safe operation
- Scrubbing systems for hazardous substance release control etc.
- Surge and overflow tanks associated with Hazardous substances.

Safety Monitoring and Emergency Communication Systems

Plants will consider safety systems associated with the detection of our people's response to release of hazardous substances as Safety Critical. These include for example
- Early gas and smoke detection systems in case of releases of hazardous substances e.g. ammonia analyzers in plant area, Chlorine dioxide analyzer.
- Pipe leak detection system associated with hazardous substances service.
- Emergency alarm systems e.g. area fire alarm, audible alarm system for emergency
- Emergency radio system for on-site and off-site communication
- Low O2 detectors for personnel protection etc.
- Fire and Gas Detection System

Notes:
- The functionality of these systems should not depend on any large control system such as a Distributed Control System (DCS). They should be simple, stand-alone systems to minimize the scope of the Safety Critical systems. The alarm signal could be input to the DCS for information and logging purposes, but this function for the DCS would not be Safety Critical.
- Systems should be considered for designation as Safety Critical if they are the primary means of communication to mitigate the impact of a release of a hazardous substance.
- One communication system shall be declared as safety critical such as radios or PA (public

address) system and ensure the applicability of all safety critical protocols on it.
- Defeat protocol is not required for radios. In case of un-availability or malfunctioning of any of radio it will be replaced immediately with spare one.

Active Mitigation Systems

Systems that need activation to reduce the potential of or minimize hazardous releases and process-related fires and explosions are Safety Critical. These include for example:

- Fire protection systems directly associated with Safety Critical Process areas, including sprinklers, deluge systems, fire water supply, storage, piping distribution systems, Fire hydrants and fire monitors.
- Safety Shower and eye washer system and its associated network.
- Breathing air systems e-g Self Contained Air Breathing Apparatus, CCR Breathing Air Systems.
- Fire extinguisher, foam trolleys
- Fire trucks
- Explosion suppression systems and fire door
- Area ventilation systems to dilute hazardous substances,
- System to provide positive pressure to exclude hazardous substances from an area or for electrical classification purpose
- Safety Programmable Electronic Controller (PEC) system.

Note: Defeat protocol is not required for fire extinguisher, foam trolleys and SCABA. In case of un-availability of any of this equipment it will be replaced immediately with spare one. In case of unavailability of spare, notice to be placed referring the nearest possible location.

Passive Prevention and Mitigation Systems

Area will consider systems that do not need activation but can reduce the potential of or minimize hazardous releases and process-related fires and explosions are Safety Critical. Examples are:

- Secondary containment systems e.g. dykes and drainage systems, sumps, process sewers or collection systems and jacketed piping systems intended to contain hazardous substances in case of release.
- Fire walls

Service and Utility Systems

Service and utility systems that help for the safe shutdown of unit during operational upset are Safety Critical. These include for example:

- Backup power supply system providing emergency electrical power to SC equipment in case of power failure to enable or maintain process safety in the event of loss of normal power.
- UPS for powered Safety Critical Controls
- Diesel generators.

Note: Normal intended of service and utility system (as described above) is to provide fail-safe conditions to the processes in case service or utility loss occur and hence, may not be designated as Safety Critical. The examples below are provided for guidance in cases where the service or utility failure does not result in a fail-safe condition.

- Nitrogen supply systems providing inert blanketing for process safety purposes – In cases where a high pressure N_2 tank or liquid N_2 is used as backup in the event of failure of pipeline N_2 supply, the N_2 tanks (high pressure or liquid) and their associated piping to the process and controls would be Safety Critical, but the normal N_2 supply pipeline would not be Safety Critical.
- Instrument air supply – The above example for nitrogen would apply in a similar manner to instrument air supplies and any emergency or backup air supply systems for Safety

Critical equipment where air is needed for the equipment to continue to operate safely.

Note: Most Safety Critical controls do not need power to perform their safety function.

Safety Critical Defeat Protocol Guidelines

The aforementioned-eight categories are summarized in table below and requirement of defeat protocol is also mentioned. Defeat protocol contains a detail set of actions required for defeating or bypassing any safety critical security.

S #	Category Name	Equipment Name	Description / Justification	Defeat Protocol Required?
01	Equipment Containing Hazardous Substances	Equipment (Vessels, heat Exchangers, Stand-by Filters, Columns, Reactors etc.)	Wetted Parts of HHP Loop	No
		Machines (Pumps, Compressors, Turbines)	Wetted Parts of HHP Loop	No
		Piping System (Pipe lines, NRVs etc.)	Wetted Parts of HHP Loop	No
		GLGs (Glass Level Gauge)	Wetted Parts of HHP Loop	No
		Inline Instrumentation (e-g Thermo-well, Flow Transmitters)	Wetted Parts of HHP Loop	No
		Standby Equipment (e-g Filters, Pumps etc.)	Wetted Parts of HHP Loop	No
02	Containment Control	PSVs	All PSVs of Plant irrespective of service	Yes
		PSDs	All PSDs of Plant irrespective of service	Yes
		Cathodic Protection System	Cathodic protection of all hazardous chemical containing underground Piping	No
		Vent Systems	All conservation and emergency vents and related piping that contains or can potentially contain hazardous substances like NH_3, H_2, CH_4 etc.	No
03	Shutdown Controls	Emergency stop switches	Shutdown switches installed at Auxiliary / Local panel for the safe shutdown of Plant / process identified as SC	No
		Sensors, Final Control Element	Sensor to Final Control element of HHP loop actuating the shutdown interlock / logic for safe shutdown of process Temperature, Vibration tripping Probes of Compressor, Turbines actuating the safe shutdown of machine	Yes
		Over speed Trip	Mechanical/Electronic OST's for all Steam and Gas Turbines up to Final control elements e.g. T&T etc.)	Yes
04	Controlled Release System / Equipment	Liquid Separators, Scrubbers	Equipment installed at the vent of HHP Loop for controlling / mitigating the effect in case of release	No
		Flare system	Equipment installed at the vent of HHP Loop for controlling / mitigating the effect in case of release	No

05	Safety Monitoring and Emergency Communication System	Emergency siren	Emergency Hooter System for communication of level-I / II	Yes
		Radios	Walkie Talkie Base Station for communication i-e b/w FCC & MCC	No
		F&G System	Fire and Gas Detection System in process areas	Yes
06	Active Mitigation System	Fire Water System	Fire water piping, Fire water Pumps, Fire hydrants, Fire monitors etc.	Yes
		Fire Deluge System	Fire Deluge Systems available on Lube oil circuits of compressor etc.	Yes
06	Active Mitigation System	Safety Shower & Eye Washer	Safety shower and eyewash fountains located in the area including related piping and pumps.	Yes
		B.A Systems	Breathing air cylinders and B.A System in control room building, MCC building	Yes
		SCABA, Scott Air Packs	SCABAs, Scott Air Pack, Full / Half Face Mask.	No
		Wind Socks	Wind Sock in Ares at different locations	No
		Fire Extinguisher / Foam Trolleys	Fire Extinguisher of CO2, DCP types and Trolleys of CO2, DCP and Foam Types	No
		Fire Trucks	Fire Truck available with ERT Team	Yes
		Explosion Doors	Explosion Doors available at plant e-g at ammonia furnace	No
07	Passive Mitigation System	Dykes	Dykes areas under the vessel containing hazardous chemicals (HHP loop vessels)	No
08	Service and Utility System	UPS System	Backup power supply system providing emergency electrical power to SC Equipment in case of power failure	No
		Diesel Generator	Backup power supply system providing emergency electrical power to SC Equipment in case of power failure	Yes

Benefits

Clear ownership of the system defined in new guidelines to improve the stewardship and SC system. The rationalization of SC system as per above criteria optimize the overall number of safety critical equipment at all units by 34%. The numbers are even more promising with reference to safety critical defeat-able devices. The Mechanical Integrity, Quality Assurance and Quality Control and Preventive Maintenance of overall equipment of site also improved resulting in optimization of resource utilization. Dedicated teams of each unit are established for implementation of system as per revised guidelines. Detailed training sessions are conducted of nominated individuals by core team to ensure proper system understanding and implementation with same understanding at whole site.

Conclusion

Implementation of PSM system at FFCL has clearly demonstrated an uplifting of the process safety standards and morale of employees with a reduced number of injuries and serious process safety incidents, improvement in reliability and productivity of plant as shared above in stats.

The term "Safety Critical" is very misleading and there is not a consensus on the formal definition in the process industry, a general theory can be gathered from the review of various definitions. Identifying these safety critical components can be an intimidating task considering OSHA's lack of concrete guidance on the matter. So other tools and references can also be applied.

Layers of Protection Analysis (LOPA) is an efficient tool for estimating frequency, probability, and severity of an incident scenario. The LOPA technique also lists the safeguards necessary to reduce the risk of catastrophic events, so using it as guidance can also help in identifying safety critical equipment and FFCL aims to use this technique results as baseline in future revision cycle of SC system.

Since a process is always evolving, the Safety Critical Components Manual should be evergreen document. It should be updated and reviewed as necessary to ensure that the information is correct and up-to-date.

Optimizing Safety, Reliability and Cost of SIS and Non-SIS Equipment

This paper will describe methods by which operating companies can optimize Safety Instrumented System (SIS) design and test methods to achieve specific safety or financial goals. Doing so requires understanding of traditional SIS performance metrics like probability of failure on demand (PFD), combined with economic metrics of installed cost, maintenance cost, and cost due to spurious trip. Using principles of reliability theory and a cost prediction model, it is possible to optimize for lowest lifecycle cost, lowest installed cost, highest onstream time, maximum safety within a specific cost constraint, or other objectives. This paper will also discuss extensions of this method to other types of (non-SIS) equipment, by which improved safety and higher uptime can be achieved.

Peter G. Hereña
Baker Engineering and Risk Consultants, Inc. (BakerRisk®)

Introduction

Operating companies must balance the need to protect people against the ability to keep the plant running. Additional equipment may be necessary to protect people and assets, but that same equipment can fail in ways that forces production offline. On the other hand, a worse problem is when equipment is removed or never installed to protect people, causing a riskier workplace for production personnel. In the world of safety systems, these two competing objectives are described as the SIL (Safety Integrity Level) and STR (Spurious Trip Rate) of the system. The SIL is a measure of risk reduction provided by a safety instrumented function (SIF), whereas STR is the frequency at which the safety function causes an unnecessary shutdown. SIL is measured in increments between 1 (the lowest performance level) and 4 (the highest performance level) as noted in Table 1 and is often associated with probability of failure on demand (PFD).

The international standard IEC 61511[1] describes the full safety lifecycle for a safety instrumented function, but for the purpose of this paper the key steps are as follows: For each hazard where a safety instrumented function is necessary, a target SIL is required. This is often assigned by use of layers of protection analysis (LOPA) or another qualified method. Then for each SIF, once the target is SIL is known, the safety function may be designed to achieve that SIL.

SIL	Unavailability (PFD)	Availability (1-PFD)	Risk Reduction Factor
4	$\geq 10^{-5}$ to $< 10^{-4}$	0.9999 to $\leq$ 0.99999	> 10,000 to $\leq$ 100,000
3	$\geq 10^{-4}$ to $<10^{-3}$	0.999 to $\leq$ 0.9999	> 1,000 to $\leq$ 10,000
2	$\geq 10^{-3}$ to $< 10^{-2}$	0.99 to $\leq$ 0.999	> 100 to $\leq$ 1,000
1	$\geq 10^{-2}$ to $< 10^{-1}$	> 0.90 to $\leq$ 0.99	> 10 to $\leq$ 100

Table 1. SIL and Relation to PFD

There are many different configurations of safety instrumented functions that can achieve a given SIL. The decision levers for designers include:

- the type of equipment used,
- the level of redundancy provided,
- the diagnostic capability,
- the type of actions performed,
- the frequency of testing,
- and the degree of maintenance.

Different configurations can minimize PFD for a given SIL rating (improve safety performance), minimize the STR for a function (improve uptime performance), or strike a balance between the two.

A method has been described by which total lifecycle cost of a SIF can be calculated.[2, 3] The calculations are based upon simplified equations established by reliability analysis techniques.[4, 5] This method factors in the annualized installed cost, testing cost, maintenance cost, and spurious trip cost. The method can be extended to other types of costs, or more precise cost accounting methods as desired. Some of the key equations from paper reference 3 are noted below:

Installed cost depends on the number of devices, the cost per device and the expected lifetime of the device.

$$C_{inst} = C_{device} \times N / L$$

Where:

C_{inst} is the installed cost of the function
C_{device} is the installed cost of the device
N is the number of devices
L is the lifetime of the device.

Testing cost takes into consideration the amount of labor time required to perform a test of the SIF.

$$C_{test} = TI \times t_{test} \times R$$

Where:

C_{test} is the testing cost of the function
TI is the test interval of the component
t_{test} is the time required to test the component
R is the labor rate of the technician(s) performing the repair.

Maintenance cost takes into consideration the amount of labor needed to replace or repair failed equipment.

$$C_{maint} = \lambda \times t_{repair} \times R \times N$$

Where:

C_{maint} is the maintenance cost of the function
λ is the total failure rate of the device (safe and dangerous failures)
t_{repair} is the repair time for a failed component
R is the labor rate of the technician(s) performing the repair
N is the number of devices

Economic impact of a spurious trip can be modeled by a very simplified model of assuming that a spurious trip causes a permanent loss of revenue according to the following formula:

$$C_{trip} = STR \times MTTR_{process} \times P$$

Where:

C_{trip} is the spurious trip cost of the function
STR is the safe/overt/spurious failure rate for the component or group of components
$MTTR_{process}$ is the mean time to restore the process to the point of producing product
P is the revenue rate of product streams from the process

System Configuration Choice	Annualized Installed Cost USD/yr	Test Cost USD/yr	Maintenance Cost USD/yr	Spurious Trip Cost USD/yr	Total Cost USD/yr
1	$833	$150	$9	$788	$1,780
2	$1,667	$300	$18	$40	$2,024
3	$1,667	$300	$18	$1,577	$3,561
4	$2,500	$300	$26	$40	$2,866

Table 2. Lifecycle Costs for 4 SIF Configurations

Table 2 shows an example from the referenced paper, which describes the cost of four standard system configurations, each of which would be capable of SIL 2 effectiveness, and their total lifecycle costs. From the information in this table alone, the selection of system choice #1 would seem to be the most effective.

The above method could be applied for a single SIF, a thousand SIFs, or more, each time providing a total-cost-optimized solution. However, it is limited in that it only optimizes by total cost. Operating companies may wish to optimize on a different variable, for example such as annualized installed cost, and examine the different lifecycle cost that arises from the points close to the installed cost minimum. This is often the case in new project planning, where installed cost may have priority in order for the project to move from one planning stage to the next.

What is therefore needed in this circumstance is to apply a flexible deterministic optimization approach, which includes several steps:

1) Identify design variables
2) Describe an objective function
3) Describe constraints, and
4) Assign variable boundaries

Design variables are variables selected because they are important in establishing their connection to overall system functionality. Design variables could require optimization, or could be included in one or more system constraints, or could be an output from a system which simply is being monitored. For our SIS lifecycle optimization approach, the design variables have largely been described in the above equations, though they could be extended for additional factors.

An objective function is a function to be minimized or maximized (optimized). The objective function in the above example was the sum of annualized installed cost, test cost, maintenance cost, and spurious trip costs. But for SIS lifecycle optimization, the objective needs not optimize on any cost. A flexible optimization approach can instead establish an objective function, the sum of spurious trip rates for all system components, which in turn minimizes the spurious trip rate of the entire system.

Constraints are additional system values or sets of equations which place boundaries on plausible answers. For example, if spurious trip rate is to be minimized, then there may be boundaries placed on total lifecycle cost – an annualized cost greater than $10,000 per year for a single safety function is not likely to be implemented.

Variable boundaries are a form of constraint, but ones that are localized to input variables (variables that are used to calculate the values of other variables). They bound known input variables to real-world values. In the spurious trip rate example, allowing the universe of all known devices in industry would take a long time to solve, and could lead to an infeasible solution, therefore the devices could only be limited to the catalog of those devices still being manufactured, and from a subset of manufacturers for which the operating company has an existing service contract.

There are two types of optimization models – deterministic and stochastic. Deterministic uses *a priori* knowledge of the system to bound the outputs. Stochastic inserts a random element factor to account for real-world situations that might be impossible to account for using a set of equations. The method in this paper used a deterministic routine because the cost and reliability equations are well established. One possible application exception could be if the spurious trip cost is highly complex or depended on many substates, then a stochastic method (not described here) may be a better approach. But this is not often expected to be the case for most operating companies. The optimization software used in this paper is the publicly available "pyOpt", which uses the Python programming language[7].

In the example described above, cost functions were used to describe total lifecycle cost for a project containing 50 SIFs. Then the following constraints were added to the model:

- Desired spurious trip rate (STR) less than 0.2 /yr (greater than every 5 years)
- Equipment (and their spurious trip rates) bounded per the site's established vendors
- Equipment anticipated to be in production for at least the next 10 years
- Equipment specific values (cost, spurious trip rate, total failure rate, repair time) bounded by 3rd party published values
- Equipment test intervals bounded between 1 year and 5 years
- SIFs bounded to their associated SIL targets
- Spurious trip cost constant for all trips
- Annualized installed cost less than 75% annualized total lifecycle cost

With this model, a trend of total annualized cost for various spurious trip rates was established (Figure 1). The minimum annualized cost of approximately $15,000 per year could be achieved at a spurious trip rate of approximately 0.4 / yr (once every 2.5 years). However, that solution is not tolerated by the first constraint of our model, which was a spurious trip rate of less than 0.2 / yr (greater than once every 5 years). If that constraint is to be satisfied, the lowest lifecycle cost would be $20,000 per year. Another way to look at the data is the "cost" of reducing the STR to cover the constraint is $5,000 per year relative to the more globally optimized solution.

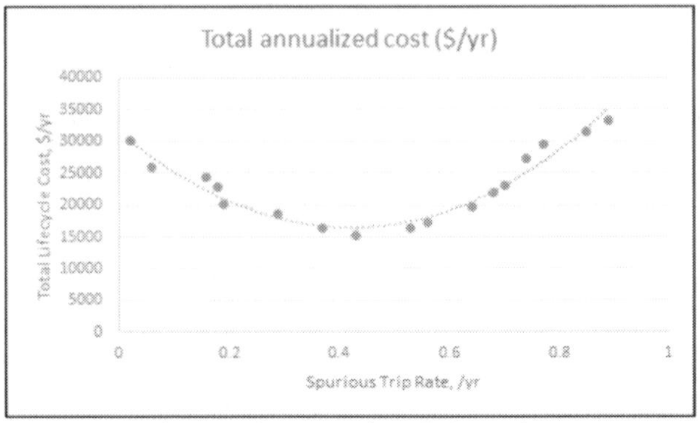

Figure 1. Total Annualized Cost for varying Spurious Trip Rates

Conclusions

Approaches have been established to minimize total lifecycle cost for safety systems. This paper describes methods by which those concepts can be extended to prioritize or optimize variables other than cost. After the objective function is established (the thing we wish to optimize in our design), the equations from prior approaches can be merged with a set of simple constraints, and evaluated through optimization software, to yield a range results of interest. From that range an optimized solution can be picked. This optimization model approach will allow for a broader range of decision options compared to previously discussed methods.

References

1. *IEC 61511 Functional Safety- Safety Instrumented Systems for the Process Industry Sector* (IEC, 2003).
2. Shah, Jatin and Hereña, Peter, The High Cost of Poor SIS Lifecycle Planning (CCPS, 2013).
3. Hereña, Peter and Magraw, Robert, Cashing in on the IEC 61511 Value Proposition (ASSE, 2014).

4. *ISA-TR84.00.02-2002 – Part 2, Safety Instrumented Functions (SIF) – Safety Integrity Level (SIL) Evaluation Techniques Part 2: Determining the SIL of a SIF via Simplified Equations,* (ISA, 2002).
5. Smith, David, *Reliability, Maintainability, and Risk*, 7th Edition, (Burlington: Elsevier Butterworth-Heinemann, 2005).
6. Gruhn, Paul and Cheddie, Harry, *Safety Instrumented Systems: Design, Analysis and Justification*, 2nd Edition (Research Triangle Park: ISA, 2006).
7. pyOpt open-source software website: www.pyopt.org

Damage to Mol Sieve Filters and Fouling of Purifier Following Bed Desiccant Change

Yara Pilbara Fertilisers is a 2200 MTPD KBR purifier plant commissioned in 2006. Drying the synthesis gas is critical to safe and efficient operation of the purifier and synloop. When desiccant is replaced, we should expect a little bit of dust. Something happened after a planned desiccant replacement where the dusting did not stop. The excessive dust blocked the downstream filters and fouled the downstream purifier as well. The problem was difficult to identify, and it took three shutdowns to identify the cause of the issue.

Alan Smith, Rajan Sinha, Neel Lalchan
Yara Pilbara Fertilisers

Introduction

The molecular sieve dryers protect the Purifier in the KBR plant by removing moisture, they consist of zeolite beads which are susceptible to dusting when agitated. Moisture and dust carryover into the purifier requires complex procedures to remove dust and derime (passing through hot gas) the fouled system. It is imperative that all aspects of design, installation commissioning and operation of the molecular sieves minimizes and protects from dust formation and carryover. This paper describes what happens when those protections fail during and after replacement of desiccant and how to handle fouling of the purifier when these failures occur.

Process Description

Molecular Sieve Drying

The methanator effluent is cooled to 4°C (39.2°F) in ammonia chiller 130-MC. The condensed water is separated in a knockout drum, 144-MD, the chilled gas then goes to the molecular sieve driers, 109-MDA / MDB, each containing APG III zeolite bead desiccant. After drying in the mol sieves, the total of water, CO_2 and NH_3 content in the syngas is less than 1.0 ppmv. Each mol sieve vessel has an outlet filter, 154-MDA/B installed to remove any dust particles above 1 micron using pleated fibrous filter elements.

Cryogenic Purifier

The purifier consists of a cold box with plate fin heat exchangers to cool and liquefy the inlet gas. The gas at dew point (-65°C (-85°F) or less) enters the purifier rectifier column, 137MF where the nitrogen and methane are separated to control the H: N ratio in the synloop ratio at 3:1.

Mol Sieve Vessels

The internals of each mole sieve vessel are listed below from the bottom to the top:

- Outlet collector lined with coarse 4 X 4 mesh
- ¾'' ceramic balls at the bottom, then
- A layer of ½'' ceramic balls, then
- A layer of ¼'' balls, then
- 18 X 18 screen followed by the desiccant
- 18 X 18 mesh on top of the desiccant than
- ¼'' ceramic support balls then
- 4 X 4 coarse mesh
- Hold down grating

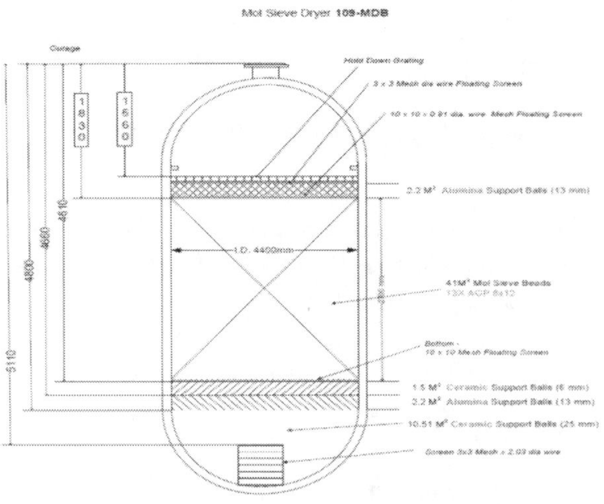

Figure 1. Molecular Sieve Schematic

Molecular Sieve Filters, 154-MDA/B

This filter consists of a filter housing vessel with 31 pleated filter elements measuring 36 inches in height and 4.5 inches in diameter per filter element. The filters are designed to capture particles above 1 micron.

Operation of the Mol Sieves

The molecular sieves are designed for a 24-hour in-service cycle followed by a regeneration cycle of 20 hours. There is a standby time of around 3-5 hours to allow for any short maintenance work if required. Dry waste gas from the purifier is normally used for regeneration and cooling of the driers. When waste gas is unavailable during start-up, the regeneration gas is obtained from a slipstream of the in-service molecular sieve.

Incident Description

The APG III desiccant installed in 2011 performed well through its installed life. The first replacement was planned in the 2018 turnaround (TAR) as a precautionary measure to ensure the next TAR cycle is met without any process problems.

The scope of the TAR included desiccant replacement as well as vessel inspection and filter element replacement. A specialized catalyst unloading contractor was appointed to conduct the replacement works. The desiccant was replaced along with the corresponding filter elements. No abnormalities were identified during the inspection of the vessels or replacement of the desiccant.

The standard purging and pre-startup checks were done prior to restarting the unit during which the beds behaved normally. There were no signs of damage, blockage or maldistribution prior to putting the new desiccant and filters into service.

The differential pressure (DP) across the filter of mol. sieve B rapidly increased to its maximum range when putting into service for the first time. The DP across the bed ‘A’ filter was normal. When the bed ‘A’ was switched to bed ‘B’, the front-end pressure increased by around 200 kPa (29 psi). Trends below show the difference of filter DP for mol sieve A and B.

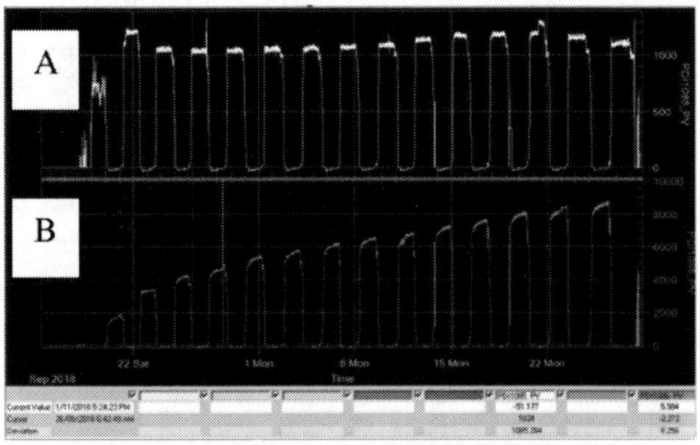

Figure 2. Filters 154MDB DP Trend

The sequence of events following the above observations are summarized below in Figure 3.

Sequence of Events

Date	Activity /observations
July –Sep. 2018	Desiccant replaced in turnaround for both 'A' and 'B' vessels
03 - 08 Sep. 2018	'B' Mol sieve put into service, Mol sieve 'B' filter DP high and increasing. 'A' mol sieve is normal
08 Sep. 2018	The plant stopped, mol sieve filters inspected, cleaned with nitrogen and reinstalled
10 - 14 Sep. 2018	Plant restarted, purifier cooldown took longer than normal. Purifier after switching from mol sieve 'A' to 'B'. Plant tripped, and left offline to investigate
14-19 Sep. 2018	Purifier derime and back blowing conducted, dust observed in the purifier shown in Figure 6. New Mol sieve filter elements installed.
19-Sep. to 27 Oct.	Mol sieve B DP still increasing online continuously. DP predictive model to plan the unload of desiccant. Continued operations with reduced plant rate while drier B in service.
28 Oct - 03 Nov. 2018	Stopped plant to inspect and replace 'B' mol sieve Desiccant in outlet piping: bottom inspection and top inspection ok. Desiccant not replaced, outlet spool cleaned and put into service
03 Nov. - 20 Jan. 2018	Plant online, mol sieve filter DP reduced but still increasing very slowly
20 Jan. - 26 Feb. 2019	During next available opportunity, the desiccant was removed and missing overlap of bottom 18X 18 Mesh identified as the cause.
28 Feb. 2019	The plant was restarted after replacing 'B' mol sieve desiccant, support balls, and mesh.

Figure 3. Sequence of Events

Downstream effects (19th Sep to 27th Oct)

The difference in the DP of mol sieve 'A'' and mol sieve 'B' resulted in cyclic plant operation during each 24-hour service cycle. The plant front end operated at higher pressure, often within <5% of the pressure relief valves (PRV) set points. Downstream, the pressure was lower and resulted in a varying load of the synthesis compressor and cycling of the purifier waste gas flow and levels.

The purifier cold box started showing signs of poor performance as soon as the mol sieve B was put in service. Operators had difficulty maintaining the liquid level in the purifier rectifier bottoms, which affected the synthesis loop ratio. Initial analysis of the dew point confirmed that moisture and carbon dioxide was not the cause of the purifier problems. The only other cause of fouling was thought to be dust breakthrough which can occur after desiccant replacement. The increased filter element DP confirms that excessive dust was present, however, the problem was worse in the B mol sieve vessel and not in the A vessel as indicated by the high DP of the filter elements.

The purifier problem worsened within two days of operation, the purifier rectifier bottom level and the H: N ratio to synthesis loop became too difficult to control. The H:N ratio fluctuations resulted in the loss of reaction in the ammonia reactor and synthesis compressor instability. The synthesis compressor was manually stopped to prevent damage to the compressor from surging caused by gas composition changes.

The plant was shut down for inspection and replacement of the mol sieve filters a total of three times between August and October 2018, and again in January,19.

Mole Sieve Filter Inspection and Cleaning in October 2018

Inspection of filters could not be done online during the 3-hour standby window because of passing inlet motor operated (MOV) valves. When the plant was shut down on 08 September for other issues, the opportunity was taken to inspect and replace the filter elements.

Dust, desiccant and ¼" ceramic balls were found in the filter housing of both 'A' and B mol sieves, the 'B' vessel had approximately three times as much as the A mol sieve. The 'B' filters had significantly more blockage and were saturated with dust.

Desiccant beads forced their way into the filter mesh and tore the outer layer of the filter cartridges. The filters, themselves were not damaged or collapsed and there were no visible breaches. Fine dust (smaller than 1 micron) came through the filters when they were knocked.

Replacement filter cartridges were not on site because new elements were used in the TAR. With a lead time of around 6-8 weeks, the only option to reduce downtime was to clean and reuse the elements. The filters were cleaned using a vacuum cleaner and a backflow with nitrogen.

Figure 4. Damaged and Blocked Filter Elements

Pipe and Vessel Inspection

Boroscope inspection was completed on the elbow below the filter inlet spool: spool removal was not done due to the time limitations. Ceramic balls were observed in the outlet piping of the mol sieve leading to the inlet of the filters.

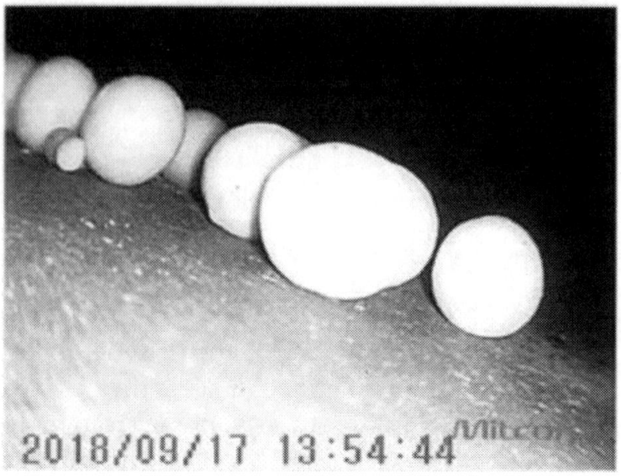

Figure 5. Boroscope Image of the Outlet Pipe Spool

Boroscope inspections were carried out on the outlet piping and top cover of the 'A' and 'B' bed by removing the thermowells and looking through the flanges. ¼" balls and desiccant were found in bed B outlet piping, only ¼" balls were found in the 'A' bed outlet piping. The time constraints did not permit the tedious exercise of removing the balls and desiccant. It was accepted that the filter elements would capture and prevent any particulates that slipped through

Purifier Fouling and Back Blowing

Fine dust, less than 1 micron passed through the filter elements and accumulated on the plates of the purifier inlet exchanger 132-MC. This reduced the heat exchange, leading to poor purifier performance that eventually caused the rectifier bottoms level to decrease to zero and lead to problems in the loop ratio control.

Dust migration into the plate fin heat exchangers of the purifier cold box cannot be removed by derime (passing through hot gas) only. Dust contamination is only removed by pressurizing then depressurizing the purifier with gas. This is known as back blowing; the process is described as follows.

Purifier Back Blowing

Before the back blow the purifier was derimed by passing through hot "moisture free gas". This removed any frozen moisture in the exchangers from the system. The purifier should always be adequately derimed prior to completing a back blow to ensure back blow works effectively.
Back blowing involves installing a sacrificial sheet flange made of layers of gasket material to act as a bursting disc. once the gasket is installed, the pressure in the system is gradually increased until the gasket ruptures. The contaminating particles will be blown out due to the sudden flow of pressurized gas to the atmosphere through the ruptured gasket.
Yara developed a procedure for the back blow based on advice from the OEM of the purifier, the plant licensor and previous back blow learnings from 2009. The optimum pressure for back blowing, gasket thicknesses, and the expected number of blows was predicted.
Knowing this information beforehand significantly improved the efficiency and effectiveness of the back blowing.
The blows and observations are summarized in the Table 1

Gasket Thickness mm(in)	Rupture Pressure Barg (PSIG)	No. of Blows	Comments
1 (0.04)	0.2 (3)	1	Burst pressure too low, no dust
3 (0.12)	1.0 (14.7)	1	Dust observed
5(0.20)	1.4 (21)	1	Large amounts of dust
6(0.25)	2.1 (31)	15	Large amounts of dust
10(0.40)	2.3 (34)	5	Used for final blows until no more dust emerged

Table 1. Details of Gaskets and Burst Pressure for Backblowing

Figure 6: Image of Dust from Purifier Back Blow

Mol Sieve 'B' DP Continues

While the back blowing was taking place, the 154-MDB filter was inspected and internal filter elements were replaced.
The B bed was then put back into service on September 19th. The DP on the filter started at 1,000 mmWC and was 1,700mmWC after the first 24-hour run. When the bed was put back in service after regeneration, the DP was 3,000mmWC. The DP continued to rise during normal operation and with a step change after each regeneration.

Mitigation Steps

- Online filter back blows with Nitrogen were attempted. Nitrogen was introduced at the bottom of the filters and a valve opened at the top 2'' to vent the dust and nitrogen through the filters. This was not very effective: the OEM suggested that a sudden change of velocity was required to get the dust out of the elements and the 2'' connections did not allow enough flow to achieve this.
- The mol sieve sequence was adjusted to maximize the online time of the 'A' bed and minimize the on-stream time of the B bed. Engineers calculated the moisture load and determined the safe operating time of one bed with a safety margin for this case. More than 32 hours of service life was achieved with the A desiccant bed while the B bed was around 24 hrs.

- Lower plant rate and lower adsorption time maintained for Bed B (83% plant rate, 24 hr adsorption time) in comparison with Bed A (90% plant rate and 32 hours adsorption time). Operations personnel got familiar with the process within a few days and were comfortable with the situation and aware of the required emergency actions if problems occurred.
- Venting upstream the mol sieve was done to reduce the inlet pressure and DP to ensure dust migration was minimized. This was a last resort and 10% venting was found to reduce the DP increase rate by over 50 %.

Planning for Replacement

The filters could not be replaced online, and eventually, the DP would be too high to economically run the plant. The other factors to be considered in the plan were:

- Lead time of desiccant
- Availability of specialized catalyst contractors
- Increasing back pressure of the front end of the plant
- Spares for passing valves
- Additional costs to expedite services and material, including air freight costs for desiccant from the USA.

Process engineers developed a predictive linearized model (modified cubic spline) of the 154MDB DP to predict the time available before the filter housing collapsed, the predicted and actual DP matched well allowing effective planning of desiccant change out.

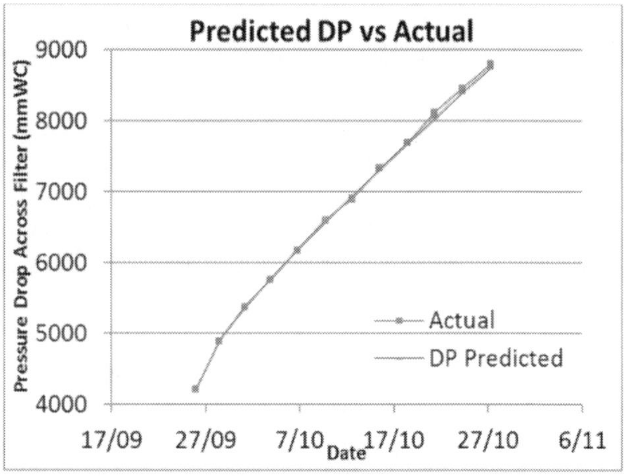

Figure 5. Predicted vs Actual DP

The plant team developed an action plan to ensure that the mol sieve DP didn't increase to catastrophic levels (>10,000 mmH_2O) according to the predictive model. Job planning was done in line with the DP trend.

Due to the delivery time of desiccant and availability of specialized catalyst loading contractor catalyst replacement contractor, the 109MDB could not be changed out until 28th October.

It was also imperative to plan the shutdown based on the status of the bed: a regenerated bed was preferred to minimize ammonia and other contaminants.

Second Inspection of Bed B

The decision was made to change out 109MDB on the 28th of October. The mol sieve B operated for 14 cycles up to 28th October the DP reached as high as 8,600 mmWC (1 month from last filter element replacement).

Top of Bed

The top manway of 109MDB was opened and the grating and top of the bed found to be in good condition. The hold-down grating was removed to inspect the 18 X 18 fine mesh above the desiccant, the mesh was checked found to be in good condition and desiccant and balls were not mixing.

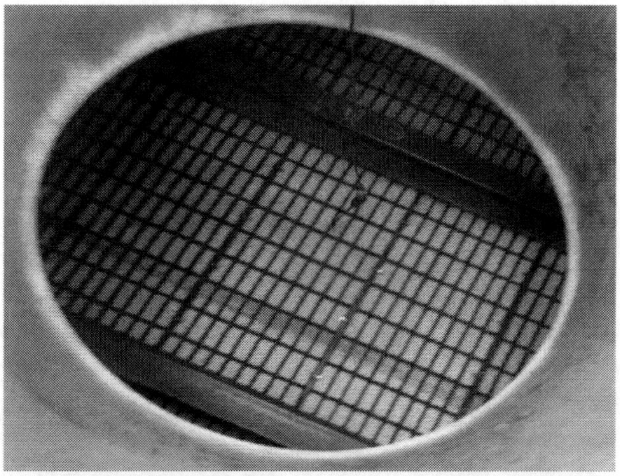

Figure 6. Hold down Grating at Top of Vessel

Mol Sieve 'B' Outlet Collector

The desiccant and support balls were found at the outlet collector when the bottom spool was removed for inspection. The mesh and collector were in good condition. The support balls on the other side of the collector were not broken or mixed with desiccant.
Pictures of re-loading the 'B' vessel during the turnaround were not available, the lack of records was explained by a camera malfunction during loading of the 'B' mol sieve. In the interest of time, the loading without photographic evidence was accepted. Photographic records of the 'A' mol sieve were available and appeared to meet the technical requirements and it was inferred the 'B' vessel was loaded in the same manner.
It was concluded the problem was not severe and may have been residual leftover material from the loading as the outlet spool was not inspected or cleaned during the TAR.

Figure 7. Desiccant and ceramic balls at the Outlet Collector

DP Still Increases

After restart, the bed DP was continued increasing after cleaning the outlet collector and spool at a lower rate of 50mmWC per cycle instead of 400mmWC. This was determined to be acceptable for short term but needed to be rectified to ensure long term sustainability of the plant. It became clear that there was a problem inside the bed.
Fortunately, the plant had to be stopped in January 2019 for repairs to other sections of the plant, and the bed was unloaded, inspected, and luckily the cause was finally identified and rectified.

Root Cause

Third Time Lucky!

The mol sieve B was opened in Jan 2019 to inspect the internals to determine the cause of the increasing DP for the third time.
Adequate time was available to completely unload and reload the bed, all material and resources were available.
When the bottom mesh was inspected, there was a gap between the screen and the vessel wall. The desiccant was passing through this void between the bottom screen and the vessel wall and desiccant along with the support balls entering the outlet spool. The backflow regenerations caused the

beads to constantly move and rub against each other causing excessive dusting that blocked the filter elements over time.

Figure 8. 18X 18 Mesh showing space at walls

Figure 9. Mixing of Desiccant and Balls Directly Under the Screen

The bottom mesh was incorrectly installed. The overlap of the 18 X 18 screen and the vessel wall was insufficient. The vessel drawings indicated a minimum of 25mm was required.

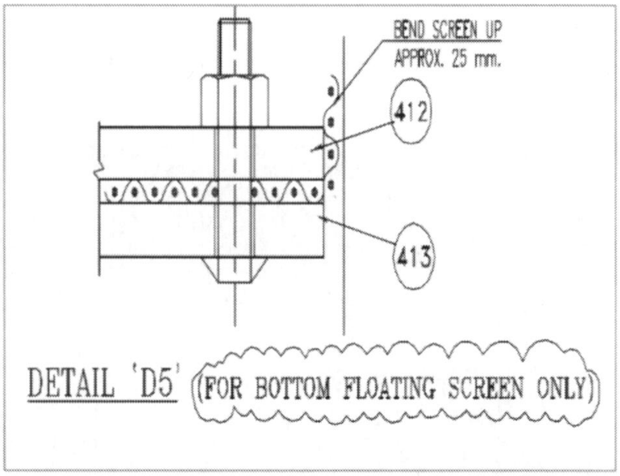

Figure 10. Drawing showing 25mm Overlap

Figure 11. Fine Mesh overlap with vessel wall after loading

Figure 12. Installed Rope Packing (Steel Wool)

Conclusions

Mol Sieve Desiccant Replacement

Based on the findings the following suggestions are proposed.

- Remove, inspect and clean the outlet spool on each mol sieve vessel whenever replacing desiccant.
- Ensure the loading is done according to the OEM requirements and in line with technical drawings.
- Loading procedures must include all aspects of the loading, this should include critical quality aspects like overlap, stitching requirements, outages etc.
- Ensure photographic or video records are obtained before proceeding to the next step.
- Install manual isolation valves in series with the inlet and outlet MOV to enable online filter replacement.
- Change filter elements to metal screen type instead of pleated to enable reuse of filters if they become blocked.
- Install rope packing beneath the fine mesh below the overlap a second layer of protection.
- A de-dusting procedure with piping provision is recommended.
- Ceramic balls are susceptible to breakage, install high grade alumina balls instead of ceramic balls
- Filters can be upgraded to use a mesh type with quick release housing cover to enable reuse if they become blocked.
- When having plant problems, take sufficient time to find the root cause otherwise the problems might come back after plant restart.

Purifier Back Blow

- Backblow the purifier at each TAR
- Backblow the inlet pipe spool piece as well as cold box inlet.

- Keeping records and reports of abnormal situations can be helpful if similar problems re-surface in the future.
- Always consult with equipment OEM or plant licensor even if the problem was experienced before. This is to ensure that new, better ways of working can be applied.

Pre-Commissioning, Commissioning, and Initial Start-up Experience of Gresik Ammonia Plant

In early 2018, Gresik completed the construction and started up its 2000 MTPD Ammonia 1B plant based on KBR's Purifier™ Technology. The plant is located close to Gresik city in East Java province of Indonesia. Commissioning and start-up was successfully completed in July 2018, with supply of ammonia to its newly constructed 1725 MTPD Urea Plant. The Performance Test was successfully completed with net energy consumption well within the guaranteed number. This paper presents an overview of the project and its journey through various stages with focus on pre-commissioning, commissioning, and start up. It also shares the problems encountered and key lessons learnt during the project phase.

Sebastian Nababan
PT. Petrokimia Gresik

Rohit Khurana
KBR

Introduction

This paper shares the Petrokimia Gresik's successful commissioning experience of a 2000 metric tons per day Ammonia Plant completed in July 2018. It provides the roadmap of the project starting from operator's team build up, training phases including training at the existing Ammonia plant, Operator Training Simulator (OTS) and then the field execution of pre-commissioning and commissioning activities.

Prior to Ammonia 1B, PT. Petrokimia Gresik operated an ammonia plant along with a Urea plant. The existing 1350 MTPD (1488 STPD) ammonia plant is based on KBR's conventional process (Former MW Kellogg). The new Gresik 1B plant utilizes KBR Purifier™ Ammonia Technology. The plant is designed for a nominal capacity of 2000 MTPD (2200 STPD) Ammonia with maximum capacity of 2200 MTPD. The 1725 MTPD (1900 STPD) Urea plant, utilizing Toyo's ACES 21™ technology was also constructed.

Project Overview

Since established in 1972, PT Petrokimia Gresik (PG) has played important role in supporting the Indonesian Government program on agricultural production with its fertilizer products. PG is one of largest fertilizer producers in Indonesia with total 5 million metric tons of fertilizer produced in 2017, mostly NPK based fertilizers. To create a more competitive overall production cost, and meet increasing fertilizer demand, PG decided to build a second Ammonia-Urea complex (Ammonia-Urea II Project), 660.000 TPY Ammonia and 570.000 TPY Urea Prill.

The Project is located in Gresik, East Java, close to the existing Ammonia-Urea 1A plant.

The project was completed in 42 months after passing through many phases and challenges, with 12.941.017 man-hours without a single fatality and no lost time injury (LTI).

Project Execution

The Gresik 1B Ammonia Urea complex was constructed on EPC Lump Sum basis by a consortium which consist of two main contractors, Wuhuan Engineering Company (WEC) as the Consortium Leader, and PT Adhi Karya Tbk (ADHI) as the member. WEC had the overall responsibility starting from Process License to Mechanical Completion and Performance Guarantee Test Run of Ammonia and Urea Plants, while ADHI had the responsibility for Cooling Water System, Demineralized Water Plant, Waste Water Treatment, and Product Handling Facilities.

Other facilities outside of this Project scope included: Raw Water, Liquid Ammonia Storage, Off-site Piping, Start-Up Boiler and Gas Turbine Generator and were supplied by independent utilities companies.

The ammonia process was licensed by KBR. Proven state of the art technologies are used in this plant to ensure that the plant can be operated safely and be most energy efficient.

From the very first stage to the end of the project cycle, engineering documents were reviewed by the PG engineering team. Any discrepancy from the Project Specification was resolved by mutual agreements among Owner, Licensor and the Contractor even though the Project was EPC Lump Sum. With this procedure, extensive experiences from Contractor, Licensor and Owner were applied to the Project at the earliest stage. Upon Mechanical Acceptance of each discrete part, the PG operation teams took the responsibility for operation under main contractor's direction and coordination. This strategy served well for PG personnel to have the deep insight of technical and operational aspect of the Project, also gave enough hands-on experience for the operation and maintenance team to operate the Plant safely and smoothly after handover from the Contractor.

Project Timeline

The Project Contract was signed on December 2014, with original duration of 34 months to Plant Acceptance, however due to Government Regulation changes, weather, and some utilities supply issues, it was revised to 42 months. BED (Basic Engineering Design) was finalized in June 2015. Construction work started as soon as contract was signed.

After mechanical completion, Steam blowing was started from July 2017 and ended on January 2018. The Reformer was fired for start-up on 15 March 2018 and first drop of ammonia was produced during converter reduction on 30 April 2018. The urea plant followed with its first product on 20 May 2018. The performance & demonstration test was completed on 18 July 2018.

Milestone	Date
Contract Signed	December 2014
BED Finalized	June 2015
First Pilling at Utility part	June 2015
First Pilling at Ammonia Plant	September 2015
Steam Blowing Start	July 2017
Air Blowing	August 2017
CO_2 Removal Cold Circulation	January 2018
Ammonia Plant firing for first start-up	15 March 2018
Ammonia First Drop	30 April 2018
Ammonia Performance and Demonstration Test	July 2018

Table 1. Project Milestone for Ammonia 1B

Ammonia Process Design Features

The ammonia plant is based on the KBR Purifier flow sheet, a low energy natural gas reforming process as shown in Figure-1 and utilizes the BASF licensed OASE White (activated MDEA) solvent for the CO_2 removal process.

The plant design is based upon the natural gas composition listed in Table 2.

Components	Mole Fraction, %
CH_4	97.5
C_2H_6	0.75
C_3H_8	0.37
C_4H_{10}	0.17
C_5H_{12}	0.21
CO_2	0.5
N_2	0.5
Total Sulphur ppm	25
Mercury content $\mu g/Nm^3$	89

Table 2. Natural Gas Feed Composition

All the components of the ammonia plant are well proven technology features. All process equipment is single train with all compressors as centrifugal compressors driven by steam turbine.

The unique design features of the Gresik Ammonia-1B Purifier plant are:

- Mild conditions for primary reforming via excess air to the secondary reformer.
- Mercury guard bed to remove mercury from the feed. Primary reformer with induced draft and forced draft fans each have steam turbine as main driver and stand by motor driver.
- KBR Proprietary Floating Head Waste heat boiler design
- KBR's nonmetallic mixing chamber (no metallic mixer or burner) in the secondary reformer
- Low Temperature Shift Guard Bed. The guard bed prevents the main catalyst bed from poisons.
- BASF's OASE (previously named aMDEA) two stage process for Carbon dioxide removal
- KBR Purifier for adjustment of synthesis gas hydrogen: nitrogen ratio and removal of all methane and the majority of the argon present.
- KBR horizontal ammonia synthesis convertor with 3 catalyst beds and 2 interchangers
- KBR unitized chiller
- Front end hot vent and a back end flare
- Use of KBR's operator training simulator
- KBR's Remote Performance Monitoring

Plant Performance

The plant performance guarantee test was conducted in July 2018 over 7-day test run period. The plant demonstrated that all performance guarantees including capacity, product specification and energy consumption were all met.

Table-3 shows the comparison of the ammonia plant performance against the design. The "measured" values were average measured data over the test run period, while the "expected" values are from the design issue of the process release. The energy value is in terms of Gcal per metric ton of ammonia on a higher heating value (HHV) basis.

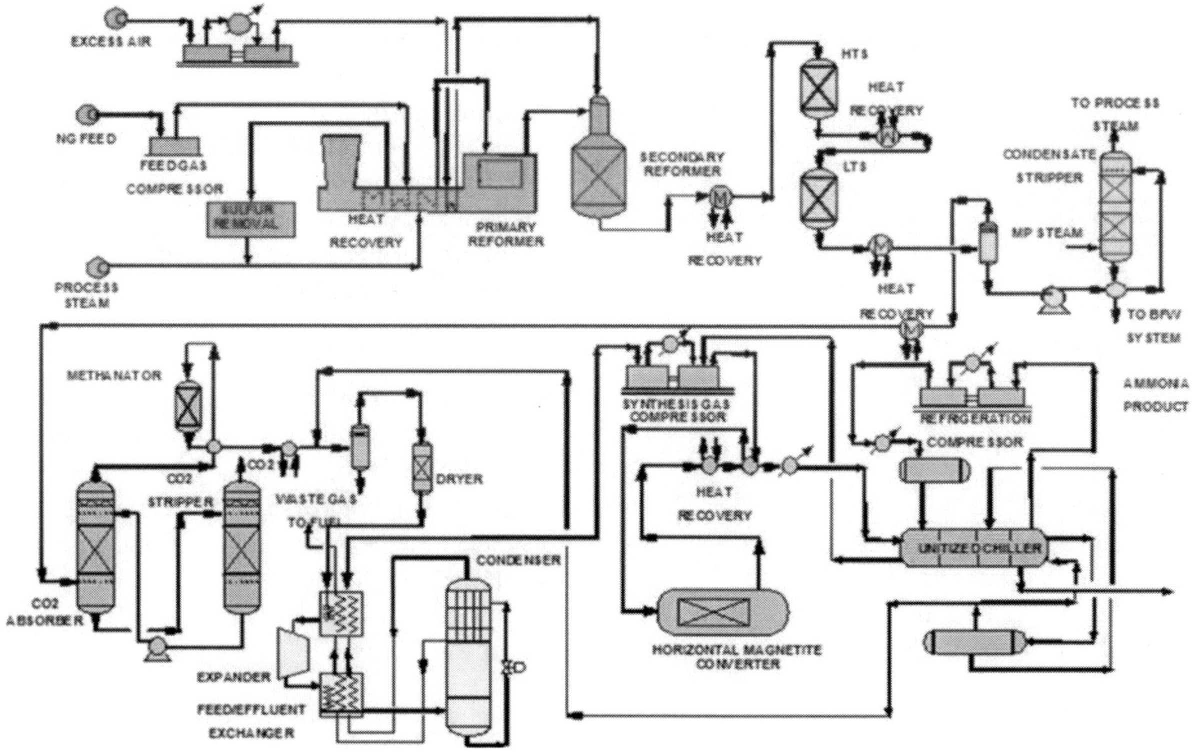

Figure 1. KBR Ammonia Purifier Process

	Expected	Measured
Ammonia Production		
Capacity MTPD	2000	2109
NH3, wt %	99.5 min.	99.86
H2O,wt %	0.5 max.	0.07
Oil, ppmw	5.0 max	0.07
Net Specific energy (HHV) (feed+fuel-steam credit+power)		
(Gcal/MT, HHV)	7.938	7.598
(MMBTU/MT)	31.5	30.17
CO_2 product		
CO_2, drymol%	99.0	99.99
H_2, drymol%	0.8	ND (Not Detected)
N_2 & inerts, dry-mol%	0.2	0.01
Stripped Process Condensate Quality		
Ammonia (ppmw)	10	4.47

Table 3. Comparison between the expected and the average measured values over the Test Run

Pre Commissioning, Commissioning and Start up

As per the LSTK contract, the main responsibility of all the activities lies with the contractor. However, all the procedures had to be approved by PG. Every pre-commissioning activity was closely monitored and supervised by PG. During Start-up Three operation teams from Contractor, PG, and Licensor worked closely to ensure the smooth and safe running of activities.

Contractor Team Role

The main contractor role was to

- Coordinate and direct all the work for the Pre-commissioning, and Commissioning.
- Coordinate all activities related to vendor and licensor
- Execute the work together with PG

Language barrier was a challenge as most of operation experts from the contractor spoke limited English. To bridge the language barrier, it was mutually agreed that contractor would assign at least one English speaking engineer in the DCS room.

PG Pre-Commissioning and Commissioning Team Role

Unlike the conventional LSTK projects, where contractor provides a complete team for commissioning and start up, the contract specified that PG would use their own resources for commissioning together with Contractor supervision. The team of relatively inexperienced operators endured all challenges remarkably which saved millions of dollars which are normally paid to the contractor for commissioning services. An organization and communication chart (Figure 2) was established to ensure a smooth running of Pre-commissioning and Commissioning.

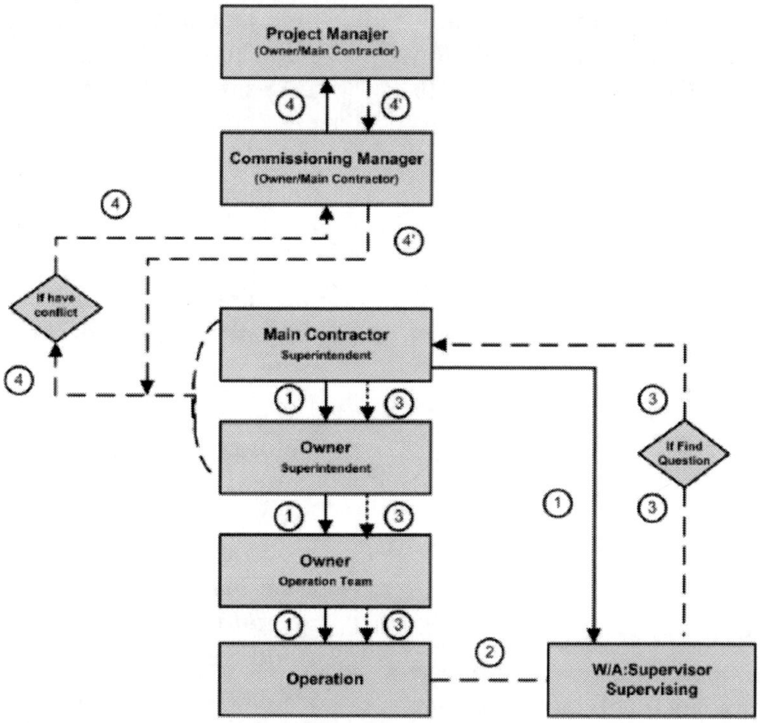

(1) Means : it is the route to pass the Commissioning and Start up order

(2) Means : Supervisor of W/A will survive the operation

(3) Means : If the Supervisor of W/A find some problem , they will give new order

(4) Means : It is the route to solve the conflict. If there are some conflict between Main Contractor's superintendent and owner's superintendent

(4') Means : It is the route to give the final order to solve the conflict

Figure 2. Organization and Communication Chart of Commissioning

The average age of PG operation team was 30 years old, which is relatively very young, with 48% of the team were below 25 years old. The preparation for operators started simultaneously with the start of the Project. New recruits placed at Ammonia 1A until the start of pre-commissioning of Ammonia 1B. During this time, operators were trained about the Ammonia-1B, and Pre-commissioning, Commissioning, and Start-up activities which would become their responsibilities.

Operator Training Simulator

Operator Training Simulator (OTS) played important role in accelerating operator understanding of the new plant. The interface and simulation were very similar to the real DCS and process. In many cases, potential problems in an actual control or interlock were identified during OTS testing. Ideally, the OTS should have been ready during construction phases, when the operation team would relatively have less activities. However, in this Project, the OTS came late due to delay in finalization of DCS engineering. The OTS was ready almost at the same time as preparation for pre-commissioning. Soon after the OTS was ready, operator started the OTS training. Total 1,840 OTS training hours (average 43.8 hrs/person) were gained.

Licensor's Role

To ensure project success, right from the start, emphasis was placed on working as a combined project team and ensured by involving all the parties at all stages of the project.

KBR, other than providing the Basic Engineering Package and Proprietary Equipment for Gresik 1B ammonia plant, was also involved in crucial stages of the project. This not only included engineering design reviews, P&ID reviews, 3D Model reviews (60%), and HAZOP but also inspection of critical equipment along with the PG. KBR & PG also conducted the P&ID walk-throughs and interlock tests together to ensure that the plant was built to satisfy all the technical requirements. Training & pre-commissioning advisory services were also provided as needed by KBR.

KBR on-site representatives monitored the startup and commissioning right from the first burner light up till the Performance Guarantee test run was completed.

Pre-Commissioning Activities

Pre-commissioning of the project is a gigantic task, it was well understood right from the beginning that success of the project would lie with a 'once through' approach for pre-commissioning, as saving of some hours by skipping steps during this activity can cost days during commissioning. Detailed procedures were developed by the contractor and approved by PG. Looking at the tight deadlines, PG maintenance team also joined the contractor's team. This also helped in getting 'on the job training' for the PG team prior to taking the full responsibility of the plant after the hand-over.

Before each specific pre-commissioning activity started, a briefing meeting was conducted between PG team and contractor team to refresh the procedures and discuss any safety concerns. Separate meeting was also arranged for briefing the operation team, to make sure that the team understood the procedures, the objectives, and safety precautions needed.

Prior to do a Pre-commissioning activity on a specific system, a walkthrough by PG, contractor and safety officer ensured the readiness of the system with respect to safety. For example, before do steam blowing to point A or B, walkthrough check would be carried out a day before. For every first-time walkthrough, each team leader led the walkthrough to make stringent safety requirements, which would become the benchmark for other operators to follow.

Line Checking

Operators did conformity check on piping construction by comparing it to Piping & Instrumentation Diagrams (P&IDs). The checks were performed at an early stage which helped in reducing rework on the piping, early recording of any concerns and modifications prior to hydrotest. This activity also helped the operator to understand the Plant piping lay out faster.

Water Flushing

Cooling water system, demineralized water system, part of CO_2 removal OASE system were cleaned using water flushing techniques. Special attention was required for temporary electrical equipment and power connections to be safe and protected from potential water flooding. Proper and adequate drainage systems were ensured to be available prior to flushing.

Good flushing resulted in lower frequency of strainer cleaning. Harder part was to flush the cooling water system, due to its huge system, and big size of pipe (72") that made it harder to achieve high flushing velocity. Frequent open flushing would result in huge waste of water, it was decided to perform closed circulation of the cooling water system after the open flushing. Mud collected on the cooling water basins were cleaned manually.

Figure 3. Cooling Water Flushing

Steam Blowing

Steam blowing was the most energy intensive in the pre-commissioning activity. It took almost six months to properly finish all the steam blowing points.

Due to the nature of the activity, special concerns on safety were always reminded to all PG and Contractor teams. Extensive coordination were made to all stakeholder such as, offsite steam supplier, subcontractor that worked closely to blowing area, nearby plants and most important nearby residents. Because of good communication made to prior steam blowing, there were no major complaints from the resident. Steam blowing was performed at daylight only and on some special occassion such as religious holidays, the steam blowing was stopped. During steam blowing, the safety officer routinely checked conditions (such as unsafe conditions, noise, complains) not only inside project area, but also the surrounding plants and residential area and closely communicated with steam blowing operators.

An efficient scheme of steam blowing was needed to decrease the amount of steam wasted during steam blowing. Inefficient steam blowing would have caused more time to blow, thus wasting lot of steam and delaying other work. Some of the major highlight of steam blowing were:

- Complete steam layout were made, and blowing out point were named. Every valve was identified on the layout with tag named on site.
- Biggest headers for steam blowing were prioritized. To clean the biggest header, it was required to blow a number of blow out points (BOP) simultaneously to create lower backpressure to increase steam velocity and *blowing factor* (K) on the pipe largest-size section.
- Temporary pipe size to silencer and silencer size significantly influenced the real calculated K as undersized temporary pipe

would create higher backpressure and lowered the steam velocity

- To optimize the blowing time, BOPs were grouped based on their lay out, and blown at the same time. After blowing for some time, the blown-group was switched to new BOPs group, and this sequence was repeated throughout the day. This reduced the need for opening closing BOPs valves frequently and also fluctuation on boiler load.

Some of safety precautions made for the steam blowing activity:

- A day before a BOP be blowed, a safety clearance check by Contractor and PG operation and safety officer was required
- Insulation must be provided for piping that is close to personnels working area and temporary power cabels.
- All operation personnel were frequently reminded of weak points of the steam piping, which are temporary lines. A dedicated emergency vent (manual valve) was provided for operator to release pressure in case of pressure increases.

Despite strict safety precautions in place, a near overpressure incident on a medium pressure steam (MS) line was experienced, however it was quickly resolved by an operator without major incident by using the emergency vent. The investigation revealed that miscommunication during switching of BOP resulted in closing of wrong valve. This incident happened at *critical-time* (near end of shift, lunch break, praying time, etc). After the incident, some additional precautions were made:

- Transmitters and venturis were installed after some blows as it was expected that the line was already clean enough. Soon the PSVs and vent lines were blowed, and PSVs & control valves reinstated. With these installations no temporary line was left on MS lines, and operator operated and coordinated much of the activity from DCS.
- It was emphasized that during operation of BOP valves, both PG and Contractor operation must be present and accompanied by an English speaking engineer from contractor.

Transient operation (start/stop blowing or switching BOPs) shall be avoided at *critical time*.

Figure 4. Steam Blowing on Reformer Side

Air Blowing

There were three methods of air blowing commonly used in pre-commissioning of an ammonia plant.

- Pressure shock blowing,
- Intermittent blowing using CO_2 Absorber as storage
- Process air compressor

Commissioning of process air compressor was delayed, as steam blowing for steam system was still in progress. Pressure shock blowing was an option, however Contractor and PG were of same view that pressure shock blowing have the highest risk activity in the precommissioning. It was decided that maximum pressure for shock blowing would be 4 barg (60 psig) to reduce the safety risk. Pressure shock blowing activity progress wereslow. For every pressure shock blowing cycle, the outlet area was cordoned off & extra personnel were placed to guard the area and make sure no person enters dangerous zone, loud sirens were sounded prior to pressurize the pipe to blast.

At the same time, air blowing using CO_2 absorber as storage was also prepared. 3 X 2500 Nm^3/h (1555 SCFM) Plant Air Compressor were used to pressurized the CO_2 Absorber. Temporary lines were constructed across the plant. The temporary lines were also hydrotested and equiped with PSVs for safety reason. This scheme served as blowing method for major portions of the pipe, while process air compressor was mostly used for dedusting of catalyst.

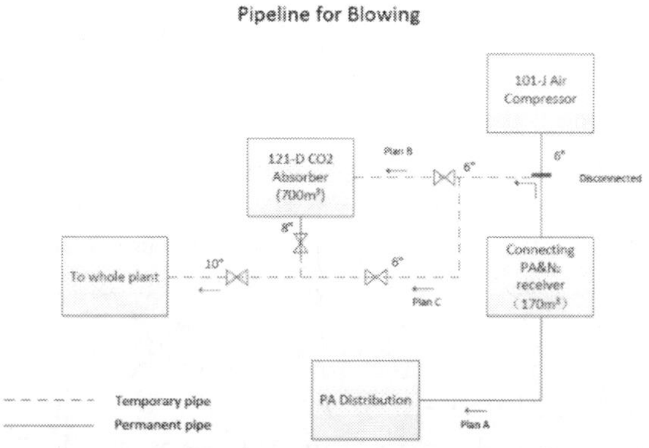

Figure 5. Scheme of Air Blowing using Plant Air Compressor and Process Air Compressor

A dynamic process calculation was carried out to study the effectiveness of blowing with CO_2 absorber as storage method. It was shown that for blowing the 16" line suction of feed gas compressor, using CO_2 absorber storage will take approximately 1.4 minute duration (Absorber pressure decrease from 8 kg/cm^{2}g (114 psig) to 6 kg/cm^{2}g (85 psig)). The velocity during blowing can reach 202 m/s (660 ft/s) at maximum compared to 9.42 m/s (31 ft/s) during normal operation. It would need 25 minutes to repressurize the CO_2 absorber again and repeat the cycle.

CO_2 Removal system cleaning

The OASE System was properly cleaned by means of mechanical cleaning, water flushing, and degreasing. The objective was to remove oil, grease, mill scale, rust, or other contaminants from the system that may contribute to foaming of the OASE solution. The time used in properly cleaning the OASE systems pays large dividends later on when process gas is admitted to the system for removal of CO_2. Therefore, special efforts were made to have the system as clean as possible before start-up. Licensor's extensive experiences in pre-commissioning of OASE system cleaning were shared with Contractor and PG team, and applied wherever it was possible.

The cleaning process was already started since the construction of CO_2 removal piping, with the *built-it-clean* philosophy on the piping construction. Construction worker dilligently cleaned every pipe sections, and Precomm team checked and recorded the cleanliness of every section. Only after thorough inspection, the pipe section was then fitted and welded to the piping.

A temporary demineralized water pipe provided good pressurized water jet to flush every chamber on CO_2 removal towers. Rust and mill scale that could not cleaned by water jet, were brushed.

CO_2 removal towers were pressurized for preparation of cold water circulation. The towers were pressurized using plant air instead of nitrogen which eliminated the safety concerns with risk of asphyxiation (leakage risk of nitrogen) of many construction workers arround the area, after cold circulation cleaning, the water was drained, and the towers were opened for inspection. No anomaly on the tower internals was found, however all carbon steel walls rust appeared back as it was before. Apparently, that was due to the usage of plant air for pressurization combined with warm water and humidity. It was decided to repeat the manual cleaning of the carbon steel wall, and redo the cold water circulation proces with pressurization using nitrogen.

Chemical Cleaning

Chemicals were used for degreasing of the CO_2 removal system and for chemical cleaning of pipes and passivation of BFW line. A Job Safety Analysis was performed for this activity. Leakages were identified first with fresh water circulation before the chemicals were loaded. During chemical circulation, operators were dilligently patrolling the circulations line, checking every flange and instrumentation point, even in the most remote location. Once small leaks were identified, the construction team immediately repaired the leak. Discharge of chemical waste was coordinated with environment department and chemicals vendor to ensure that the procedure and execution is complied with regulation.

Commissioning Activities

Prior to start of the commissioning, PG dedicated a special team to do *hammer test* (quick check on bolt tightness) on all bolt flanges; loose bolts were immediately retightened. One fire incident happened from an untight flange leak in the CO_2 removal Process Gas line; the fire was immediately extinguished by PG operator.

Ammonia-1B PG organization chart was designed to be 4 shifts for normal operations, however, during Commissioning until Plant Performance Test, it was decided that the team is splitted into 3 shifts with these objectives:

a. More people in one shift, will ease the operator work load.
b. Each people will be exposed to 12 hours learning for each shift, this is expected to accelerate the operator learning curve, compared to normal 8 hours shift.

A Pre-start-up Safety Review (PSSR) was finished before start of Commissioning (NG feed in) to make sure that every step before commissioning was completed and the plant was ready to receive Natural Gas. Step by step commissioning process were explained to the team with the help of Licensor Commissioning personnel. Licensor personnel were always present at the plant and guided the commissioning process until the finish.

Nitrogen Circulation & First Firing

Natural gas compressor was used for nitrogen circulation for warming up of front end circuit. Reformer was fired after the nitrogen circulation was established, and reformer and High Temperature Shift Converter temperature were increased slowly before being switched to steam heating. Low Temperature Shift Converter was reduced using hydrogen taken from Ammonia-1A interconnection line, and warmed up using dedicated blower and heating circuit.

Gas Introduction

During steam heating, it was found that both feed gas control valve and feed gas shutdown valve were leaking. To prevent coking of the reformer catalyst, the drain valve between feed gas control valve and shutdown valve was opened and directed to safe location. After steam heating, natural gas was fed to the reformer after reaching the desired outlet temperature of 700°C (1292 °F). Operators performed routine checks on tube color, and looked for any hotspot sign on tubes. Reformer tube skin temperatures were check by special high temperature thermal imaging. After reduction of reformer catalyst, process air was introduced. HTS was reduced and desulfurized without any issues, and LTS dedusting followed soon. The CO_2 removal section was commissioned smoothly. For each step, operator and safety officer kept on inspecting flanges for any leaks.

Methanator and Molecular sieve drier

Prior to methanator heating, the heating system was rechecked by operators. Many leakages were identified in the steam line to methanator trim heater, and methanator heating was put on hold until the leakages fixed. The area was to be cleared of personnel during methanator heat-up until effluent bed temperature above 205°C (400 °F). The methanator pressure was maintained as low as possible during heat-up for deterring the reaction rate during the carbonyl formation period as well as to reduce any presence of leakage. Above 205°C as the possibility of carbonyl formation was now remote, the methanator was pressurized, and checked by field operator and safety officer for leakage.

After the methanator, process gas was introduced to molecular sieve dryer. Dryers were regenerated two cycles each before starting synthesis gas to the Synthesis loop.

Converter Catalyst Reduction

Flanges of synthesis loop were leak tested at every step of pressure increase until it was ready for the catalyst reduction process.

The Start up heater was lit for ammonia converter catalyst reduction. The reduction process was carried out under direction and supervision by catalyst vendor. Reduced water was sent to the Urea Plant to be used during Urea Plant Start-up. The Hydrogen to Nitrogen (H/N) ratio at synthesis loop was maintained at around 2.3 – 2.6 during reduction as the purifier was bypassed, the reduction process was finished in 80 hours. The ammonia unit soon produced full capacity of amonia after purifer unit was put on line.

Purifier Start-up

To save time, Purifier unit was bypassed during ammonia converter reduction process. During this time, the purifier and expander cold box were dried out using nitrogen. High consumption of nitrogen led to longer time for this process. Once ready, the purifier was started-up smoothly without any major issues. Thereafter ammonia production rate was increased to 100% load for the performance test.

Challenges Faced

Though the journey of this project was very smooth, there were some challenges during the course of start-up and commissioning.

ID/FD Fan Clutch Issues:

Just about a week after lighting up the Primary Reformer, the ID fan clutch failed. A few days thereafter the FD fan clutch also failed. Both the clutches were de-coupled and removed to the workshop for repairs. Bearing damage was suspected initially. However, after opening the clutches, in addition to bearings, torque transmission parts were also found damaged. The clutches appear beyond repair. The damage on the clutches were attributed to poor storage conditions. Fortunately, as the plant had a dual drive for both ID and FD fan, the clutches issue did not hold up the startup, otherwise it would have caused major delay.

Reformer Hot Air Door Damage:

Metallic sheets were provided as heat shields to protect the hot air door cables and limit switches. However, during one of the trip events, the heat shields were inadequate cables and the limit switches were burnt from the hot gases. . To avoid repetition of this damage, opening of these doors were extended away from the combustion air duct by installing an extended duct away from hot areas.

Expander Oil Leak

After few days of operation with Purifier online, the plant was shut down for some days due to external factor. During the restart, it was observed the inlet guide vane of the expander would not open beyond a particular value of around 40-45%. Several attempts were made to operate the valve but that did not improve the condition. Expander vendor was immediately called to the site. After opening the expander box, a few drops of oil were noticed. Due to the ingress of the oil at sub-zero temperature ~ (-)130 °C, frozen oil cause the linkage between the actuator stem and guide vane assembly broken. The plug-in unit (spare unit in the form of complete modular expander) was installed to restart the plant as early as possible.

Converter Inlet Flange leak

Converter inlet piping break flange developed heavy leak at a pressure of 130 kg/cm^{2}g (1850 psig). Despite tightening to maximum hydraulic pressure using bolt tensioner, the leak could not be stopped. Eventually a vendor was called in to clamp and seal the leak on-line. Daily routine check were programed for the leakage flange.

Expander duct flange leaks:

Prior to purifier start up, high pressure leak test at 30 kg/cm^{2}g (426 psig) was performed for the Purifier piping in the expander box and no leaks were detected. During Purifier cool down, at a temperature of about -100°C (-148 °F) at the inlet of expander, the expander duct sample point started indicating explosives. Purifier shut down was taken to identify the leaks. After tightening at cold conditions, very small leakages were still found on three points, and boxes were installed on the suspected flanges to identify leaks. A tubing vent was provided for each box for venting outside of the expander duct. Purifier was again cooled down. Three flanges had minor leaks as detected from the boxes installed. However the expander duct itself were free of hydrogen and no explosive detected. It was decided to keep operating the purifier with the leaks till the next turnaround. As a safety measure, nitrogen purging to expander box was increased, steam hoses were provided at the flange box tubing vents, and explosive meters monitored the hydrogen content on a daily basis.

Synthesis Gas Compressor Startup-Recirculation in Loop

During the synthesis gas compressor first startup, it was observed that recycle stage (final) was surging as soon as it was started. Root cause analysis pointed out that some hydrotest water left in the synthesis loop which converted to ice at negative temperatures. To take corrective action, the operating parameters in refrigeration system was modified to ensure that the gas is always above 0 °C (32 °F). This also ensured melting of ice to water which was removed as ammonia-water mixture.

Lessons Learnt

1. Good team work between Owner, Contractor, and Licensor team is a success key for smooth and safe Pre-commissioning and Commissioning.
2. Owner commissioning team involvement during construction phase offers the best oportunity for personnel training and familiarization of the new plant.
3. Resources, tools, or facilities for pre-commissioning should be identified and prepared during engineering stage. Extra nitrogen supply for Pre-commissioning and commissioning must be prepared in advance.
4. Special care must be made to ensure no hydrotest water is left on synthesis loop and refirigeration loop.
5. OTS - Operator Training Simulator – OTS is a powerfull tool for training of operator, the delivery should be as early as possible. Developing of OTS should be as similar as possible with the real DCS, interlock, and control philospphy.

Summary

The Gresik 1B plant startup experience highlights the importance of a robust equipment preservation program, thorough plant pre-start-up checkout, and effective pre-commissioning practices.
Despite the challenges faced, the plant successfully met all its energy consumption, production, and product quality guarantees.

References

PT Petrokimia Gresik: "Final Report of Ammonia Urea II Project", Gresik, Indonesia 2018

KBR: "Operation Philosophy Manual", July 2015.

Project Records, including daily, and weekly minute of meetings

Inspection for High Temperature Hydrogen Attack

Inspection for High Temperature Hydrogen Attack (HTHA) has proved problematic. In the last few years, the industry has adopted the Advanced Backscatter Ultrasonic Technique (AUBT) as a standardised inspection technology for HTHA. This approach involves not only a prescriptive assessment of ultrasonic data, but also on interpretation of that data to identify other possible sources of signal. In the event that the received UT pattern is not strongly representative of HTHA, alternative reasons for that signal are provided. The final result is an ultrasonic determination of the worst case scenario and a qualitative determination of the probability of that result. Recent experiences have highlighted shortcomings with the AUBT approach and that modifications and enhancements to the methodology can significantly increase the likelihood of detection of HTHA.

David Keen
Incitec Pivot Limited

Chris Jones
Quest Integrity Aus Pty Limited

Charles Thomas
Quest Integrity NZL Limited

Introduction

Managing the risk of damage due to high temperature hydrogen attack (HTHA) has presented a major problem to the synthesis gas industries. The reasons are largely twofold. Firstly, the likelihood of HTHA occurring has proven difficult to predict. Secondly, detection of HTHA by inspection is problematic. Given the potential for extreme consequences in the event of a catastrophic release of highly flammable gas, as illustrated by the Tesoro Anacortes Refinery incident in 2010 [1], the need for a multifaceted approach based upon an integrated risk management plan is imperative. The consequence of HTHA failure is illustrated by Figure 1, which shows the hydro-treater exchanger that failed at the Tesoro Anacortes Refinery and resulted in seven workers losing their lives. This photograph is reproduced but remains a poignant reminder of the potential for catastrophe represented by unchecked HTHA damage.

This paper describes recent experiences relating to damage found in equipment that may have been considered safe using traditional approaches for managing the risk and highlights the inspection techniques used to detect that damage.

Figure 1. The damaged heat exchanger at the Tesoro Anacortes Refinery [1]

High Temperature Hydrogen Attack

The process of HTHA requires the dissolution of atomic hydrogen. This is normal and all ferritic steels operating in hydrogen at elevated temperature and pressure will dissolve hydrogen. The higher the temperature and pressure, the greater the amount of dissolved hydrogen. Another manifestation of this phenomenon is the frequent need to "bake-out" or heat treat steels to remove dissolved hydrogen before attempting to weld the steel. The dissolved hydrogen is reactive. It will react with any free carbon and with internal carbides, both of which have significant influence on material strength. This reaction generates methane gas. The more stable the carbides in the steel, the less likely the reaction is to occur, and hence more alloyed grades are more resistant to HTHA. Methane is not soluble in steel and forms in small pockets and around internal inclusions. The pressure in these pockets can be extremely high and localised high stress levels are generated leading to the formation of internal fissures and micro-cracks. Near-surface pockets of methane can lead to blistering occasionally leading to visual warning of damage. The formation of micro-fissures and the loss of carbon results in serious reductions in strength, ductility and toughness. The reduction of these material properties will ultimately result in the component being unable to withstand the normal service loads and catastrophic failure will occur. The visible and serious consequence of HTHA is illustrated in Figure 1.

Figure 2 shows an example of the micro fissuring that is caused internally by HTHA. Ideally, this should not occur at all. If a risk exists, a reliable inspection technique is essential.

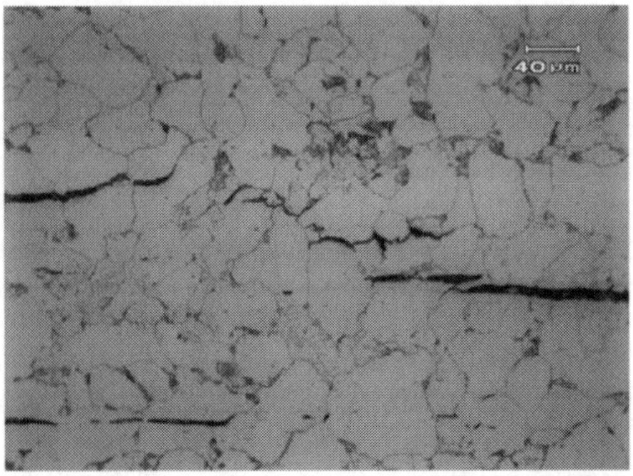

Figure 2: Internal fissuring caused by HTHA

Management of HTHA

The approach to HTHA management used by Quest Integrity and Incitec Pivot Limited (IPL) is summarised in Figure 3, which places emphasis on determining risk and acting to mitigate that risk. The ultimate goal is typically the elimination of risk by replacement with a modified design. Of prime importance are the identification of at-risk plant and the level of that risk. For many years, significant faith has been placed in the Nelson curves [2] as described in API 941 to predict the likelihood of HTHA. These curves are based on reported performance and have been subject to a number of significant changes as new experiences come to light.

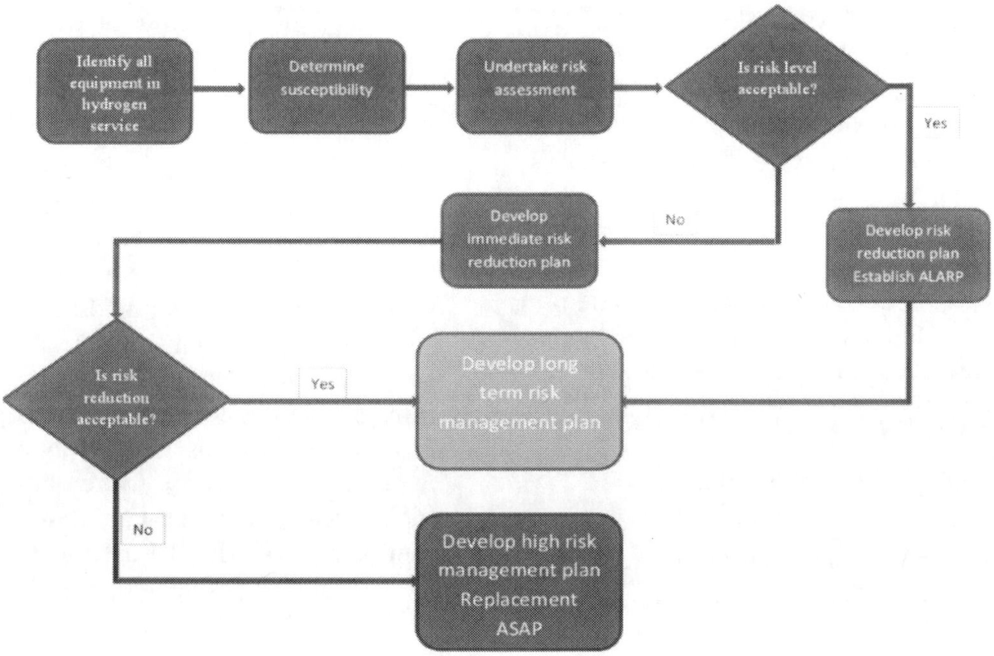

Figure 3: An approach to management of HTHA.

Figure 4 reproduces the Nelson curve from API 941 and this set of curves represents the core information in the API standard. In the 1990s, there was a growing number of cases of HTHA damage occurring in Carbon-½Molybdenum steels. The reaction was to remove the Carbon-½Molybdenum curve from the API 941 guideline. This placed a large number of items of Carbon-½Molybdenum equipment in limbo with the recommendation that the material should be treated as carbon steel as far as its resistance to HTHA was concerned. A large number of such items of equipment already in service and designed to the original Carbon-½Molybdenum curve now operated above the carbon steel curve.

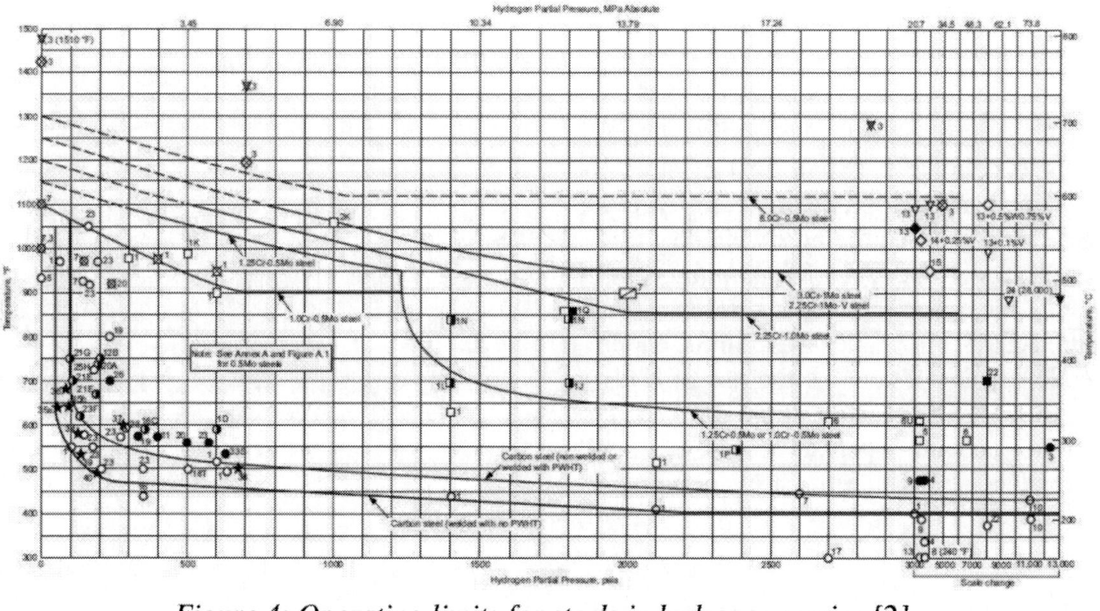

Figure 4: Operating limits for steels in hydrogen service [2].

An attempt was made to rationalise the experience with Carbon-½Molybdenum steel by providing different risk factors [3] for annealed and normalised material. The implication is that the metallurgical condition of the material, in this case the form and stability of the carbides, was significant. There is however, no reason to believe that such an effect is unique to Carbon-½Molybdenum steel or indeed parent metal as opposed to weld and HAZ material. An attempt to quantify risk factors using a parameter Pv dependent on pressure, temperature and time in service was included in the 2008 version of API 581 [3]. The Pv parameter approach has now been removed from the 2016 version [4].

The second major change has occurred in the wake of the Tesoro Anacortes Refinery incident and the subsequent Chemical Safety Board report into that incident [1]. The investigation concluded that failure of the exchanger occurred due to HTHA in a carbon steel pressure shell that was operating below the Nelson curve. Significantly, however, the exchanger had not been post weld heat treated and all the observable damage was present in the heat affected zone. API 941 contains significant qualitative advice concerning the influence of stress on the risk of HTHA and the 2016 edition contains the first attempt to quantify this for carbon steels by introducing a new curve for carbon steel that has not been post weld heat treated.

As in the case of metallurgical condition, there is no reason to believe that the effect of stress is unique to one particular grade of steel. However, the codes will likely require post weld heat treatment for alloyed grades As API 941 points out in some detail, the experience for steels that operate within the ASME allowable stress limits, and below the Nelson Curves, is that the risk of HTHA is negligible. The influence of stress does need to be taken into account when severe stresses such as weld residual stress, thermal stress or high piping loads, for example, exist. Knowing of the existence of such stresses is an essential part of any risk management plan.

Despite the above comments, API 941 and the Nelson curves remains the principal guide to the risk of HTHA. It is an experience-based document and continuous modification in the light of new information is to be expected. By themselves, the Nelson curves do not guarantee freedom from HTHA. As noted above, the API 581 [3] approach to establishing risk levels using a single quantifiable parameter Pv has been removed from the recommended practice. The 2016 version [4] has adopted a very conservative approach particularly for carbon and carbon-½Molybdenum steels. For these steels, the current version of API 581 assigns high susceptibility for any component operating above 177°C (350°F) with a hydrogen partial pressure exceeding 0.345MPa (50 psia). This is very conservative relative to the API 941 Nelson curves for both post weld heat treated, and non-post weld heat treated carbon steels. It is acknowledged however that this is an interim measure while a more quantitative method is under development.

For low alloy steels, calculation of risk factor has also been simplified. It acknowledges that the Nelson curves do not represent an absolute measure of risk and assign risk level based upon the margin by which any component operates below the Nelson curve (Figure 5).

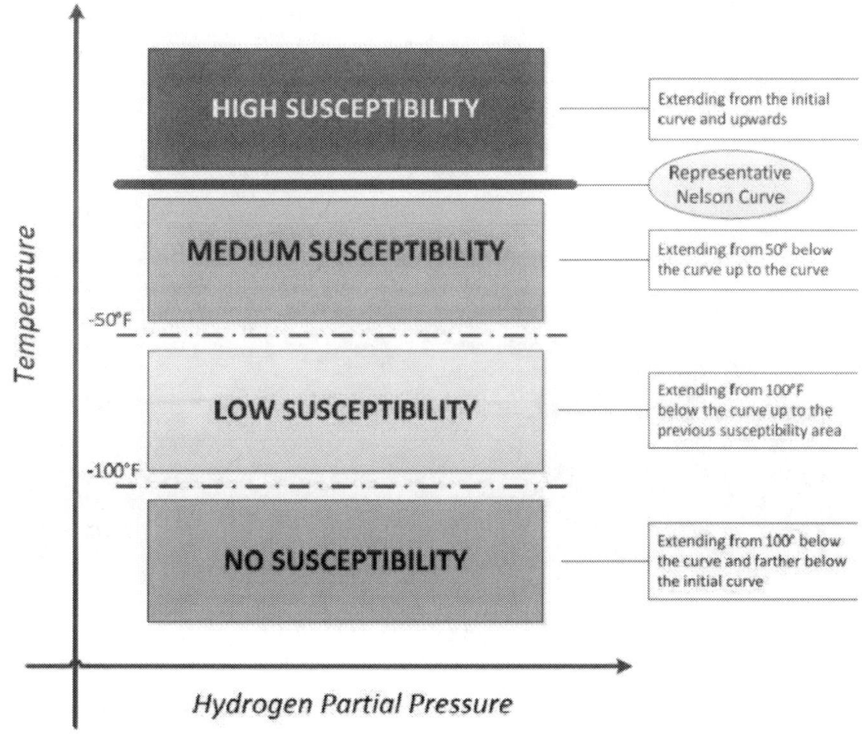

Figure 5: Risk assessment in an HTHA environment, API 581 [4].

This high level of conservatism on the one hand and scepticism around the reliability of the Nelson curves on the other hand has led some owners to develop their own methodologies. An example approach is summarised below.

Step 1: Identify Equipment in Susceptible Service. All equipment in hydrogen service is identified and categorize as:

- High Susceptibility – operating on or above the Nelson Curve.
- Low Susceptibility – operating below but within a 50°F and 50 psi below the curve.
- No susceptibility – operating more than 50°F and 50 psi below the curve.

Step 2: Undertake risk assessment. For all equipment identified as being susceptible to HTHA, a risk assessment is undertaken. Using standard risk methodologies, a risk matrix can be generated. The factors to be taken into account for HTHA include:

- Temperature.
- Hydrogen partial pressure.
- Material of construction.
- Exposure time.
- Stress.
- Cladding.
- Prior failures.
- Inspection history.
- Heat treatment.

Consideration of these factors and the consequences of failure result in a risk ranking using a matrix similar to that shown in Figure 6. The text against the various risk ratings provides user guidance, which will be unique to each organisation and is developed as a fundamental aspect of its pressure equipment integrity management processes. In the case of IPL, any component with risk assessed to fall within the red zone (upper right corner) is defined as "unacceptable to the board".

Such a risk ranking would typically result in a requirement to redesign and replace as soon as reasonably (safely) possible.

Likelihood	1	2	3	4	5	6
Almost certain	16	22	27	33	34	36
Likely	11	17	23	29	32	35
Possible	7	12	18	24	29	33
Unlikely	4	8	13	19	25	30
Very Unlikely	2	5	9	14	20	26
Almost impossible	1	3	6	10	15	21
	Consequence					
Financial Health and Safety Environment Business Interruption Reputation	Insignificant	Low	Moderate	High	Serious	Extreme

Figure 6: Example risk matrix

Step 3: Based on level of risk, develop an equipment targeted risk management plan. A key and primary objective would typically be to reduce the risk for those items of equipment in which the risk is assessed to lie within the high risk (red) area of the risk assessment matrix (Figure 5). This may be achieved in the short term to allow operation to the next scheduled outage by reducing operating conditions, usually temperature. Long term solutions involve:

- Development of robust Inspection Test Plans (ITP). This may include destructive testing.
- Enact inspection at scheduled outage.
- If HTHA found, consider equipment replacement with upgraded metallurgy.

Step 4: Integrate with Asset Integrity Management Plan. It is critical for the ongoing management of risk that knowledge and operational limits be understood and acted upon with the wider organisation. Key issues that must be addressed include:

- Knowledge, training and knowledge management. It is essential that the significance and risk of HTHA and indeed all potential damage mechanisms are understood throughout the organization.
- Establishment of Integrity Operating Windows (IOWs). The risk of HTHA is strongly influenced by operating conditions. Establishing and implementing IOWs is required to ensure continuing operation. This should include reporting protocols so that the impact of any excursion can be assessed.
- Plant Standard Operating Procedures (SOPs). The impact of HTHA should acknowledge the significance and risk of HTHA and the control measures that are in place to manage that risk.
- Management of Change (MOC). Robust MOC procedures are required to ensure that any changes to operating conditions do not impact negatively on the reliability of the plant.

Inspection

The risk management plan for each item of equipment that is deemed susceptible to HTHA inevitably calls on inspection to determine the presence and extent of HTHA. The ultimate control of HTHA is to replace equipment at risk. This is usually a costly solution and justification would normally require evidence that HTHA damage had indeed occurred. API 941[2] contains commentary on the various candidate inspection techniques and concludes that no single inspection technique is ideally suited to detection of HTHA damage particularly in its embryonic stage when micro fissuring is just beginning to form.

Ultrasonic inspection (UT) has been found to have the best chance of detecting HTHA. Four generic types have been used and these are as illustrated schematically in Figure 7.

- Backscatter techniques.
- Velocity methods.
- Attenuation.
- Spectral Analysis.

Backscatter ultrasonic testing involves irradiating the test material with moderately high frequency ultrasound and detecting the backscatter noise that is reflected back from HTHA. This can be used to detect HTHA and also to provide an estimate of the concentration and through thickness extent of damage.

Velocity methods make use of the fact that the altered microstructure from HTHA produces material with different shear wave ultrasound velocity than the unaffected material, whereas low frequency compression wave ultrasound velocity is relatively unaffected.

The back-wall reflection is essentially a wall thickness test and the extent to which this signal is attenuated by the presence of HTHA can be used as a measure of the extent of damage. An extension of this technique is spectral analysis in which the dependence of attenuation on the frequency of the ultrasonic signal is analysed. Higher frequency ultrasound is more extensively attenuated.

No single version of these techniques is ideal on its own. As such, a methodology has been developed that uses all four of these techniques. This is the Advanced Ultrasonic Backscatter Technique or AUBT. In this methodology, backscatter methods are used as screening tool to locate damaged areas and based upon the backscatter pattern observed, one or more of the other UT techniques are employed to quantify the nature and extent of HTHA damage.

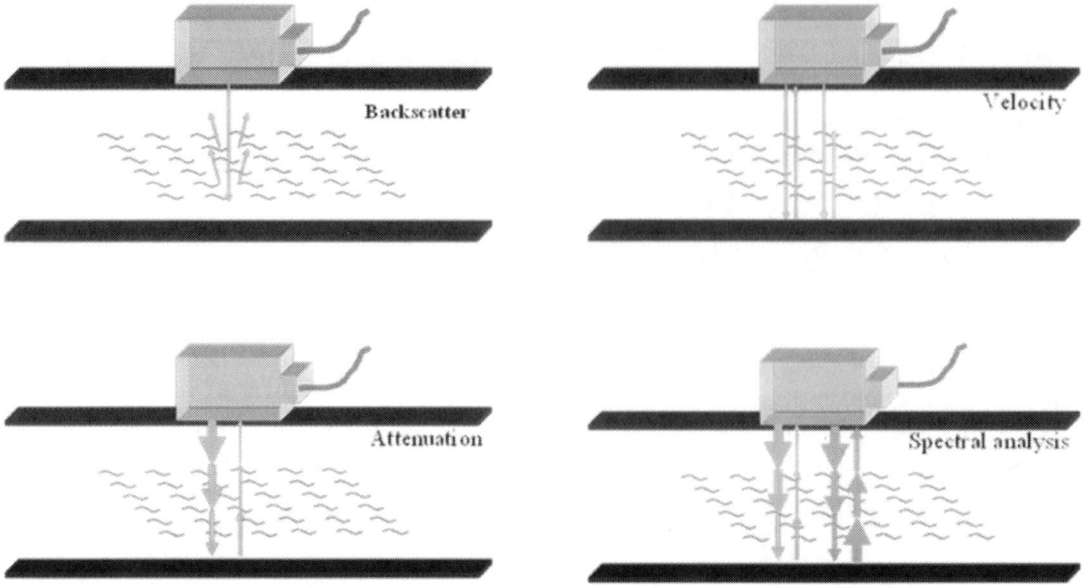

Figure 7. Schematic illustrations of UT techniques for HTHA.

The AUBT technique has been used extensively in the integrity management and inspection program at Incitec Pivot Plants (IPL). The results are summarised in Table 1:

Equipment items assessed susceptible to HTHA	**75**
Equipment items inspected for HTHA	**36**
Equipment items confirmed to have HTHA	**22**
HTHA identified by AUBT	**1**
HTHA identified by AUBT with qualified and experienced operator	**18**
HTHA confirmed without inspection	**2**

Table 1. Summary of HTHA inspection results at IPL.

The analysis of the history of inspection for HTHA at IPL has indicated the strong dependence on the knowledge and experience of the operator undertaking the inspection for HTHA. Of prime importance is the ability to interpret the findings of any of the various UT methods and understand which to subsequently use to build as complete a picture as possible of the extent, type and location of damage. AUBT inspections that have been contracted using inexperienced inspection personnel have been largely unsuccessful. Those inspections that have been undertaken by inspectors in which the owner had confidence have proved very successful. In this context, "qualified" means proving to the satisfaction of the plant owners that the inspectors and the procedures that they will employ have a very high chance of detecting HTHA. It does not mean that the inspector has a certificate that he has attended an HTHA training course. The complexity of HTHA inspection methodologies has highlighted the need for each operator/owner to critically assess the capabilities of individual inspectors as part of their QA/QC approvals and not rely upon a third-party certification.

Given the absence of any single and reliable UT technique that can be relied upon to detect HTHA, the findings at IPL has been that the experience and the knowledge of the inspector is paramount to the success of the inspection program as illustrated by the summary of results listed in Table 1. In the absence of the requisite knowledge and experience, only one case of HTHA was detected by AUBT as against 18 with an experienced operator. The importance of a robust asset integrity management program was noted previously and a number of the activities that needed to make such a program were highlighted. The results of the HTHA inspection program at IPL firmly emphasise the importance of other inputs to such a program including:

- Approval procedures for service providers.
- Critical review and statistical analysis of inspection results.

In the case of HTHA, it is strongly recommended that plant owners develop their inspection program for HTHA in partnership with their selected inspection company and maintain that partnership as long as is possible.

Closing Remarks

The experience of managing HTHA at IPL has highlighted the importance of a robust asset integrity management program. HTHA cannot be effectively managed in isolation by simple inspection and a large number of factors come to bear to maximise the likelihood of detecting and controlling the risk that this damage mechanism presents. Operating such a program is a continuous activity. The linkage between risk level and decision making particularly as that relates to capital expenditure is well established at IPL. The need to ensure that all staff appreciates the risk and that controls need to be in place is ongoing with IOWs, SOPs and training programs being developed to reflect that risk.

Inspection for HTHA is an essential part of the asset integrity management program and it has been found that such inspection is a particularly difficult task requiring significant expertise. Successful detection and quantification of HTHA requires the ability to select the most appropriate technique for any given inspection challenge and optimise that technique. The skill and experience of the inspection personnel is paramount and must be demonstrable to standards embedded in company Pressure Equipment Integrity Management (PEIM) procedures.

References

1. US Chemical Safety and Hazard Investigation Board Report 2010-01-I-WA, May 2014. (http://www.csb.gov/tesoro-refinery-fatal-explosion-and-fire/).
2. API Recommended Practice 941. Steels for Hydrogen Service at Elevated Temperatures and Pressures in Petroleum Refineries and Petrochemical Plants. Eighth Edition February 2016.
3. API Recommended Practice 581. Risk-Based Inspection Technology. Second Edition September2008.
4. API Recommended Practice 581. Risk-Based Inspection Technology. Third Edition April 2016.

Yara's Experience with Reformer Revamp at its Ammonia Plant in Saskatchewan

Yara Belle Plaine's Ammonia plant in Saskatchewan has undergone a series of revamps increasing its capacity from the original 1500 tons/d to current 2125 tons/d. The Primary Reformer firebox was expanded in 2009 to extend the tube and burner rows which created unforeseen challenges such as failure of inlet pigtails and flue gas mal-distribution. This paper details the benefits, challenges and lessons learned from this revamp of the Primary Reformer and describes many improvements that were made in 2018 which have been successful in improving plant safety, reliability and efficiency.

Waseem Haider, P.Eng
Yara Belle Plaine Inc.

Thorsten Brakhane
thyssenkrupp Industrial Solutions AG

Mike Hubbell
Johnson Matthey

Introduction

Yara owns and operates a fertilizer complex in Saskatchewan, Canada consisting of Ammonia, Urea and Nitric acid plants. The facility was commissioned in 1992 and has been in operation since then. The ammonia plant was designed by thyssenkrupp Industrial Solutions (tKIS), then Uhde, with a nameplate capacity of 1500 tons/d which increased to 2125 tons/d as a result of major expansions completed in 1997 and 2009. The expansion projects included the additions of a pre-reformer, booster air and synthesis gas compressors, interstage chillers along with a significant revamp of the primary reformer.

Background

Original Design of Primary reformer

- Primary reformer at Yara Belle Plaine Inc. is a UHDE/Foster Wheeler design down fired balanced draft unit. Originally there were 240 tubes in five rows for a total absorbed radiant heat duty of 85 MW which was supplied by 108 gas-fired burners in six rows. Outer row burners operated at 50% capacity compared to the inner row burners. Total heat load of convection section was 64 MW.
- Feed steam inlet manifolds and pigtail support beams were of suspended design using constant load hangers. Ball bearing supports compensated for vertical movement. Figure 1, 2 and 3 provide the principal sketches.
- Collection headers and transfer line were of classical UHDE design which decouples high temperature and pressure bearing service. The connection of the reformer tube outlet to the manifold header is refractory-lined, see figure 4 and 5, thus the skin temperature profile of external pressure bearing parts decreases very rapidly to a temperature suitable for carbon steel. Internal outlet tubes can freely expand downwards within the outlet manifold nozzle and need to withstand only the temperature but no additional forces. Prevention of hot bypass streams is achieved by installation of gas barriers.

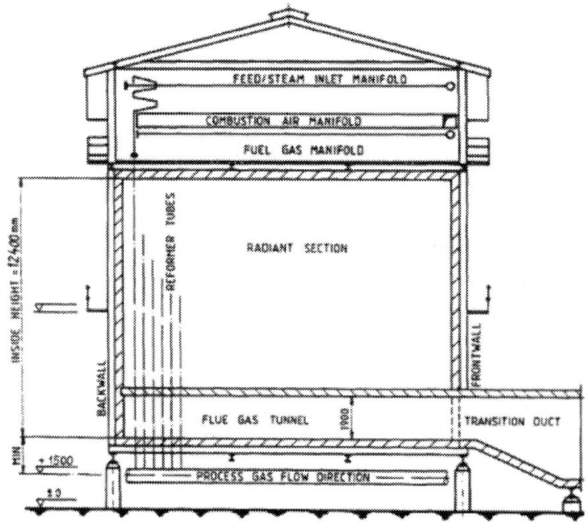

Figure 1. Principle Sketch of reformer

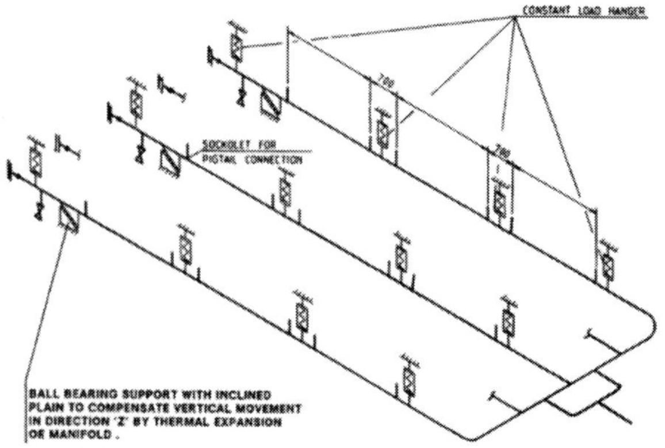

Figure 2. Principle Sketch - Inlet manifold

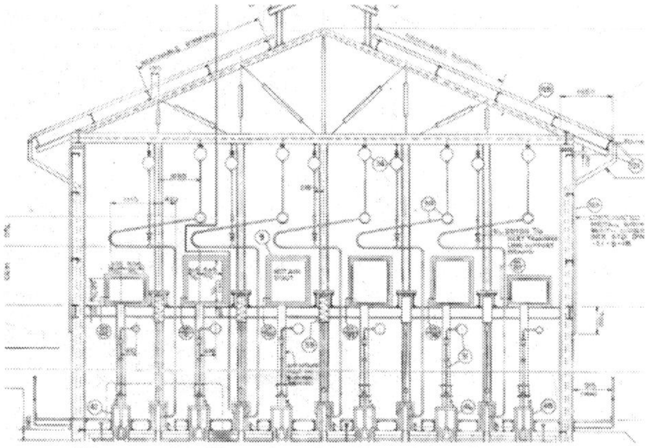

Figure 3. Sketch of inlet pigtail support

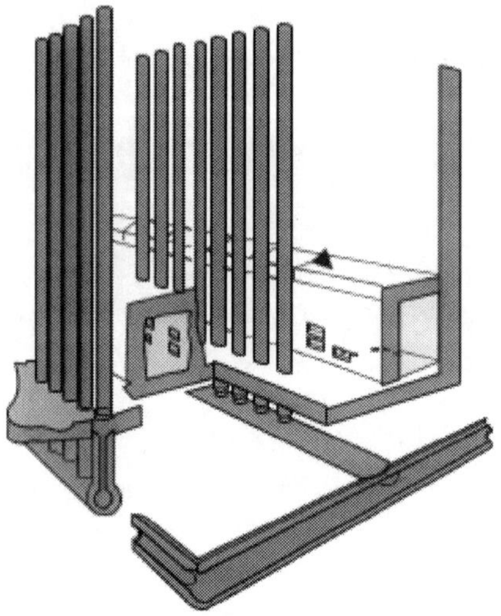

Figure 4. Principle Sketch - Outlet manifold

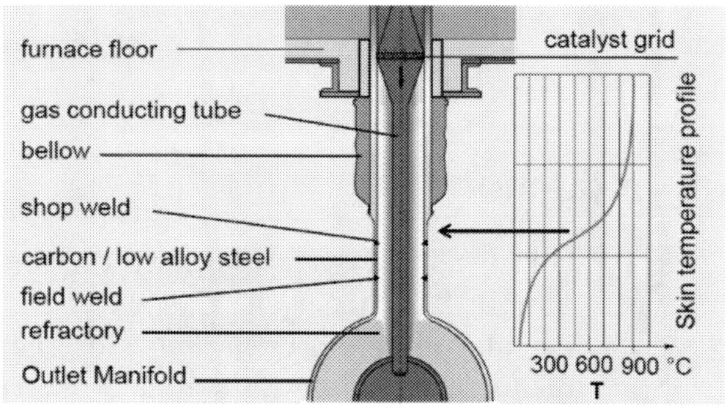

Figure 5. Tube to manifold – Temperature profile

History of Primary Reformer

- 1992 Year of Commissioning
- 1993 During a startup, one reformer tube failed, and thirteen others overheated due to moisture accumulation and freezing in pigtails. Negative slopes on pigtails were identified to be the cause of this failure. Affected tubes were disconnected and subsequently replaced.
- 1997 First expansion of Ammonia plant to increase nameplate capacity from 1500 tons/d to 1850 tons/d. This included adding a pre-reformer, a larger process air convection coil and molecular sieve dryers on make-up gas to the synthesis loop. The ID fan rotor was upgraded. The outage also included maintenance work on the Inlet feed/steam headers. The feed/steam line from final convection coil outlet to the inlet of the individual distributors was replaced due to cracking caused by embrittlement.
- 2003 Aging of reformer tubes originally installed in 1992 necessitated re-tubing. All tubes were replaced in two phases during 2003 and 2004.
- 2009 The second expansion of the Ammonia plant was undertaken to increase the nameplate to present day 2125 tons/d. This major expansion included installation of booster air, synthesis and CO_2 compressors together with an expansion of the primary reformer among other things. The reformer was expanded along its axis as shown in figure 6. Eight tubes were added to each row which increased the total to 280 tubes. Three burners were added to each row increasing the total to 126. Post this expansion, the reformer operated with a radiant and convection section duties of 116 and 93 MW respectively. The inlet and outlet manifold system were extended keeping the original design principles.

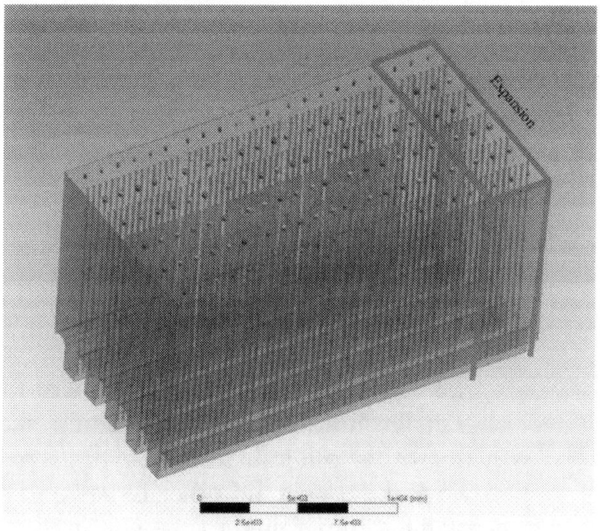

Figure 6. Representation of 2009 expansion

- 2010 As a direct outcome of the 2009 extension of the inlet manifold system, multiple pigtail failures were encountered between 2010 to 2014. A total of 16 pigtails were replaced, most as a preventive measure after cracks were observed. A root cause analysis showed stress induced cracking. Figure-7 shows the location of pigtail failures.

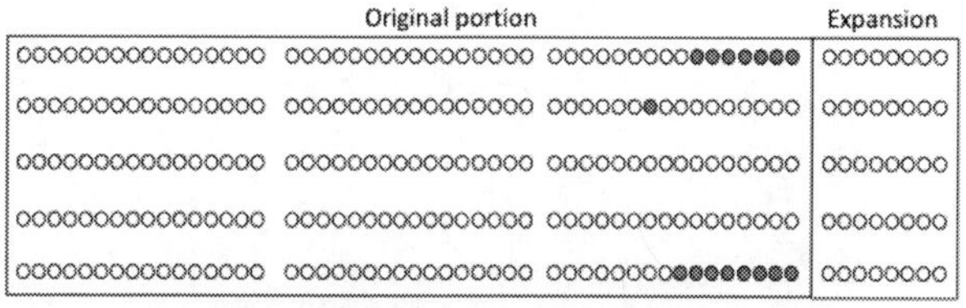

Figure 7. Map – Pigtail failures

- 2015 The inlet and outlet manifold systems were replaced on the primary reformer. There were a number of changes to the design of the inlet manifold system to address the issues encountered post 2009 expansion. This included change in diameter of pigtails, metallurgy upgrade, heat treatment together with changes to support system. The outlet manifold system consisting of five collection headers and transfer line was also replaced due to aging with a newer design. The new refractory concept was a two layer castable concrete comprising of a 50mm hot-face (Gouda Curon 180 T Sp) and 75mm insulating layer (Gouda Golite 135 R). There was no high-alloy steel liner within any header or transfer line in the new design.

- 2016 The primary reformer experienced hotspots/bulging on the newly installed outlet manifold system a short time after 2015 start-up. Yara made the decision to replace the entire affected section of the transfer line using corresponding section of the old transfer line. After repair work there were no other issues until a planned turnaround of the site in 2018. Root cause analysis of this failure identified higher velocities in the outlet manifold system to be a major factor. Double layer refractory design decreased the inner diameter reducing cross sectional area to about 79% in each collection header and to 88% in the transfer line whereas the plant capacity had increased by 41% of original design. These increased velocities are thought to have provided enough driving force for bypass flow leading to hotspots and subsequent bulging.

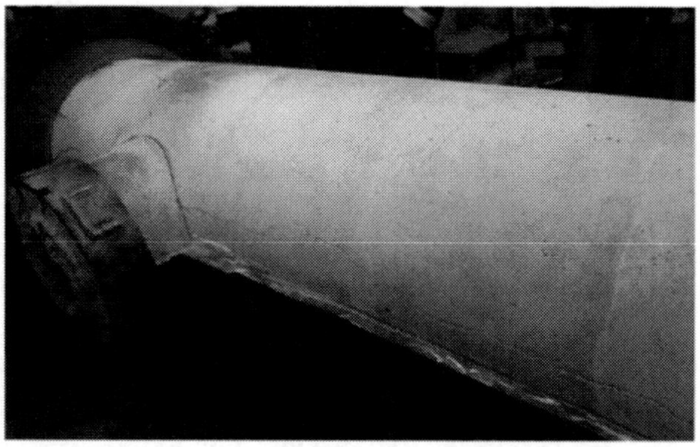

Figure 8. Transfer line section with longitudinal cracks

Value and Challenges from 2009 Expansion

Analysis of the past fifteen years of operating data shows that the 2009 expansion brought an average increase of 10% to daily production under normal operating conditions, and half of this gain got translated into annual production levels in the initial five years. Downtime encountered due to issues such as pig tail failures contributed to the five percent that was lost.

Pig tail failures from 2009 to 2015 were the biggest issue introduced by the 2009 expansion. The issue was ultimately fixed in 2015 when the entire inlet manifold system was replaced including the supports. Original pigtails were ¾" Sch 10 321H. In 2005 the size was increased to 1" Sch 40 with a small portion of ¾" left on each end. Two investigations were conducted into pigtail failures and both concluded that the cause of failure was stress induced cracking or stress relaxation cracking. In 2013, in order to better understand the stresses in inlet distribution system, and after the 2nd pigtail failure, Yara completed a detailed stress analysis. Figure-9 provides some insight into the problem. The high stress area included the expansion section and twelve pigtails into the original section where most of the failures occurred.

The second biggest challenge was the thermal imbalance as the expansion section ran considerably cooler than rest of the reformer. Figure-10 shows a 3D plot of tube metal temperature profile in the reformer. Modelled methane slip numbers corresponding to this temperature profile range from a minimum of 10 to a maximum of 17 mol% dry. Average methane slip ran from 13 – 14%.

Thirdly, combustion air distribution issues required modifications and adjustments to address some of the thermal imbalance problems. Furthermore, capacity limitation on combustion air fan, especially during summer months, directly limited plant rate.

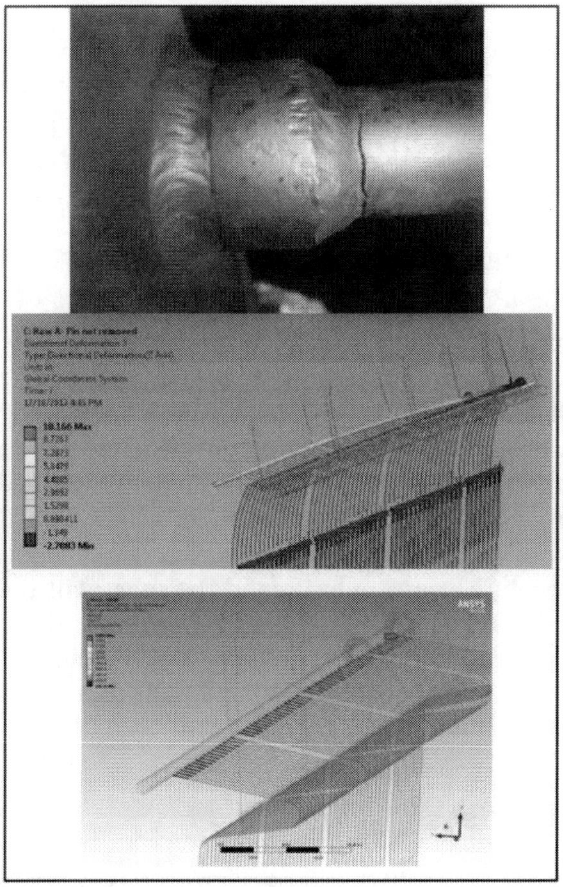

Figure 9. Overview of stresses introduced

Finally, Erosion and refractory damage in various parts of the reformer have required maintenance every turnaround. The problem has been more pronounced in the transition duct connecting the tunnels with the convection section.

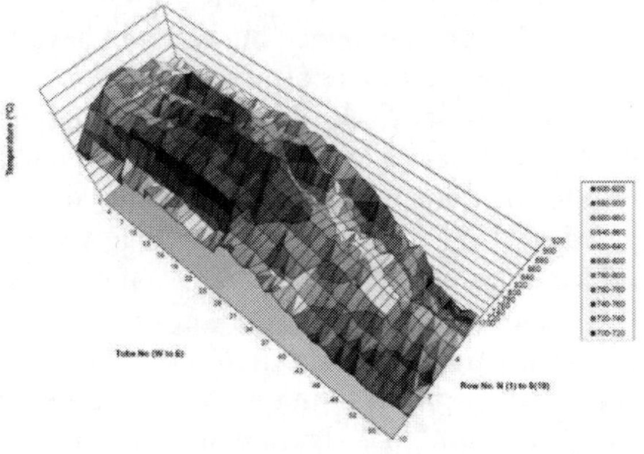

Figure 10. Tube metal temperature map (From JM's Reformer Survey 2017)

2018 Reformer Project

Yara completed investigation work on the transfer line failure in 2016 and 2017. In-house engineering assessments were also completed on some of the other issues such as thermal imbalance in the reformer, burners, combustion air system, remaining life of tubes and flue gas extraction tunnels. Yara subsequently engaged tkIS to complete a detailed engineering study on the reformer to address some of the identified issues. The following deliverables were chosen based on in-house assessments and investigation of the transfer line failure.

1) Design of outlet manifold system
2) Assessment of the existing combustion air supply system and basic design
3) Specification of new burners
4) Economics of primary reformer tube replacement
5) Assessment and recommendations on flue gas extraction system including tunnel wall openings and impact of these upgrades on convection section

Outlet Manifold System

Several design changes and options were discussed between Yara and tkIS to address some of the problems identified in the investigation of the transfer line failure. Finally, the following main features were agreed and have been safely and successfully implemented:

- The new manifold shall comprise, like the original one, an alloy shrouded hot face with one layer insulating castable.
- Inside diameters of the liner at header and transfer line have to consider the higher capacity of the plant and reduce the velocities to an acceptable limit.
- The pressure parts had to comply with the latest edition of API941 – Nelson Curve, therefore the shell material was upgraded from ASTM A 516 Gr. 65 to A387 Gr. 11.

Finally, as the insulating concrete of the original system was no longer available on the market, two alternatives were evaluated. Firstly, an insulating concrete based on light fire clay with very low thermal conductivity. Secondly, a bubble alumina concrete with less than 0.7% SiO2 but significantly higher thermal conductivity which required a larger lining thickness as well as a larger shell OD. There were several design challenges as the manifold had to be installed underneath the existing reformer, i.e. a tight fit between the manifold and the steel structure and the existing foundations of the reformer box. In addition, the existing centerline had to be maintained as the transfer line is directly connected to the secondary reformer inlet. Any increase of the transfer line OD would also require the addition of a reducer at the inlet to the secondary reformer.

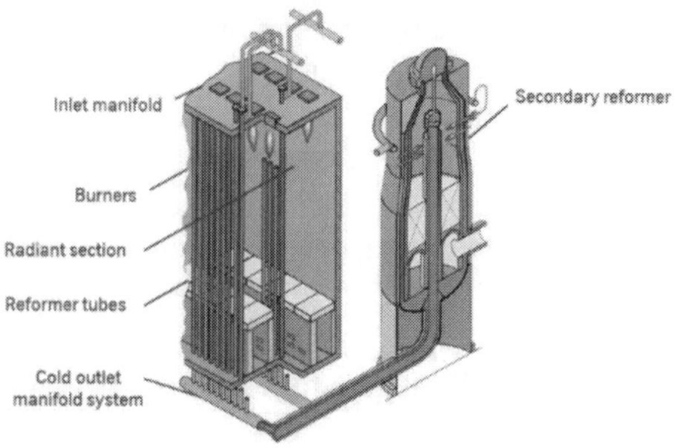

Figure 11. Representation of outlet manifold

Figure 12. Pre-fabricated outlet manifold

To evaluate the material behavior after casting, dry-out as well as welding and PWHT of the pressure shell, real scale test pieces were fabricated with both options. Based on the results, Yara and tkIS decided to select the bubble alumina-based material despite being more expensive - due to the advantages of constant product quality (as made from artificial aggregates); very low shrinkage during dry-out and last but not least, the excellent track record in all tkIS ammonia plants since 2010.

Reformer Tubes

The exchange of the reformer tubes included a redesign considering the latest available reformer tube metallurgy. The use of modern micro alloy material reduced the minimum sound wall to 8 mm (0.351 in) in the new catalyst tubes, but fabrication constraints limited the tube wall thickness to 10 mm (0.394 in). The outer diameter was kept the same, and as a result the new catalyst tube inner diameter increased from 110 mm to 112 mm (4.331 in to 4.409 in).

Burners

Prior to the 2018 reformer project, two different types of burners were installed, and both types had in addition two versions with lower capacity for the outer row burners. The main (original) part of the reformer was equipped with AIROIL DFR burners, 72 inner and 36 outer row burners with outer row burners at 52% of the capacity of the inner row burners. The AIROIL DFR burners are combustion air staged burners and have been extensively used in reformer applications in the past. In contrary to the DFR burners, the burners in the extension area were fuel staged. The basic design data are listed in below table and compared to tkIS' recommendation for a new set of reformer burners.

For evaluation of the burner performance, tkIS conducted a process simulation of the reformer using tkIS' internal simulation tool (taking into account the actual plant load data submitted by Yara). The convection bank was also simulated in order to ensure the required pre-heating / superheating of the different media and to assess the impact on pressure drop and velocities.

		Reformer Burners prior revamp				New Reformer Burners	
		Main box area		Extension area		complete reformer	
		Inner Row	Outer Row	Inner Row	Outer Row	Inner Row	Outer Row
Burner Type		DFR 35	DFR 20	IPSF-35	IPSF-20		
		air staged		fuel staged		air staged	
Heat release, normal (100%),	[MW]	1.744	0.907	1.694	0.879	2.17	1.48
relative to inner row	[%]		52		52		70
Max. heat release as per burner curve	[MW]	1.92	1.0	2.12	1.10	2.33	1.63
Number of Burners		72	36	12	6	84	42
Excess Comb. Air	[%]	10		15		14	

Table 1. Comparison of reformer burners

This simulation of the reformer showed that a total heat release of 240 MW (incl. sensible heat of comb. air and fuel) was required. At 0.15 MPa (g) fuel pressure – which is the maximum operating point as per AIROIL's burner curve – all the installed burners had a heat release of only 202.3 MW, so especially the DFR burners were not rated for the actual plant load.

It was recommended to adapt the reformer burners to the actual required firing. In order to increase the reformer tube temperatures in the outer rows, tkIS recommended to increase the capacity of the outer row burners to 70% of the inner row burners (from 52%). This relation is according to tkIS' actual design approach for reformers and has been successfully implemented on all tkIS reformers in the last 20 years.

Yara worked with Zeeco in Broken Arrow, OK for engineering and fabrication of new burners for the reformer. As the reformer was a critical path item in the 2018 turnaround and given the expected congestion on top of the reformer deck, Yara proceeded with a partial replacement of burners only replacing the outer row burners and expansion section burners. It was Yara's assessment that most of the performance issues will be addressed this way and the rest of the burners can be replaced during the next turnaround.

Figure 13. Burner at Zeeco's test facility

Flue Gas Collection

While the combustion air manifold is responsible for uniform distribution of oxygen to the burners, the flue gas collection via flue gas tunnels is responsible for a uniform flow within the radiant box of a top fired reformer. A uniform flow is required, within certain limits, in order to avoid issues like flame impingement on the reformer tubes or the walls and (together with the firing) to ensure an even heat distribution across the reformer tubes. The imbalance of the reformer box could be seen also in the temperature differences in the flue gas tunnels. A dt of approx. 66K shows an improper heat distribution inside the radiant box.

The layout prior the 2018 project contained two undesired features. The distribution of the ports across the outer rows was constant along the tunnel wall, which didn't consider the pressure drop inside the tunnels, in the direction of flue gas flow. The inner row tunnels are fed from two sides, while the outer row tunnels from only one side. Moreover, the outer tunnels had only approximately 30% of the open ports compared to the inner tunnels, thus the flue gas flow had a strong drift to enter the inner tunnels.

The revised design for inner and outer tunnels has taken into account the increased firing rate of the outer row burners (70%) and the high flue gas velocities at the end of the tunnels (due to the increased capacity compared to original design). To force flue gas entering the outer tunnel rows in the same magnitude as the firing rate of the outer row burners, the number of ports inside the outer tunnel walls had to be increased from 124 to a total of 229.

Figure 14. Flue gas tunnel with open ports for collection

Combustion Air Ducts

Basis for re-design of combustion air ducts was agreed between Yara and tkIS to be the amount of combustion air required as per process simulation at 240MW. The connection of the expansion section to the main ducting especially needed to be re-worked to achieve an even distribution of combustion air to the burners. Installation of completely new ductwork was deemed necessary. The design of the combustion air duct considered:

- The required burner capacity and its corresponding air flow properly distributed throughout the combustion air manifold
- The combustion air duct velocity profile in order to mitigate the pressure drop to the best possible scenario. As the capacity of the outer row burners was increased to 70%, the combustion air flow needed to be adjusted accordingly. However, given the connection of the outer distribution duct at the main header and the limited space in the penthouse, the first duct section could not be enlarged to 70% of the cross section of the inner duct. In order to limit the extra pressure drop – and with this any imbalance in combustion air distribution – the first section of the outer distribution ducts had to be designed as a rectangular box in order to reduce the velocity as soon as possible to the same magnitude as for the inner ducts. Figure-15 shows how the air ducts have been re-shaped to achieve a constant or slightly decreasing velocity profile along the combustion air headers.

Yara chose to proceed with duct work modification in the outer rows only for the reasons already shared under burners. Remaining work will be completed in the next site turnaround.

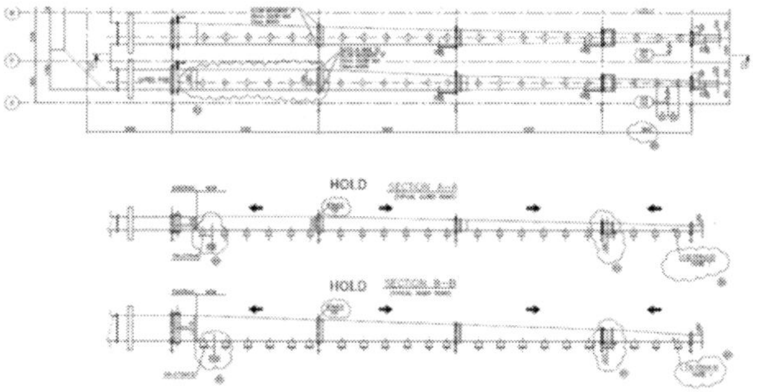

Figure 15. Combustion air duct work modifications

Combustion Air Fan

The FD fan and its driver were found to be under rated for the required combustion air flow. A tkIS reference plant operating at 2125tons/d was also using a larger fan than what was installed at Yara Belle-Plaine. Upgrade of the combustion air fan and its motor was recommended by tkIS based on process simulation. Yara subsequently worked with the OEM of the FD fan and was able to meet the requirement by using a larger diameter impeller in the old casing. Motor upgrade was also implemented as part of the project.

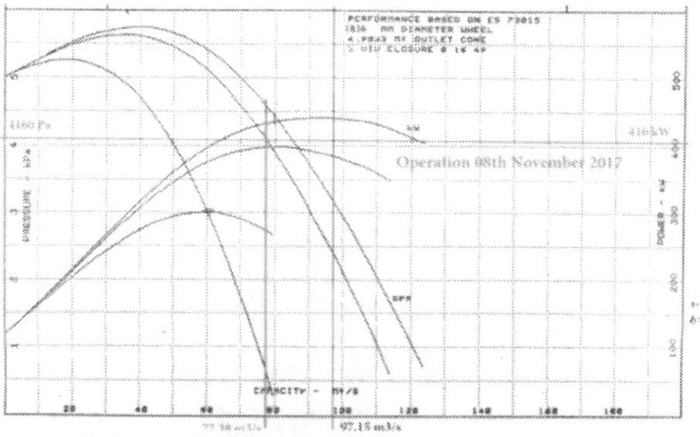

Figure 16. Expected operating point

Catalyst and Pre-Loading of Reformer Tubes

Yara Belle-Plaine and Johnson Matthey agreed to a combination of KATALCO 25-4Q over KATALCO 57-4XQ for this project. The top catalyst layer is promoted with potash. The combination of potash and pellet size yields a very high activity catalyst which will inhibit carbon formation throughout the life of the charge and extend the life of the tubes by lowering the tube skin temperatures. The bottom layer catalyst was chosen for a few reasons. Since most of the reforming reaction occurs in the top portion of the tube, a high activity catalyst is not as critical in the bottom portion of the tube. KATALCO 57-4XQ has a slightly lower nickel content and no potash. Additionally, the pellet size is bigger than the KATALCO 25-4Q which means the loaded density is less, effectively lowering the active surface area of the catalyst. Yara opted for this option as it results in lower pressure drop and more throughput while still nearly achieving reaction equilibrium

Yara chose to preload the new reformer tubes before installation. Since reformer work was the critical path item for turnaround 2018, doing so reduced the overall timeline by at least four days. For the purpose, a dedicated tube stand was built. 22 tons of steel and approximately 1700 man hours were required for building this stand. Loading of new tubes including the dP measurements were completed prior to plant shutdown. Old tubes did not need to be unloaded and were placed on the stand after removal as the stand was built with some extra space for placing old tubes.

A final dP measurement was repeated after installation and before startup. No tube needed to be offloaded. It is Yara's experience that preloading of reformer tubes makes economic sense during re-tubing of reformers as it reduces the overall timeline of outage and the additional cost is easily recovered i.e. one and a half day of net production gain in this case.

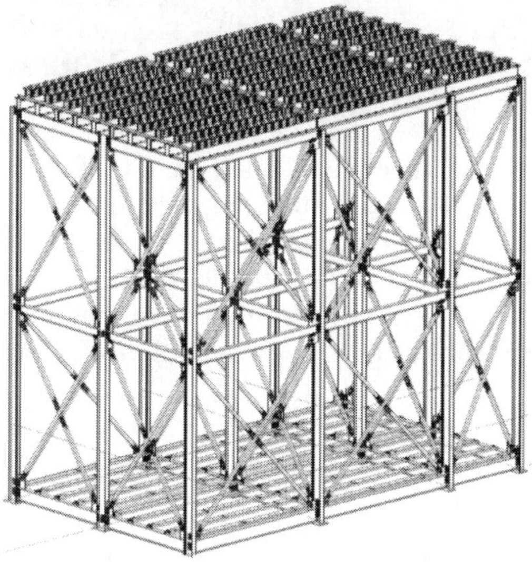

Figure 17. Principle sketch of Reformer tube stand

Figure 18. Reformer tube stand

Figure 19. Flange numbers were used as a reference

Outcome – 2018 Reformer Project

The project was completed within the allocated budget and timeline. Collaboration of teams within Yara and with other organizations including tkIS and JM has helped to resolve some of the long-standing issues of thermal imbalance in the reformer. One year of operation since completion of project has shown significant improvement in overall balance of the reformer and data supporting this assessment is presented below in the form of a comparison of tube metal temperature (TMTs) plots.

The first observation from these plots is that the replacement of burners in the outer rows and changes to the air ducts erased the discrepancy seen in 2017 when tubes in outer rows were on average 30°C colder than the inner rows. Increasing the temperatures in the outer rows to match the inner rows helped improve the reformer balance and lower methane slip by 0.3 mol% dry start of run.

The second observation from Figures 20 and 21 is that the east end of reformer is still colder than the west end but the balance from south to north is now better throughout. This was expected as Yara proceeded with partial replacement of burners and deferred the inner row burners for next turnaround. Same applied to the combustion air duct work. The improvement in reformer balancing can also be seen by comparing the tube metal temperature frequency charts in Figures 22 and 23 presented below.

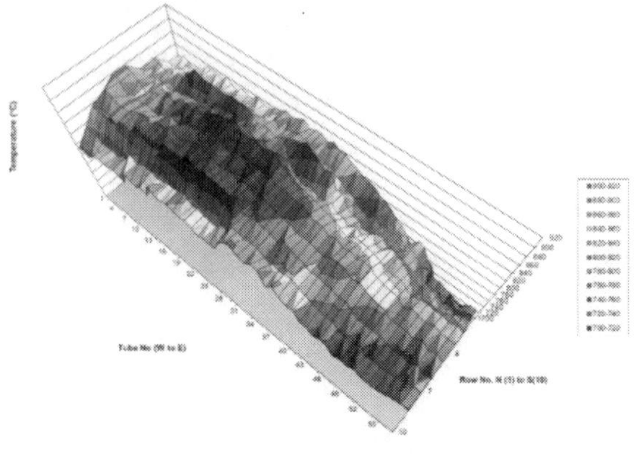

Figure 20. Corrected TMTs – 2017 (From JM's reformer survey 2017)

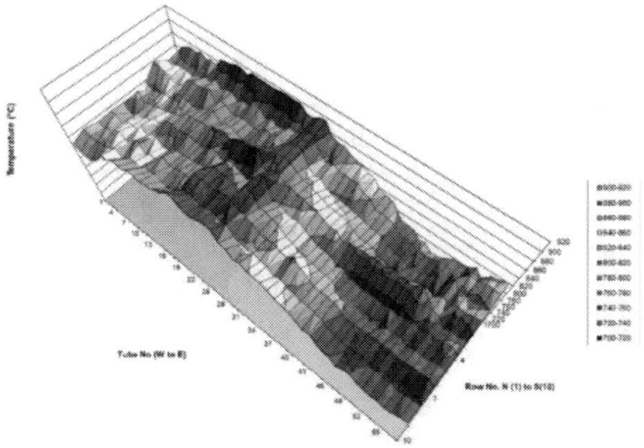

Figure 21. Corrected TMTs – 2019 (JM's reformer survey 2019)

Most of the tubes that were less than 800°C in 2017 are now greater than 800°C after the upgrades. These tubes were in the cold zone of the northeast corner seen in Figure 20. The few tubes that were greater than 905°C were eliminated which will help preserve the life of the tubes. The frequency plot of tube metal temperatures is presented below.

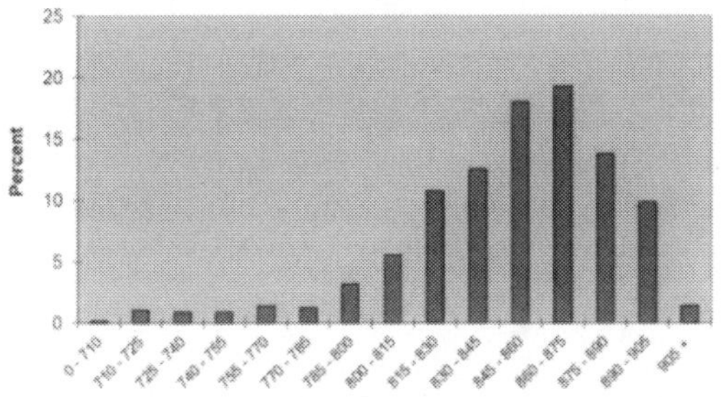

Figure 22. Corrected TMTs (°C) –2016/17

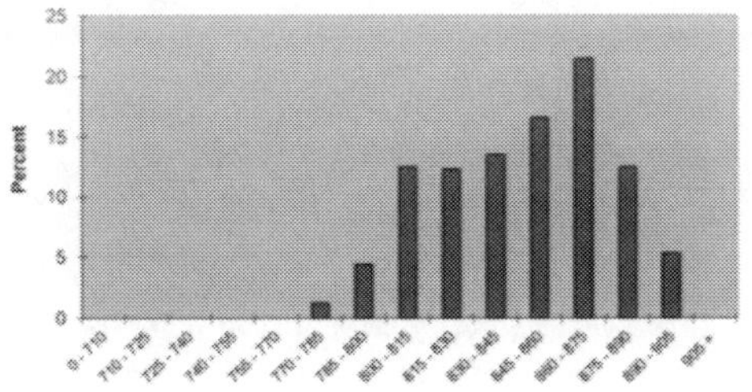

Figure 23. Corrected TMTs (°C) –2019

Conclusion

The primary reformer is often a capacity bottleneck in Ammonia, Methanol and other related plants. This case serves as a credible reference of expanding a top fired methane steam reformer casing along its axis which could, in certain cases, be more economical than adding a row of tubes – a method that has been employed more often. While there have been challenges, and it has taken several years to fix some of them, it is achievable and should be considered as a possible alternative to adding an additional row of tubes when assessing the economics of expansion in methane steam reformers.

Damage Mechanism Reviews and Mitigation Methodology for Ammonia Equipment

This paper presents a methodology for conducting damage mechanism reviews (DMRs) that can help operating companies maintain equipment integrity by providing a foundational basis on which to manage and mitigate common damage mechanisms that may affect plant equipment. The DMR methodology allows ammonia-related industries to draw on recent API recommended practices to provide a holistic management resource for the plant's Mechanical Integrity Process Safety Management programs. The assessment methodology is demonstrated for typical process unit equipment in an ammonia plant. BakerRisk's Asset Mechanical Integrity Plan identifies the potential damage mechanisms based on known process parameters, selects the appropriate inspection methods, and communicates relevant factors and possible mitigation options as a resource for the site.

Daniel J. Benac, P.E., CFEI and Dorothy Shaffer
Baker Engineering and Risk Consultants, Inc. (BakerRisk®)

Reliability in operating companies depends heavily on the mechanical integrity of the equipment and piping.

In an ideal world, materials would be chosen that would be impervious to deterioration effects from process conditions. In the real world, this is either impossible or cost-prohibitive, so various types of equipment damage continues to occur. For the refining industries, American Petroleum Institute (API) has identified some common damage mechanisms for typical refinery and chemical process units along with proper inspection techniques in developing API RP 571, *"Damage Mechanisms Affecting Fixed Equipment in the Refining Industry,"[1]* followed by the soon to be released new edition of API RP 970, *"Corrosion Control Documents."* [2]

The ammonia-related industries can utilize these resources and methodology with specific industry knowledge to manage and mitigate the consequences of the knowns and unknowns in their Asset Mechanical Integrity (MI) programs. An example of damage mechanism reviews for ammonia equipment is presented in this paper along with an example of BakerRisk's Asset Mechanical Integrity Plan as a preferred option for generating an overall foundation document and plan for improving equipment reliability.

Damage Mechanism Review (DMR):

It's easier to find something if you know what you are looking for. While the MI programs for some process areas of a plant are easily managed, others are complex and can be affected by changing operating conditions. For the ammonia equipment and piping industries, potentially life-limiting damage mechanisms include high temperature mechanisms, high temperature hydrogen attack (HTHA), stress corrosion cracking (SCC), general corrosion, and brittle fracture, to name a few. [3, 4, 5, 6, 7]

Awareness of mechanisms that can damage equipment is essential in developing the solution to inspect the equipment properly, mitigate damage, and prevent failure. The pertinent damage mechanisms are also input for evaluations regarding fitness-for-service, and rate of attack needs to be understood to determine the remaining life. For a proper risk-based inspection (RBI) program or during a hazards analysis, the appropriate mechanisms are identified so that the probability of failure can be determined when addressing reliability issues. While API RP 571 is targeted for refineries, it does provide a good resource for other industries in identifying the characteristics of damage mechanisms and their proper inspection techniques.

API RP 571 provides general guidance as to the most likely damage mechanisms affecting common alloys used in the refining and petrochemical industry and is intended to introduce the concepts of service-induced deterioration and failure modes. These guidelines provide information that can be utilized by plant inspection personnel to assist in identifying likely causes of damage, to assist with the development of inspection strategies, and to help identify monitoring programs to ensure equipment integrity.

Identifying the damage mechanisms by the site's MI subject matter expert (SME) is not enough for proper management, as MI interacts with many plant functions such as operations and equipment design. A damage mechanism review (DMR) requires a multi-disciplined team that includes a process engineer and the inspection/reliability SME working with the damage mechanism SME. Operations personnel are also utilized for operating conditions, such as startup and shutdown.

Reviewing trended failures or life-limiting damage or near misses should be included in the DMR. [8]

API RP 970 outlines a document with information on the process unit, damage mechanisms, and inspection plan. The format of the Asset Mechanical Integrity plan draws on variety of additional sources — such as a facility siting study for consequence evaluation — to provide broad perspective that generates a process safety foundation for operations management. Team input is recommended to provide a foundational document for the site that represents plant functions that can impact the mitigation plan.

The Asset Mechanical Integrity plan includes the following:

- A description of the unit.
- Process flow and corrosion loop diagrams
- A list of probable damage mechanisms and failure modes, along with information on each.
- Guidance for operations and process support, such as operating conditions that may impact damage mechanisms and walk-around field observations such as vibration or insulation damage.
- Equipment limits/applicable integrity operating windows (IOW's).
- Recommended inspection practices.
    - The critical damage mechanism circuits, their potential for failure, and the failure consequence.
    - Critical operational, process, or inspection mitigating factors, and the responsible parties for taking action.
    - Guide to the report's location for additional operational, process, safety, and inspection actions and input.

Developing an Asset Mechanical Integrity Plan:

The process of analyzing damage mechanisms and creating an equipment and piping plan begins with identifying a process unit of interest for damage mechanism assessment; in this example, a Selexol unit was considered that removes and concentrates CO_2 and H_2S discharged from a petcoke gasifier.

Unit Description: Selexol is an acid gas removal solution comprised of dimethyl ethers produced from polyethylene glycol. It is a physical solvent rather than a chemical absorption solvent. Figure 1 shows a Process Flow Diagram for the unit. The consequence of failure is high on an unmitigated basis due to the H_2, CO_2, CO, and H_2S content of the gas. The green highlight is the Flash Gas Type circuit. The blue highlight is the Lean Cold Selexol circuit. Both are discussed later in the paper.

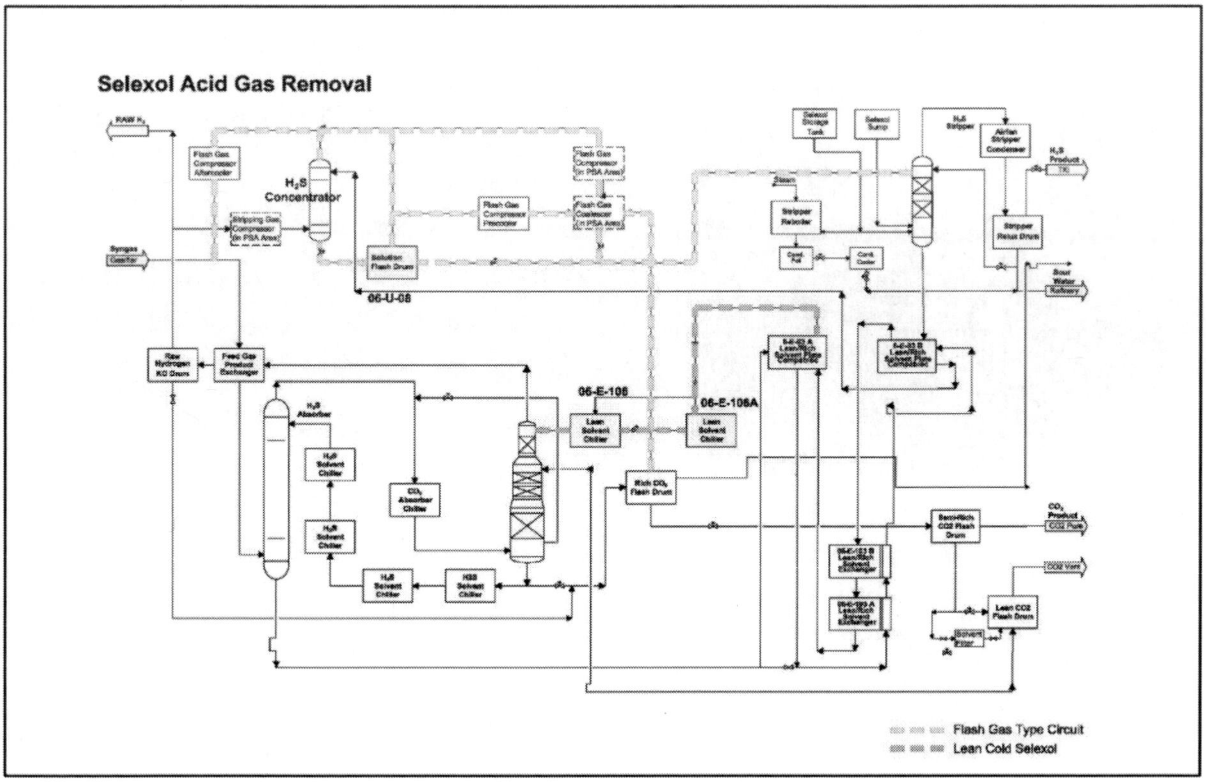

Figure 1. Process flow diagram for the Selexol Unit

In the Selexol unit, the cooled gasifier discharge is treated with chilled Selexol solvent in several absorption and flashing steps to clean the feed gas. The CO_2 and H_2S streams are concentrated and the Selexol is regenerated in a steam-driven stripping tower.

The equipment parameters and operating conditions are compared to the damage mechanisms that could potentially cause life-limiting damage. In the damage mechanism review, the scenarios / mechanisms considered were given a qualitative likelihood rating based upon the known conditions. The following rating was used:

- *No likelihood to occur:* Conditions are not present or possible.
- *Low likelihood to occur:* Condition is theoretically possible; but judged unlikely to occur.
- *Medium likelihood to occur:* Condition is judged likely to occur in the lifetime of the plant.
- *High likelihood to occur:* Condition has been observed in the facility or a similar facility.

As part of a damage mechanism and mitigation review, equipment in two different damage circuits denoted with different damage characteristics were examined: the H_2S Concentrator Circuit and the Lean Selexol Chilling Circuit. The exchanger (06-E-106) and the Solution Flash Drum (06-U-08) were assessed.

Lean Selexol Solvent Source:

The 06-E-106/106A Lean Solvent Chillers, shown in Figure 2, are the final exchangers in chilling the unit's stripped Selexol solution to 15°F (-9 °C) prior to feeding the top section of the CO_2 absorber 06-U-107. The chillers can be operated in series or in parallel.

Figure 2. Lean solvent chillers 06-E-106 / 106A

Like most acid gas removal solvents, lean Selexol is relatively noncorrosive. Unlike potassium carbonate or amine based systems, Selexol has no added water, which potentially reduces some active corrosion mechanisms.

Selexol leaves the bottom of the H_2S Stripper 06-U-105 which uses a steam-driven Stripper Reboiler 06-E-08 to bring the solution temperature up to 301°F (149 °C) at 23 psig (0.16 MPa) in order to remove H_2S from the Selexol stream. This stripping process should maintain the "Lean Solvent" relatively free of water and dissolved gases, including CO_2 and H_2S. The flow then passes through a series of Lean/Rich Solvent Exchangers (06-E-93B plate exchanger, 06-E-103A&B) prior to the 06-E-106 A/B Lean Solvent Chillers.

The Selexol enters the tube side of the horizontal exchanger and is chilled by ammonia refrigerant on the shell side. The final temperature is controlled by the temperature on the shell side for a given flow of Selexol.

Table 1 summarizes the materials of construction and operating conditions for the 06-E-106 exchanger. These materials of construction parameters and conditions were compared to the possible damage mechanisms that could potentially cause damage. Only the most reasonable damage mechanisms were considered based on experience and known parameters that could limit the life of the cooler. These damage mechanisms are described further in Appendix A.

Condition	Description
Date fabricated (U1)	1998
Time in service	Since 2000
Materials of Construction	Shell: SA 516 70N (0.75-inch thick) (19mm) PWHT Head: SA 516 Grade 70N (11/16-inch thick) (17mm) PWHT Channel: SA 516 Grade 70N (1.125-inch thick) (29mm) PWHT Nozzles: SA 105/SA106 Tubes: SA334-1 welded Tube Sheet: SA350-LF2
Insulation	Yes, vessel painted
Design Temperature	Shell: 200 to (-20) °F (93 to -28 °C) Tube: 150 to (-20) °F (66 to -28 °C)
Design Pressure	Shell: 250 psig (1.7 MPa) Tube: 700 psig (4.8 MPa)
Typical Temperature	Shell: 6°F (-14 °C) Tube: 75°F inlet to 40°F outlet (24 to 4 °C) (Outlet 15°F on original PFD)
Typical Pressure	Shell: 18 psig (0.12 MPa) Tube: 630 (4.3 MPa) inlet-600 outlet, psig (4.1 MPa)
Process Stream	Shell side: Ammonia refrigerant Tube side: Selexol 100%
Inspection/ Repair History	See Inspection Report
Process Upset (thermal shock)	As this exchanger is preceded by several other exchangers, hydraulic shock should be low.

Table 1. Materials of Construction and Operating Condition for Exchanger (06-E-106)

Table 2 summarizes the potential damage mechanism assessment. Damage mechanisms that could apply are listed with their probability of occurrence.

Scenario/ Mechanisms	Carbon Steel Shell (Probability)	Carbon Steel Tubes (Probability)	Comments
Welding/Fabrication failure from repairs	Low	Low	No repairs, welds PWHTed
General wall thinning (Acid attack)	Low	Low	Verify corrosion rates
Flow-assisted Corrosion (FAC)	Low	Low	Flow rates are from 3 to 5ft. (0.9 to 1.5 M) /sec– not high enough to cause metal loss.
High Temperature Degradation (creep, HTHA)	No	No	Low to ambient temperature operations - not a concern
Low Temperature Exposure (brittle fracture)	Low	No	Low pressure and normalized and PWHTed material should not present a concern on the shell. Tubes and tube sheet are low temp materials.
Ductile Overload Failure (High Pressure)	No	No	No high pressure conditions
Ice freezing in a line (<32°F)	No	Low	Water is not normally present, excepting at washout for turnarounds.
Solution Freezing	No	Medium	Selexol with water will become slushy at 0°F (-18°C)
Corrosion Under Insulation (CUI)	Medium	No	Vessel is painted, and the carbon vessel shell operates at about 6°F, which is below the maximum corrosion rate at a temperature of 150°F to 250°F (66 to 121°C)
Internal Environmental Cracking	Low	No	Ammonia SCC possible in high hardness or non-PWHT welds.
Degradation (SCC)			Welds are PWHT. Welding Procedure OW-483 - Hardness below BHN 225.
Mechanical Fatigue (high cycle vibrations)	Low	Low	Steady state operation
Thermal or High Stress Fatigue (low cycle fatigue)	Low	Low	Equipment cycled upon shutdowns and could be subject to sudden temperature swings.
Thermal Shock	No	No	Temp -29°F (-34 °C), but at low pressure.

Table 2. Possible Failure Scenarios/Mechanisms for Exchanger (06-E-106)

- CUI is a concern because of insulation discontinuity and damage. Figure 3 shows a location where water can ingress under the insulation.

Figure 3. Potential area of CUI on lean cooler

- Ammonia SCC is also a potential concern, especially at welds. Therefore, it should be verified that welds are of sufficiently low hardness.

Brittle fracture is a potential mechanism for refrigeration systems, particularly where high pressures and cold temperatures are present. The drawings for this vessel indicated that the proper material specifications for cold service were used.

The Selexol Solvent Flash Drum

The 06-U-08 Selexol Solvent Flash Drum, shown in Figure 4, receives a high flow of Selexol with dissolved and two-phase gases after absorbing CO_2 and H_2S from the bottom exit of the 06-U-02 H_2S Concentrator operating at 245 psig (1.7 MPa) and 226°F (108°C).

Figure 4. The flash drum 06-U-08

In the Solution Flash Drum, a two-phase flow enters and the temperature reduces as the pressure drops by 30 psig (0.21 MPa), allowing gases to flash and be separated from the Selexol solvent, still rich in H_2S in the flash drum. The flash gases pass on to the flash gas system for recovery.

The Selexol solution is rich in CO_2 and H_2S, and contains water at ~0.2% (as absorbed from the main Feed Gas Stream). The Selexol leaving the flash drum is fed to the top inlet feed of the H_2S stripper.

Table 3 summarizes the materials of construction and operating conditions for the 06-U-08 Flash Drum. The parameters and conditions of the materials of construction were compared to the possible damage mechanisms that could potentially cause damage. Only the most reasonable damage mechanisms were considered based on experience and known parameters that could limit the life of the cooler. These mechanisms are described further in Appendix A.

Condition	Description		
Date fabricated (U1)	1982		
Time in service	Fabricated in 1982. Inspected before use.		
Materials of Construction	Shell: SA 516 70 (1.45 inch thick) (37mm) (PWHT uncertain) Head: SA 516 Grade 70 Nozzles: SA106B No impact testing		
Insulation	Yes		
Design Temperature	425 to (-20)°F (93 to -28 °C)		
Design Pressure	485 psig (3.34 MPa)		
Typical Temperature	200°F (93 °C)		
Typical Pressure	215 psig (1.48 MPa)		
Process Stream	H2	33.33	mol%
	CO_2	63.12	mol%
	O2/Ar	0.21	mol%
	N2	0.34	mol%
	CH4	0.39	mol%
	CO	1.59	mol%
	H_2S	0.98	mol%
Inspection/ Repair History	Clam shell repair at the bottom. Inspection in 2015 found excessive corrosion/pitting under deposits at the bottom drain valve area which was not characterized – appeared as likely acid attack.		
Excursion concerns	Water, oxygen, or chloride contamination and cooler temperatures will accelerate corrosion rates.		
Shutdown/Start-up concerns	On shutdown—cooler temperatures promote CO_2 and H_2S absorption in residual liquids. Contamination with water and oxygen in the presence of CO_2 and/or H_2S greatly increases corrosion rates for carbon steel. This could occur on shutdowns, in conjunction with a water flush.		

Table 3. Materials of Construction and Operating Condition for Solution Flash Drum 06-U-08

Table 4 summarizes the potential damage mechanism assessment. Damage mechanisms that could apply are listed with their probability of occurrence.

Scenario/ Mechanisms	Carbon Steel Shell (Probability)	Comments
Welding / Fabrication failure from repairs	Low	Original weld procedure should provide low hardness in the welds with PWHT.
General wall thinning and pitting (Carbonic Acid attack, H_2S attack)	High	1) Inspection report 15-379-bsm-V1, corrosion at the bottom 0.5"deep 2) Carbonic and H_2S acid potential 3) Under deposit corrosion, dead legs, low flow areas. 4) CO_2/H_2S corrosion-rates are high where cooler temperatures or condensation occurs. This can also impact connections; for example, static level gauges without tracing and insulation.
Flow-assisted Corrosion (FAC)	Medium	Flow rate through bottom nozzle is ~5 ft./sec. at 386,000 lb./hour without entrained gasses. Higher flows and entrained gas would increase this flow rate. Inlet flow is ~3 ft. (0.9M)/sec, excepting surges.
High Temperature Degradation (creep, HTHA)	Low	H_2 present but operating temperature 150°F (66 °C) below API 941 curve
Low Temperature Exposure (brittle fracture)	Low	Brittle fracture review performed. Shell: Minimum temperature -15 to -16°F (-26 °C). Shell material normalized. No Charpy impact testing (exempt at those temps) Nozzles: minimum temp. -150°F (-101°C). Pads: 47°F (8 °C)
Ductile Overload Failure (High Pressure)	Low	Vessel re-rated by FFS and PRV pressure changed
Ice freezing in a line (<32°F)	Low	Verify that no locations for water getting trapped after flushing for inspections.
Corrosion Under Insulation (CUI)	Medium	Insulated. The carbon vessel shell operates in the maximum corrosion rate at a temperature of 150°F to 250°F (66 to 121 °C). CUI can occur.
Internal Environmental cracking Degradation (SCC)	Low	Wet H_2S: H2 cracking a low concern in the weldments. Weld procedure A200 Hardness HAZ HB211 (HRB95), which is below HB 237.
Mechanical Fatigue (high cycle vibrations)	Medium	Piping velocities – enter in and exit piping.
Thermal or High Stress Fatigue (low cycle fatigue (LCF))	Low	Startup/shutdown cycling.
Pressure shock	Low	Sudden surge of flow may damage inlet deflector.

Table 4. Possible Failure Scenarios/Mechanisms for Solution Flash Drum 06-U-08

Review of the inspection reports and process conditions identified some of the following potential concerns:

- Corrosion was the worst on the bottom of a "drain nozzle." During inspections, corrosive attack was found on the bottom near a drain of the vessel, and at least ½-inch (13mm) of metal was removed. Figure 5 shows the attacked region. The inspection report did not identify the damage mechanism, but most likely it was an acid attack due to dissolved CO_2 and H_2S found in collected water or condensation, assisted by under-deposit corrosion.

Figure 5. Observed corrosive attack and pitting on the bottom of the flash drum

- Flow rates are from 3 to 5 ft. (0.9 to 1.5 M)/sec at the inlet and exit nozzle, respectively. Both of these are acceptable and should not accelerate corrosion. There is the possibility of accelerated corrosion occurring underneath deposits.
- Cooler temperatures at upper level gauge nozzles provide an opportunity for condensation corrosion and create localized corrosion with high CO_2 and H_2S levels absorbing on wetted surfaces.

As cooler temperatures promote CO_2 and H_2S absorption, uninsulated areas or areas of low flow have increased corrosion potential. Contamination with water and oxygen greatly increases corrosion rates for carbon steel. This could occur with process upsets that introduce more water in the incoming feed gas, or during shutdowns in conjunction with a water flush.

Asset Mechanical Integrity Plan for Selexol Unit:

After an initial Damage Mechanism Review (DMR), the Asset Mechanical Integrity plan is developed for the piping circuits with the vessels. This would include supporting detailed analysis tables, along with circuit maps, circuit/equipment-specific mitigation plans for operations/process engineering, and an inspection outline to provide guidance to the inspection group.

Connected equipment, including piping and instrument installations with similar damage mechanism concerns are grouped together in a "damage management circuit." These circuits provide a broader view of potential effects and efficiencies in the effort.

In addition to the damage mechanism input, the unmitigated consequence of failure is provided through HAZOPs or Quantitative Risk Assessments (QRAs).

Knowing the circuit and the potential failure consequences, the unmitigated failure consequence can be assessed as shown in Table 5.

Circuit	Unmitigated Consequence	Unmitigated Failure Risk
H_2S Concentrator Bottoms	High	High
Lean Selexol Chilling Circuit	Medium	Low

Table 5. Unmitigated failure consequence

Mitigation and Inspection Plans:

Critical operational, process, or inspection mitigating factors and the responsible parties for taking action are considered for the circuit.

Flash Gas Type Circuit:

Concerns and mitigation options for a Flash Gas Type Circuit, shown in Figure 1, are described below.

Concern: H_2S concentrator circuits have a high potential for heavy corrosion with pitting in areas where condensation occurs, low flow areas, or in dead legs.

Mitigation: Check deadlegs and thinner wall attachments (Action: Inspection)

Concern: Low potential for Wet H_2S cracking.

Mitigation: Verify that weld hardness is below the recommended minimum; if above, perform inspection for SCC. (Action: Inspection)

Concern: Accelerated corrosion may be linked to startups or shutdowns where acid gases cool and accumulate in Selexol or water to accelerate corrosion rates.

Mitigation: (1) Consider impact of using water flushes without controls to prevent CO_2/H_2S in the atmosphere. (Action: Operations, Process Engineering, Inspection). (2) Consider nitrogen blanket.

Lean Selexol Chilling Circuit:

Concerns and mitigation options for the Lean Cold Selexol Chilling circuit, shown in Figure 1, are described below.

Concern: Brittle fracture is a low potential damage mechanism mitigated by proper materials specifications that must be maintained. (Action: Maintenance, Engineering)

Mitigation: (1) assure pressures are low at very low temperatures. (2) Repair welds require low temperature evaluation for brittle fracture.

Concern: Medium potential for Corrosion under Insulation is possible, report insulation damage. (Action: Operations)

Mitigation: Perform visual inspection and UT of suspected areas.

Concern: Low Potential for Ammonia SCC.

Mitigation: Verify that weld hardness is below the recommended minimum; if above, perform inspection for SCC. (Action: Inspection)

SUMMARY

Damage Mechanism Reviews (DMRs) provide a systematic, proactive method for identifying potential damage mechanisms prior to failures. Having a broad understanding of these damage mechanisms and how operations or process conditions can be used to develop an Asset Mechanical Integrity Plan. When implemented, this plan aids in mitigation and inspection of damage or failures, and will produce more proactive process safety management decision-making when considering plant modifications or new equipment designs.

The development of an overall inspection and mitigation plan with input from impacted groups, such as that provided by the Asset Mechanical Integrity Plan, can provide an efficient method for conveying important actions to mitigate damage and prevent equipment failures.

It is essential that the developed DMR and the Asset Mechanical Integrity Plan be effectively communicated to plant and management staff, and the actions implemented, otherwise the high risks are not mitigated.

ACKNOWLEDGEMENTS

The authors express gratitude to Coffeyville Resources Nitrogen Fertilizers, LLC for providing process and equipment data for the Selexol Unit and allowing BakerRisk to share this paper at the 2019 Safety in Ammonia Plants & Related Facilities Symposium.

REFERENCES

1. API RP 571, "Damage Mechanisms Affecting Fixed Equipment in the Refining Industry" 2nd edition April 2011.
2. API RP 970, Corrosion Control Documents, 1st edition 2017.
3. A. H. Faraji, "High Temperature Nitriding in Corrosion in Ammonia Converters," Ammonia Plant and Safety & Related Facilities, Volume 51, pp183-191, 2010.
4. D. J. Benac, M. W. Hester, "Elevated Temperature Failure Mechanisms in Ammonia Plant Equipment – Reducing the Risk of Failure" AIChE, Safety in Ammonia Plants & Related Facilities Symposium, 2013, Vol. 54, pp. 199-211.
5. D. J. Benac, M. Gilbertson, "Multi-layered Secondary Urea Reactor Failure" AIChE, Safety in Ammonia Plants & Related Facilities Symposium, 2012, Vol. 53, pp. 29-41.
6. D. J. Benac, D. Shaffer, D. Wood "Managing Cold Embrittlement and Brittle Fracture Hazards in Ammonia Related Industries" 60th AIChE, Safety in Ammonia Plants & Related Facilities Symposium, 2015.
7. D. J. Benac, Mike Hester, and N. Cherolis, "Unexpected High-Energy Failure of Type 304L Stainless Steel Piping in the Tail Gas System in Nitric Acid Plant due to IGSCC" Journal of Failure Analysis and Prevention, April 2017.
8. D. J. Benac, D. Shaffer, J. Redman "Achieving Equipment Reliability through Understanding Precursor Failures" 62th AIChE, Safety in Ammonia Plants & Related Facilities Symposium, 2017, Vol. 58.

Appendix A - Mechanisms Definition and Key Parameters

Brittle Fracture

Definition:

Carbon steel and alloy materials can undergo a transformation from ductile-to-brittle behavior as the material temperature drops, resulting in low toughness at low temperatures and increasing the hazard of sudden brittle fracture. The ASME B31.3 for piping and ASME Section VIII for Pressure Vessels codes provide extensive rules for carbon steel materials that are subject to cold temperatures. While the codes and standards are available on cold embrittlement, additional steps may be needed to ensure codes and standards are properly utilized and addressed. Examples where cold embrittlement can be a concern:

Key Parameters:

- Piping with water that freezes
- Thick-wall pressure vessels
- Repair-welded equipment that have altered material properties.
- Cryogenic storage systems
- De-pressurized systems
- Materials used with low toughness an no guarantee for low temperature properties.
- Pre-existing cracks

Fatigue

Definition:

High-cycle fatigue (HCF) is often the result of high frequency vibrations and low stress. Low-cycle fatigue LCF is typically due to thermal or high stress cyclic conditions. Corrosion fatigue is fatigue in a corrosive environment. Fatigue is the mechanical degradation of a material under the cyclic loading. Nearly all engineering structures experience some form of alternating stress and are exposed to harmful environments during their service life. The environment plays a significant role in the fatigue of high strength structural materials like steels, aluminum alloys, and titanium alloys.

Key Parameters:

- Material, corrosive environment, cyclic stresses and stress risers are critical factors.
- Crack initiation sites include concentrators such as pits, notches, surface defects, and changes in sections or fillet welds.
- Unsupported piping.

Flow-Assisted Corrosion (FAC)

Definition:
Flow-Assisted Corrosion (FAC) is a localized corrosion process that occurs due to high flow rates and removal of protective films on metal surfaces.

Key Parameters:

- Flow rate, piping dimensions

CO_2 Corrosion

Definition:
Carbon dioxide (Carbonic Acid) corrosion is localized aqueous CO_2 corrosion of carbon and low alloy steels in an electrochemical process involving the anodic dissolution of iron and the cathodic evolution of hydrogen. Carbon dioxide in a weakly acidic gas that is corrosive when dissolved in water becomes carbonic acid (H_2CO_3). Often large shallow pits occur.

Key Parameters:

- Carbon dioxide, water concentration, process conditions

Corrosion Under Insulation (CUI)

Definition
CUI is a particularly severe form of localized corrosion. Intruding water or moisture (external) is the key driving mechanism. Special consideration must be taken during design to avoid corrosion by not permitting water to enter a system either directly or indirectly. Corrosion may attack the jacketing, the insulation hardware, or the underlying equipment. For high temperature equipment and piping, water that enters the insulation material will eventually reach a region under the insulation and come in contact with hot piping or process equipment metal walls. A saturated salt solution may then develop in this area, which becomes a problem during shutdown and reheating. Upon reheating, the wall will temporarily be in contact with the saturated solution, and stress-corrosion cracking may begin.

Key Parameters:

- Damaged or partial insulation of piping and equipment.
- Carbon steel with insulation operating within a temperature range of 0° F to 300° F (-18 to 149 °C) Maximum corrosion rate occurs at a temperature of 150° F to 250° F (65.6 to 121 °C).
- Shutdowns and Restarting.

Wet H_2S Damage (WHIP)

Definition:
"Wet H_2S Cracking" can occur in carbon steel and alloy due to the effects of aqueous hydrogen charging in "Wet H_2S" process environments. Types of damage that can occur as a result of aqueous hydrogen charging include sulfide stress cracking (SSC) of hard weldments and microstructures, hydrogen blistering, hydrogen induced cracking (HIC) and stress-oriented hydrogen induced cracking (SOHIC). The key characteristic of Wet H_2S cracking mechanisms involve charging of the steel with hydrogen at relatively low temperatures, most often as the result of corrosion process, which evolves hydrogen.

Key Parameters:

- Environmental Conditions (pH, H_2S level, containments and temperature).
- Material Properties (hardness, microstructure and strength).
- Temperature (ambient to 300°F (149 °C))
- Welded Steel.
- Carbon steels and low alloy steels above a hardness of 237 Brinell Hardness

(BH).

- Containing an aqueous phase with more than 50ppmw (weight) total sulfide content in aqueous phase.
- Containing more than just 1ppmw total surface content in the aqueous phase and a pH less than 4.0.

Ammonia Stress Corrosion Cracking

Definition:

Welded carbon steel in anhydrous ammonia is highly susceptible. Intergranular cracks can occur in heat affected zones. Impurities in liquid ammonia such as air or carbon dioxide can cause stress corrosion cracking of mild steel. Ammonia is highly corrosive towards copper and zinc alloys.

Key Parameters:

- PWHT of carbon steel eliminates susceptibility to Ammonia SCC.
- Weld hardness should not exceed 225 Brinell Hardness (BH).
- Contamination with air or oxygen increases the tendency to crack.
- Anhydrous ammonia with less than 0.2% water may cause cracking in carbon steel.

Rupture of Piping and Fire in Synthesis loop of an Ammonia Plant

The paper analyzes the root cause of a fire incident that occurred in synthesis loop of Ammonia Unit- 2 of KRIBHCO fertilizer complex. There was a localized cracking/rupture in a spool piece of 1.3-meter length of the vertical leg of pipe between two exchangers at the downstream of synthesis converter. The paper also sheds light on rehabilitation and additional reliability measures taken for sustained operations. The findings of the incident are shared to sensitize and enlighten the industry when carrying out revamp/equipment replacement projects during outages or when carrying out critical activities in synthesis loop considering potential high temperature hydrogen attack (HTHA) and compliance of material compatibility as per API RP 941 Nelson Curve, API 581 HTHA Technical Module and US CSB Recommendations.

M R Sharma - Operations Director
J Sondhi - Addl. GM (Mech. Maint.)
Krishak Bharati Co-operative Ltd (KRIBHCO)
Hazira, Surat – INDIA

Introduction

Krishak Bharati Cooperative Limited (KRIBHCO) operates one of the largest ammonia urea complexes in India (2.2 million tons of urea per year). The fertilizer complex was revamped during the years 2011 and 2012 for increasing capacity and reduction of energy consumption. The revamped plant capacities were 1890 x 2= 3780 MTPD (4170 STPD) of Ammonia and 1662.5 x 4= 6650 MTPD (7330 STPD) of Urea.

As part of the plant upgrade activities, synthesis loop Boiler feed water exchanger (123-C) of Ammonia Unit II was replaced in February 2012 (Figure 1) due to its' past tube plugging history. The elevation of inlet nozzle of the new 123-C was higher than the previous one. To accommodate the change in inlet nozzle elevation of the 123-C, a pipe spool piece of 1.3-meter (4ft) length was inserted into the vertical leg of piping (as shown in Figure 2) from 123-CA (a new exchanger installed in 1993 plant revamp) exit to 123-C inlet. The fire incident happened due to localized cracking/rupture at the weld joint of this inserted spool piece which was carrying synthesis gas at a temperature of 285°C (545°F) and pressure of 179.5 bar (2603.5 psig). The dislodged spool piece location is marked in Figure 3. The incident took place on 5th November 2017 at 10:08 AM and the fire was controlled within 15 minutes. The plant was operating normally prior to this incident.

The fire caused damage to the equipment, pipelines, instrument and electrical cables, insulation etc. in the vicinity, which were assessed and restored. The plant operations were fully normalized by 30th November 2017.

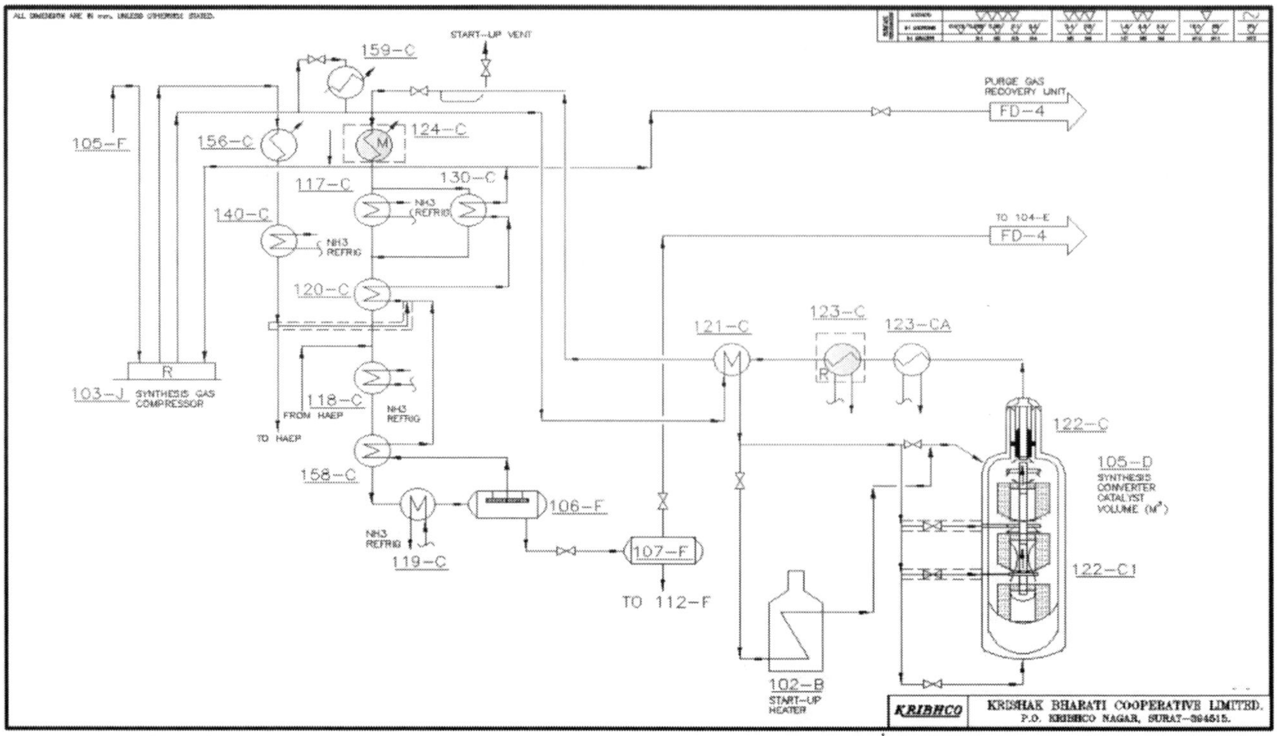

Figure 1. Flow diagram of synthesis loop between 103-J compressor 3rd stage outlet to 105-D converter

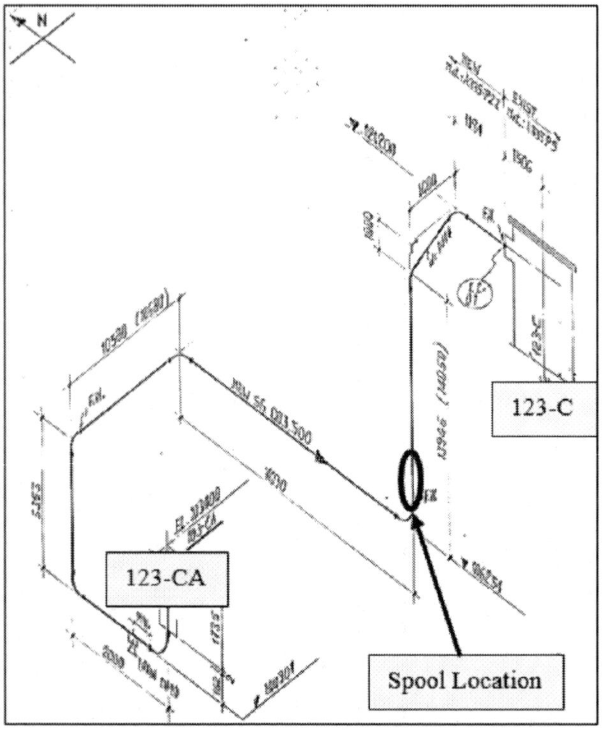

Figure 2. Isometrics of pipeline from 123-CA exit to 123-C inlet, showing location of spool piece inserted during revamp in February'2012

Figure 3. Location of failed spool segment of vertical leg of synthesis gas from 123-CA exit to 123-C inlet

Technical Data for Pipeline

Fluid Handled	Synthesis gas
Operating Pressure	179.5 bar (2603.5 psi)
Operating Temperature	285.0°C (545°F)
Gas Composition (Mole %)	H_2 – 60.0% N_2 – 18.26% NH_3 – 17.85% Ar – 3.89%
Dimensions	508mm (OD) x 50.01mm (thick)
Pipe Material	ASTM A335 Grade P22

HTHA Overview

Exposure of carbon and low-alloy steels used for piping and pressure vessel in high temperature hydrogen service can lead to onset of degradation known as HTHA or called high temperature hydrogen attack. The occurrence of HTHA primarily depends on partial pressure of hydrogen, temperature, and time with reference to material used in pipeline or pressure vessel.

When hydrogen acquires the lowest possible energy state, it will typically combine with available elements to form compounds such as water (H_20), hydrochloric acid (HCl), methane (CH_4) and ammonia (NH_3). The ability to form these compounds and its stability depends primarily on temperature and pressure. Natural gas which predominantly consists of methane, when reformed with steam, will decompose primarily into carbon dioxide and hydrogen. This molecular hydrogen at high temperature and pressure gets partially dissociated into atomic hydrogen. It is a reversible reaction and at pressure and temperature equilibrium exists between the molecular and the atomic hydrogen.

$$H_2 \longleftrightarrow H + H$$

This atomic hydrogen being soluble in steel and under the influence of certain temperature conditions and partial pressures tend to enter iron lattice as an interstitial atom. This hydrogen not only accumulates at discontinuities as molecular hydrogen, but also reacts with the iron carbide in steel at high temperature to produce methane. This phenomenon leads to "decarburization" of steel.

$Fe_3C \rightarrow 3Fe + C$ [Decomposition of iron carbide]

$C+2H_2 \rightarrow CH_4$ [Reaction of H and C to produce methane]

Methane being insoluble in iron, builds up as a gas in the small pockets at grain boundaries, resulting in the development of grain boundary fissures. The buildup of methane produces significant pressures at the grain boundaries or inclusions and will hence provide a mechanism for greater strain. The pressure built up at the grain boundary affects cohesion strength of the steel and results in "inter-granular separation". Similarly, the pressure builds up at inclusion interfaces constantly enlarges the voids. Linkage of these voids, with the assistance of grain boundary fissuring, leads to step-wise cracking and ultimately "blistering". In these instances, the material becomes irreversibly damaged which causes severe reduction in mechanical properties of the material. This high temperature hydrogen attack (HTHA) typically occurs above 204°C (400°F) at relatively high hydrogen partial pressure.

Keeping in view of the incident that occurred at Ammonia unit-II in November'2017, high pressure synthesis gas pipelines were evaluated for susceptibility to high temperature hydrogen attack. Assessment for each pipeline was carried out in accordance to the following criteria:

1. API 941 – Nelson curve for susceptibility to HTHA
2. API 581 – Pv parameter for susceptibility to HTHA
3. US Chemical Safety Board (CSB) recommendations

API 941 Recommendation

API 941 discusses the resistance of steels to high temperature hydrogen attack. According to API 941, at normal atmospheric temperatures, gaseous hydrogen does not readily permeate steel, even at high pressures. HTHA is observed in steels exposed to temperatures above 221°C (430°F).

To evaluate susceptibility of HTHA in synthesis gas pipelines, all the relevant data like operating pressure, operating temperature, fluid composition, material of construction etc. were tabulated for 42 lines in the synthesis loop from synthesis gas compressor 3rd stage outlet to converter inlet.

The minimum operating temperature on the Nelson curve is 149°C (300°F). In KBR technology KRIBHCO Ammonia plant, only 3 pipelines in the synthesis gas loop operate above 149°C (300°F). These 3 pipelines were plotted on the Nelson curve (shown in Figure 4).

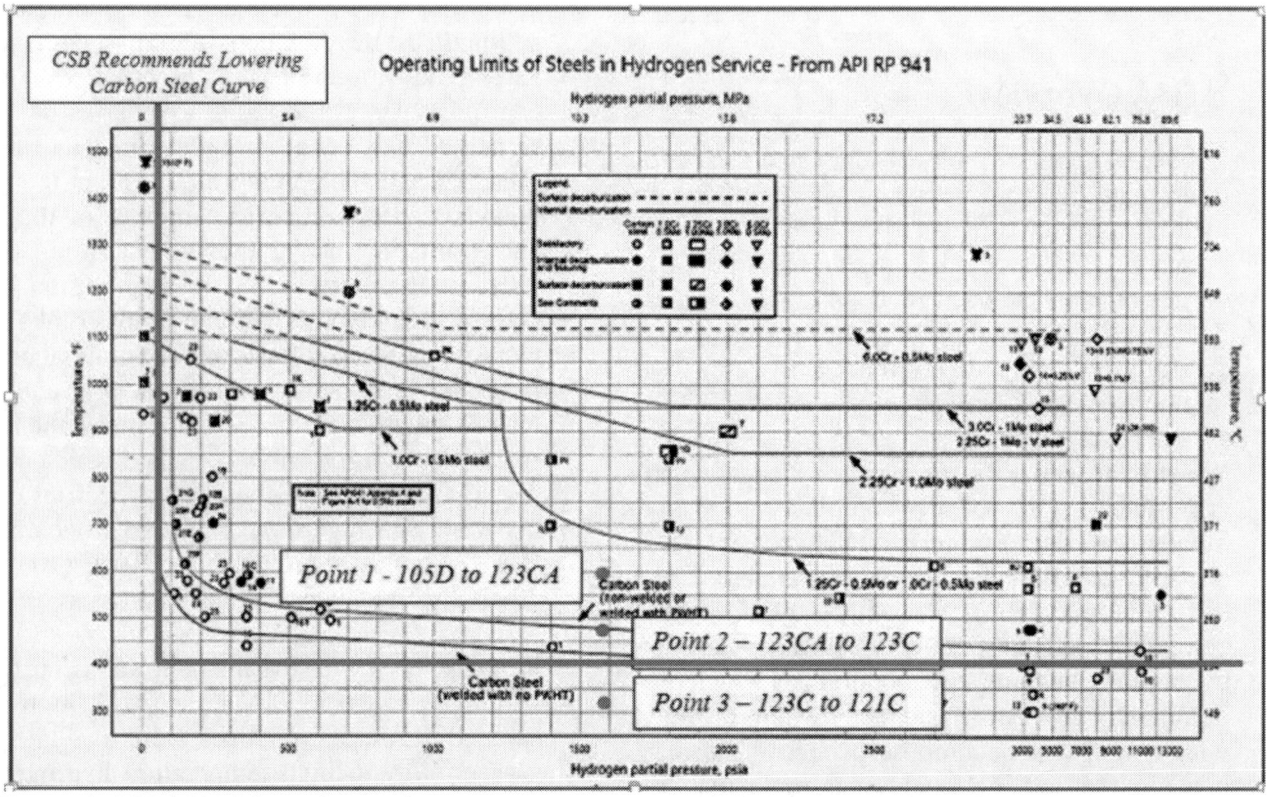

Figure 4. API 941 Nelson Curve showing operating point of three identified pipeline of synthesis loop and CSB recommendation line (red color) for carbon steel

The materials of construction for these three pipelines are as follows:

- 105-D converter outlet to 123-CA inlet (ASTM A335 Grade P22).
- 123-CA outlet to 123-C inlet (ASTM A335 Grade P22).
- 123-C outlet to 121-C inlet (ASTM A53 Gr B).

As shown in Figure 4, none of these pipelines are susceptible to HTHA as per Nelson curve for the material of construction used.

API 581 Recommendation

Susceptibility to HTHA can also be assessed through API 581 HTHA Technical Module. A single parameter Pv has been developed for carbon steel and low alloy steels operating above 204.4°C (400°F) and 5.5 bar (80 psi). It relates partial pressure of hydrogen, temperature and time.
Parameter Pv can be calculated with following formula:

$$Pv = \log (P_{H2}) + [0.000309*T*(\log (t) + 14)]$$

where:

P_{H2} – Partial pressure of hydrogen (kgf/cm^2)

T – Operating temperature (K)

t – Time (hrs.)

To evaluate susceptibility of HTHA to syngas pipelines, all the relevant data like operating temperature, hydrogen partial pressures, operating hours, material of construction etc. were tabulated. With the help of these data, parameter Pv was calculated for all the relevant pipelines. In Ammonia plant, only 2 pipelines in the synthesis loop operate above 204.4°C (400°F) as can be seen in Figure 4 above. The calculation for each pipeline is as follows:

1) 105-D converter outlet to 123-CA inlet:

   Operating pressure = 184 kgf/cm^2 (2617 psi)

   T = 619.2 K (346.2°C, 655.2°F)

   P_{H2} = 109.02 kgf/cm^2 (1550 psi)

   t = 24 year = 190080 hrs.

   Pv = log (109.02) + [0.000309*619.2*(log (190080) + 14)]

   Pv = 5.73

2) 123-CA outlet to 123-C inlet

   Operating pressure = 183 kgf/cm^2 (2603 psi)

   T = 558.2 K (285°C, 545°F)

   P_{H2} = 109.02 kgf/cm^2 (1551 psi)

   t = 24 year = 190080 hrs.

   Pv = log (109.02) + [0.000309*558.2*(log (190080) + 14)]

   Pv = 5.36

The calculated Pv value of these two pipelines are below the specified value for A335 Grade P22 grade material (as seen in Table 1).

Material	Critical Pv Factor			
	High susceptibility	Medium susceptibility	Low Susceptibility	No susceptibility
Carbon Steel	$Pv \geq 4.70$	$4.61 < Pv \leq 4.70$	$4.53 < Pv \leq 4.61$	$Pv \leq 4.53$
C, 0.5Mo (annealed)	$Pv \geq 4.95$	$4.87 < Pv \leq 4.95$	$4.78 < Pv \leq 4.87$	$Pv \leq 4.78$
C, 0.5Mo (normalized)	$Pv \geq 5.60$	$5.51 < Pv \leq 5.60$	$5.43 < Pv \leq 5.51$	$Pv \leq 5.43$
1.25Cr, 0.5Mo	$Pv \geq 6.00$	$5.92 < Pv \leq 6.00$	$5.83 < Pv \leq 5.92$	$Pv \leq 5.83$
2.25Cr, 1.0Mo	$Pv \geq 6.53$	$6.45 < Pv \leq 6.53$	$6.36 < Pv \leq 6.53$	$Pv \leq 6.36$

Table 1. API 581 susceptibility to HTHA – Pv factor table

US Chemical Safety Board (CSB) recommendation

In the aftermath of Tesoro (Washington, US) Anacortes incident (2010), US Chemical Safety and Hazard Investigation Board recommended to prohibit the use of Carbon Steel material in processes that operate above 204.4°C (400°F). The US CSB has recommended a lower curve for CS material than API 941 standard recommends (as shown in Figure 5).

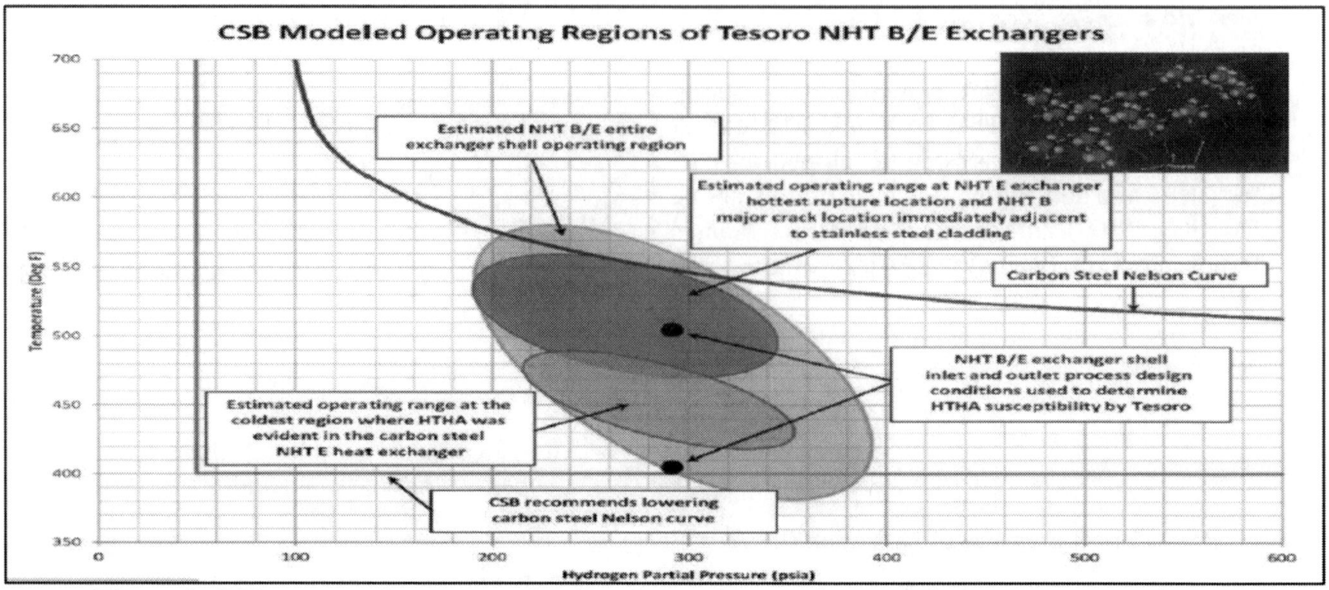

Figure 5. CSB proposed new operating line for carbon steel material after Tesoro incident

Out of the 3 pipelines subjected to further assessment, only one pipeline (123-C outlet to 121-C inlet line) was of carbon steel (A53 Grade B) having operating point of 106.5 bar (1550 psi) hydrogen partial pressure and 176°C (350°F), which also lies well below the recommended curve by CSB.

Therefore, none of the pipelines are susceptible to HTHA as per CSB recommendations due to the material of construction used and the particular process parameters operated within.

Equipment Audits for HTHA

The equipment that operates above the minimum defined temperature of the codes were also audited for susceptibility for HTHA.

- 123-CA and 123-C exchangers, synthesis gas flows into tubes at operating pressure of 179.5 bar (2603.5 psi).
- 123-CA inlet temp 346.2°C (665°F) / outlet temp 285°C (545°F).
- 123-C inlet temp 285°C (545°F) / outlet temp 176.4°C (349.5°F).
- 123-CA tubes are of A335 Grade P22 material and 123-C tubes are of A335 Grade P11 material.
- 123-CA and 123-C exchanger heads are of A335 Grade P11 material. These operating points were also mapped on Nelson curve (same points in Figure 4). Both point lies inside safe region and hence are not susceptible to HTHA.
- Converter outlet exchanger 122-C is of SS-304 material. Hence, it is also not susceptible to HTHA.
- Converter top dome is of SA516 Grade 70 material and is in contact with inlet synthesis gas at 133°C (271.4°F). Since it is well below minimum defined temperatures in the codes, the converter top dome is not susceptible to HTHA.

Thus according to above criterion, standards and recommendations, operating conditions of these equipment lie well within the safe zone and hence are not susceptible to HTHA.Repair and Rehabilitation

The following steps were taken to repair the damaged section:

- 23-meters (75 ft) pipe of ASTM A335 Grade P22 (20″ Schedule 160) was procured on emergency basis. Though the failure had occurred in a localized portion, the entire vertical and horizontal leg of the remaining pipeline was also replaced as a preventive measure to rule out any detrimental effect of stresses which might have occurred due to the fire and subsequent water quenching during fire-fighting.
- Bevel edges of all pipes and elbows after edge preparation were subjected to liquid penetrant test to check for any surface indications.
- Welding Procedure Specifications for A335 Grade P22 (P No. P5A) was duly reviewed and approved before proceeding and welder's qualification tests were performed before putting them on the job.
- Pre-heating in the range of 150-200°C (302°F-392°F) was employed. The control of the inter-pass temperature to the maximum of 200°C (392°F) was maintained to restrict grain growth and precipitate coarsening.
- After completion of welding, post weld heat treatment (PWHT) of all the welded joints including the heat affected zone (HAZ) was performed as per relevant code practice.
- The weld joints which were fabricated in Central workshop were 100% radiographed. However due to radiography source limitations, field joints were radiographed after filling up to 50% of thickness. After completion of full welding, Ultrasonic Flaw Detection of these field joints were carried out and found acceptable.
- Stress analysis of the additional load due to the insertion of the 1.3-meter length spool piece was performed, and the spring supports were found to be adequate as per analysis.
- The damaged and bent lines (small bore) near 105-D and 123-C were required to be replaced including displaced and damaged supports. These lines included:
  - Fuel Gas to Start-up Heater pipeline – 50NB and 80NB (60 meters, 200 ft)
  - Low pressure steam pipelines – 40NB (72 meters, 236 ft)
  - Synthesis Gas to Start-up Heater – 150NB (20 meters, 65 ft)
  - Instrument Air and Plant Air lines (36 meters, 118 ft)
- In situ metallography, hardness measurement and thickness measurement of equipment in the near vicinity was carried out, and no microstructure degradation/abnormality was noticed. Additionally, weld joints of nearby pipelines were evaluated by Ultrasonic Flaw Detection, and no significant defects were detected.
- Minor localized bulging of outermost layer of Ammonia Converter (105-D) due to localized heating was observed at an elevation near the failed spool piece (see Figure 6).

Figure 6. Bulging on the outermost layer of converter105-D

Assessment of the health of converter was carried out based on following studies:

- Adequacy of 105-D shell wall thickness by ignoring the contribution of the outermost layer.
- Assessment of bulge penetration into the inner layers.
- Material degradation of outer layer due to effect of fire in vicinity.

The converter was constructed as a multi-layered equipment for a design pressure of 234.4 bar (3399.7 psi) and in accordance with ASME BPVC Sec VIII Div-2, edition 1980. The required thickness of shell was calculated in compliance with above standard with respect to original design conditions mentioned in Table 2.

Parameter	Value
Total shell thickness	201 mm (7.91")
Shell thickness without outer layer	192 mm (7.55")
Thickness required as per ASME BPVC Sec VIII Div-2, edition 1980 for original design condition	196.1 mm (7.72")

Table 2. Thickness calculation summary

From Table 2 it is concluded that the available thickness of inner layers (ignoring outermost layer assuming all other layers and weld joint of equipment are intact) are:

- Adequate for before revamp operating conditions of 196 bar (2842.7 psi)
- Adequate for post revamp operating conditions of 181.4 bar (2631 psi)
- Inadequate for original design conditions 234.4 bar (3399.7 psi).

In addition to code calculated thickness, a finite element analysis (FEA) was carried out to evaluate the portion of equipment remaining after removing the 9mm thick outer layer at the bulge region. It was observed the linearized stress intensities in the local bulge affected area (without outer most layer) did not meet the requirement of protection against plastic collapse of "Design by stress analysis requirement" of part 5 of ASME Sec VIII Div. 2. Hence the design of the shell in absence of outer layer is marginally inadequate to withstand design loads as per original design.

So as a preventive measure and to meet the requirement of design condition based on above studies, repairing of the localized bulged portion of the outermost layer was carried out by putting a new patch of required dimension 4000mm X 1100mm (13.1ft x 3.6 ft)) (Figure 7).

Figure 7. Bulged portion of converter was assessed, and patch of required grade and dimension was welded

Metallography assessment of affected area was also carried out and no microstructure degradation was observed.

- One leg of the structure support of the 123-C buckled at the area adjacent to the fire (Figure 8), which was replaced as per existing configuration (Figure 9). Other bracket supports to shell weld were checked by NDT methods.

Figure 8. One leg of structure support of heat Exchanger 123-C had buckled at area adjacent to fire

Figure 9. One of the buckled leg of structure support was replaced by new as per existing design

- External visual inspection and ultrasonic thickness measurement of 123-C shell at areas adjacent to fire were checked and measurements were compared with unaffected areas and design values by NDT qualified personnel. The examination revealed no abnormality.
- The foundation of nearby equipment reinforced concrete components were assessed by rebound hammer testing (NDE method) for strength adequacy. It was found that average strength was more than the required value of $20N/mm^2$. However, as a preventive measure in Start-up Heater 102-B, jacketing of columns by micro-concreting was carried out. The foundation of 105-D was jacketed by non-shrink grout.
- The nearby damaged cables, structures, insulation, lighting etc. were replaced.

Post repairs and rehabilitation, the entire synthesis loop was pressured up to 193 bar (2799.2 psi) with process gas and held for 10 minutes to check for any pressure drop by closing the pressurization valve. The system pressure was constant signifying no leaks.

Root Cause Analysis

During the ammonia converter revamp in 1993, a new exchanger 123-CA was added between 123-C and 105-D to make the 123-C run on milder working temperatures. Hence pipe (20″ Schedule 160) was installed for taking synthesis gas from the outlet of 123-CA to the inlet of 123-C. The material of the installed pipe in 1993 was upgraded to Alloy Steel ASTM A335 Grade P22 to improve safety in operations as per API 941.

During the 2011/2012 revamp, the elevation of the inlet nozzle of the new/replaced 123-C was higher than the previous one. To accommodate for the change in the inlet nozzle elevation of the new 123-C, a pipe spool piece of 1.3-meter (4 ft) length was inserted in the vertical leg of piping from 123-CA exit to 123-C inlet. Evidently during the insertion of 1.3-meter spool piece from 123-CA to 123-C, the following unintended events happened:

Incorrect Pipe Material Used

Positive material identification (PMI) of the inserted 1.3-meter spool piece revealed that Low temperature carbon steel A333 Grade 6 (incorrect material of construction for this service) was inserted. Since the existing material was ASTM A335 Grade P22, the 1.3-meter spool piece added should also have been of same material. LTCS (A333 Grade 6) is not recommended as per API 941 and discontinued since 1991 for such service as synthesis gas at the operating temperature being handled by pipe.

- Source of Spool Piece – During the revamp, modification of the synthesis gas compressor 4th stage suction and discharge lines was being carried out at the same time. This job was in the near vicinity, which required material specification of LTCS A333 Grade 6 of similar dimensions of 20″ Schedule 160 to be used for a line that operates at a temperature of around 50°C (122°F). At this temperature, LTCS material is acceptable according to API 941, API 581 and CSB. It is evident that by overlook a spool piece from this pipe length was used in 123-C to 123-CA pipeline which required upgraded material of A335 Grade P22 for long term operations reliability.

Incorrect Filler Weld Metal

PMI of the failed weld at the welding area was carried out and compared with PMI of other healthy welds in the pipeline. The analysis of the failed weld showed that welding was done with 1¼Cr-½Mo electrode (composition like ASTM A335 Grade P11) while healthy welds in the same pipeline showed 2¼Cr-1Mo electrode (composition like ASTM A335 Grade P22).

Further PMI of welding reveals that welding electrode of 1¼Cr-½Mo material composition (equivalent P11) was used to make up for any shortfall of the 1.3-meter length of A333 Grade 6 spool piece by building up the spool piece to maintain the root gap the at site.

PMI was also carried out on the remaining line segments and elbows of the removed line wherein it was found that except for the 1.3-meter spool, the entire rest of the pipeline parent material and weld joints confirmed to A335 Grade P22.

Hence in the welding area three different metallurgies were present which is not a standard engineering practice for such critical high pressure piping.

Defective Welds

The visual observation of both welds on the A335 Grade P22 sides indicate partial or complete lack of fusion at the root pass (Figure 10 and 11).

Figure 10. Shows partial or lack of fusion at root of A335 Gr P22 pipe bottom weld with spool piece

Figure 11. Shows partial or lack of fusion at root of A335 Gr P22 pipe top weld with spool piece

Improper fusion at the root of the weld created a stress concentrator which contributed to the weld failure. It is worth noting here that the weld had passed radiography testing (RT) acceptance by the independent third-party agency (TPA) certification. The project stage site was under contractual independent third-party agency for quality supervision and compliance and all stage wide clearances were given to the TPI to avoid project delays without intervention of KRIBHCO's inspection. It seems both the contractor and the TPI agency inspector failed to identify the defects in the radiographs of the welds. The contractor did not realize the criticality of service of this pipeline.

Post incident the contractor has been de-registered and blocked from participating in tendering for KRIBHCO.

HTHA Corrosion

Visual inspection of failed spool piece revealed blisters on the surfaces. Since the operating point of this pipeline lies above the CS line on Nelson curve (as can be seen from Figure 4) and being in high temperature and pressure environment of hydrogen for more than 5 years, HTHA was evident in the form of blistering. Here it is worth noting that there was no visible effect of HTHA (like blistering, fissuring) on A335 Gr P22 side of the pipeline.

Hence it is evident that combination of above reasons contributed to the failure. Result and Discussion

HTHA detrimental effects are predominantly in HAZ areas of welds, as the 8th edition update of API RP 941 indicates, so the most likely location of HTHA attack is at or near to welds. Due to weld irregularities that were observed at the root pass of the dissimilar weld joint, it gave a preferential site for progressive HTHA damage. Stresses caused due to HTHA in weld and HAZ would have reduced the ductility and strength of the carbon steel spool pieces substantially below its design values (as is evident from test results of longitudinal specimen in Table 3).

Test	Result	Required
Yield Stress (N/mm^2)	284	240 min
U.T.S (N/mm^2)	359	415 min
% Elongation	8.85	22 min

Table 3. Tensile test result of failed spool pieces

Blisters were seen at multiple location in the failed pieces of the CS pipe (Figures 12 and 13), which visually corroborated the effect of HTHA.

Figure 12. Blister crack depth on the edge where fracture take place

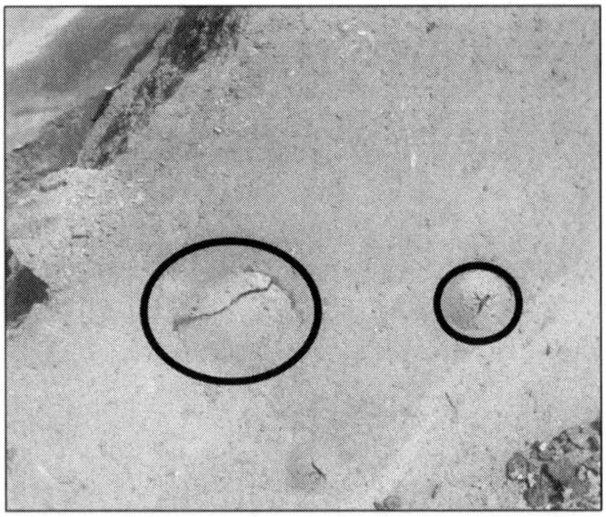

Figure 13. Blister observed on the inner surface of failed pieces indicating HTHA

In high hydrogen partial pressure and high temperature environment, metallurgical irregularities/discontinuities at the weld interface of dissimilar joint (A335 Grade P22 and LTCS) had given preferential site for HTHA, eventually causing cracking along the plane of weld. These cracking along dissimilar weld metal interface is known as dis-bonding. It is a long-term failure which occurs in welding of dissimilar joint caused by hydrogen migration even for PWHT welds.

Severe reduction in elongation and ultimate tensile strength due to high temperature hydrogen attack in the failed piece were observed.

Lessons Learned

The following lessons learned are being shared from this incident:

- Select the proper material for the operating conditions. For increased temperatures, consider the use of alloys with higher weight percentages of chromium and molybdenum.
- Use actual operating temperatures for assessing HTHA susceptibility and validate that the actual operating temperatures and pressures are below the API 941 curve by a defined amount.
- Select inspection methods and frequencies to detect initial stages of HTHA.
- Ensure that even if duly reviewed and approved written procedures are in place to provide guidance during fabrication, erection and commissioning, their implementation should be monitored. Adequate numbers of TPI personnel should be at site to take care of adherence to quality checks at various stages of work execution.
- Surprise visit/check may be carried out by TPI personnel during radiography testing (RT) to check for any intentional duplicity of similar RT passed welds re-radiographed by contractor, in lieu of weld to be radiographed by giving different joint numbers on same weld to pass the weld/repairs.
- During any equipment replacement requiring outage/modification, if the new design ratification demands any change in orientation/elevation of any nozzle, that design should be adequately flagged as a stop-job hold point by Equipment Engineer and necessary information should be passed to relevant areas such as a Piping Engineer for proper design of the job.
- If a change in elevation/orientation of any nozzle requires a change of the connecting pipeline isometric, the isometric change/modification should also be approved and flagged for implementation with adequate care and material sorting/identification must be done before start of job.
- WPS (welding procedure specification) for the specific job should be created and approved not only by TPI but also by the client.
- For any repair/maintenance activities PMI test request should be initiated for material identification in case the material is sourced from site and not withdrawn from centralized stores.

- Even from centralized stores where PMI test is subjected before acceptance in stores, any material with-drawl for critical service should be requested for PMI test before putting in service to ward off any mixing.

  The contractor works to meet the tight schedule time lines of any project/job and accepts all the clauses of the contract including compliance to requisite standards. The contractor's deputed site personnel should be competent to understand the scope and extent of standards. Works on equipment's /pipelines operating with hydrogen service should be given special attention. For such jobs the works contract's should be clearly worded such that after inspection checks carried out by TPI agency it should be counter checked and approved by Client's/Owner's personnel.

Future Work

Much research has been done into the areas of HTHA detection. In general terms, ultrasonic techniques were used to detect HTHA at the cracking stage after micro-cracks and/or blistering has occurred. After technical audit of syngas pipelines for HTHA, KRIBHCO has planned to conduct quantitative HTHA assessment in 2019 by using Advanced Ultrasonic Backscatter Technique (AUBT), Advanced Velocity Ratio, Time of Flight Detection (ToFD), Angle Beam Spectrum Analysis etc. for abnormality, if any and base line for further monitoring as the need be.

Conclusion

- The dissimilar weld joint along with stressed sites of weld imperfections gave preferential sites in HAZ and weld for high temperature hydrogen attack which eventually lead to brittle weld to weld failure and in turn causing spool piece to be torn in multiple pieces longitudinally.
- Inadequate controls of work lead to a lack of detection that the wrong material had been poorly welded into a critical service line, the material being prone to HTHA in this service.

Index Terms:

API – American Petroleum Institute
RP – Recommended Practice
HTHA – High temperature hydrogen attack
KRIBHCO – Krishak Bharati Cooperative Limited
SG – Synthesis Gas

Glossary to Terms

102-B – Start-up Heater for Converter
121-C – Converter Feed Effluent Exchanger
122-C – Converter Interchanger
123-CA – Additional Boiler Feed Water Preheater
123-C – Ammonia Converter Effluent/Boiler Feed Water Exchanger
105-D – Ammonia Synthesis Converter
103-J – Synthesis Gas Compressor
MTPD – Metric Ton per Day
PMI – Positive Material Identification
LTCS – Low Temperature Carbon Steel
NB – Nominal Bore
HAZ – Heat Affected Zone

References

[1] API *RP* 941, "Steels for Hydrogen Service at Elevated Temperatures and Pressures in Petroleum Refineries and Petrochemical Plants"- Eighth edition February 2016.

[2] API *RP* 581, "Risk-Based Inspection, Base Resource Document"- 2016 edition.

[3] U.S. Chemical Safety and Hazard Investigation Board Recommendations (CSB) - 23 July 2012 / Published online: 14 August 2012 ASM International 2012 U.S. Chemical Safety And Hazard Investigation Board (CSB) Investigation, "Catastrophic Rupture Of Heat Exchanger (Seven

Fatalities) Tesoro Anacortes Refinery, Anacortes, Washington, April 2, 2010, Report 2010-08- I-WA May 2014".

[4] ASTM Standard Number A335, " Standard Specification for Seamless Ferritic Alloy Steel Pipe for High Temperature Service" - 2017 edition.

[5] ASME code for Pressure Piping B31.3 – 2016 edition.

[6] Excerpts from "Reducing the Risk of High Temperature Hydrogen Attack (HTHA) Failures" by Daniel J. Benac • Paul McAndrew, from Journal of Failure Analysis and Prevention, Issue 6/2012. Journal no. 11668.

[7] Excerpts from proceedings at Engineering Symposium in Rochester Hyatt Regency Downtown, Rochester, NY April 18, 2017 on "Safety Culture and Safe Sealing- Anacortes Refinery Explosion" by Jim Drago, P.E. Principal Applications Engineer, Garlock Sealing Technologies, Palmyra, NY.

[8] Excerpts from thesis by Brent Matthew Saba, on "Evaluation of mechanical fitness for service of high temperature hydrogen attacked steels" Publication May 2003. LSU Master's Theses. 3883. https://digitalcommons.lsu.edu/gradschool_theses/3883.

[9] Excerpts from article on "Avoiding High Temperature Hydrogen Attack" By Baker Engineering and Risk Consultants Inc., in leaflet publication November 2015.

NH3 Terminal Case Study: Achieving SIS/SIL Compliance to ANSI/ISA61511 via SharePoint®

The first international functional safety standard (IEC 61511) for Safety Instrumented System (SIS) was released in 2003. The second edition was released in 2016 and adapted by ANSI/ISA for the USA in 2018. As of writing this paper, many American process sites are only marginally closer to a Safety Integrity Level (SIL) implementation strategy to meet all requirements of the standard. In fact, the latest release has significant changes that increases that gap between current implementation practices and full standard compliance.

These consensus standards have approximately 300 "shall" statements. The "shall" statements of IEC 61511, as well as each item under the "shall" statements, cover over 650 individual requirements for each safety instrumented function (SIF). Tracking the fulfillment of these requirements is a demanding task and collections of tracking spreadsheets have not proven effective. Attempts have been made to meet the requirements with more capable, enterprise-level software solutions, but to-date none have accomplished such a daunting task.

There are multiple reasons that can be cited for why such SIS/SIL requirements are not being met. Are we over- engineering our solutions to meet these requirements? Is the solution too complex and doesn't work well with your company's current processes and practices? Do your SIL implementation tools improve both safety and reliability?

This paper will present a NH3 terminal case study and show how full compliance can be achieved. The solution applies comprehensive, off-the-shelf safety lifecycle tools that can work with existing site practices. The process will be sustainable, simple, and most importantly, cost effective. The exciting news is that Microsoft® SharePoint® can bring all the requirements into focus and any company can obtain full ISA/IEC 61511 compliance.

Curt Miller
SIS SILverstone LLC

Introduction

The purpose of this document is to provide guidance to companies that desire compliance to the international SIS functional safety standard similar to the their achievement in process safety management (PSM). It will be based on a SharePoint® process applied to a NH3 terminal case study.

Similar to PSM compliance, the process will require discipline and management support since considerable time and resources will need to be utilized. The good news is that multiple benefits will result to justify the investment.

With the release of the updated ISA61511 [R1] standard, the PSM and SIS design worlds have come together for three main reasons. First, the new standard is an adoption of the international IEC 61511-2016 standard which had embedded hazard and risk analysis. Secondly, LOPA (layers of protection analysis) or other SIL methodology is required and will involve both PSM and SIS designers to be completed effectively. Thirdly, the important functional safety management (FSM) process is also now included. This "ISO 9000" type quality process brings planning and verification aspects to safety lifecycle practices to ensure that all the steps are completed thoroughly.

WHERE ARE THE COMPLIANCE GAPS IN THE SIS DESIGN STANDARD THAT A WELL MANAGED PSM PROGRAM CAN OVERCOME?

The ISA61511 safety lifecycle can be generally broken down into 3 phases. They include the analysis, realization, and operations phases as shown in the following Figure1.

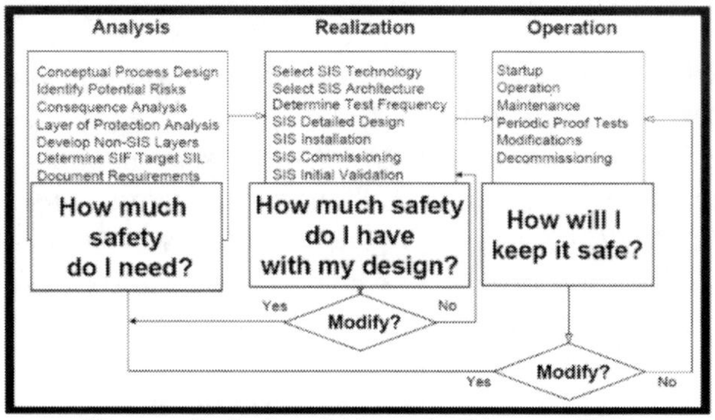

Figure 1. Safety Lifecycle Phases [R2]

With these phases in mind, we will now look at where a better PSM methodology can help ensure functional safety.

A. Functional Safety Management (FSM) – Support in All Safety Lifecycle Phases

FSM was added to the standard to overcome the systematic flaws that are responsible for the majority of SIS installation failures. Examples of systematic causes include:

- System designed could not respond in time to prevent the hazard
- Selected sensor technology does not fit the application
- Installed system could not be tested online
- Changes made to a safety function's setpoint shortened its response time

The Health and Safety Executive [R3] grouped 34 accidents where the SIS failed into the following lifecycle phase causes:

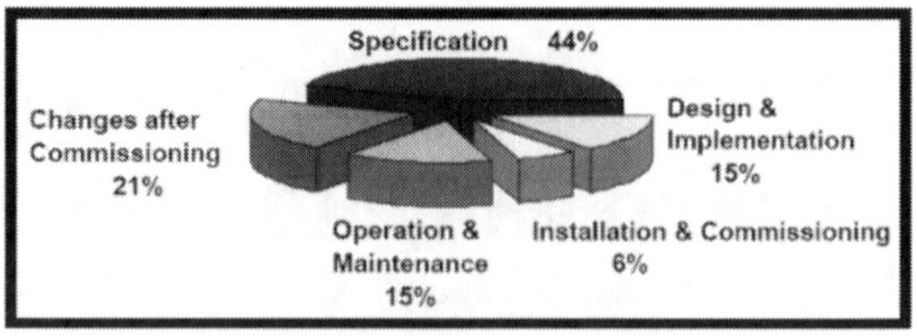

Figure 2. HSE Pie Chart Showing Lifecycle Phase % Failure [R3]

FSM was needed to provide "checks and balances" against the specification, design, and implementation glitches in the safety lifecycle process.

Polled Response on FSM Practices [R2]

Although the poll is somewhat dated, industry acceptance of functional safety management process can be seen in the following survey results:

FSM Poll Questions	Yes	No	Unsure
Which of the following do you feel would be ISA61511 compliant?			
A. Functional Safety Management procedures	40%	30%	30%
B. Project Safety Plan	64%	7%	29%

Table 1. Polled Response to Functional Safety Management Activities [R4]

Functional Safety Management Procedures

This question was asked to draw out known gaps between current and "good safety engineering practice". We can immediately see that the new requirements of ISA61511 have not been widely embraced from the unexpected low 40% favorable response to question A. FSM is the documented process that ties together the full set of safety lifecycle activities – this is a natural activity for PSM to facilitate.

Project Safety Plan

With a 64% affirmative answer to the project safety plan question, we know that most projects are following some sort of safety plan. The key question is whether the plan is sufficient. For example, does each user have a standard that they consistently follow and refine when needed to include new measures? Do their suppliers utilize it as well?

Although it will take more time to document the first time through, a planned project cycle will proceed more efficiently - resulting in a higher level of safety and lower overall costs. *(This will probably be a formalization of what your best project managers are already doing.)* An example is shown below:

CLIENT NAME :						Page 2 of 11			
Project :						PSP No.: 01		Version: 01	
CLIENT ORDER No. :				COMPANY REF :		DATE OF ISSUE:			
No.	Activity	Input	Process	Verification Record/ Output	Responsible Verifier	RESPONSIBLE			
						COMPANY	CLIENT	VERIF. TYPE	Date
1.4	CLIENT Kick-off Meeting	CLIENT Order Scope of Work	Project Management	Minutes of Meeting		H			
2.0	Produce Project Safety Plan	**TEM-PSP-V1**	**PRC-FSM-V1** Functional Safety Management	Project Safety Plan for use throughout project Project Schedule	PM	A H	A	**SM**	
2.1	Produce Project Competency Log	**TEM-COM-V1**	**PRC-FSM-V1** Functional Safety Management	Project Competency Log file for use throughout project	PM	A			
2.2	Produce Project Safety Log	**TEM-ACT-V1**	**PRC-FSM-V1** Functional Safety Management	Project Safety Log file for use throughout project	PM	A			

Figure 3. Project Execution Safety Plan

The safety plan should include:

a. Safety lifecycle steps
b. Roles, responsibilities, and competencies
c. Verification methods for each step

It is expected that PSM personnel are well equipped to support the development of the safety plan and monitor its execution during the SIS project and after its installation.

FSM Issue #1: Management Must "Buy In" or the Process Will Fail

For a manager, the compliance requirements noted in this paper should only involve a moderate investment in time and resources. However, they do require a strong and ongoing philosophical commitment since that is where the benefits really build upon each other.

As with most procedures and methods of organizing work, one can either drown in the paperwork or use it as a lifeline to cost savings and other improvements. The difference will come down to how well the process is managed. All documentation should provide a net value or else it should not exist.

Other FSM Issues Found in Functional Safety Management Implementation

Another issue when trying to implement FSM is that some contractors employed to support SIS installations do not fully understand the requirements. Hence, such projects continue to follow their normal design path. When the SIS safety lifecycle requirements are not implemented in their logical order, costly redesigns and selection of underspecified equipment may occur. In other situations, equipment may be over specified and even result in more redundancy than required.

Another issue found in meeting FSM compliance was completing the Functional Safety Assessment (FSA) after validation testing and before startup. The hardest part of fulfilling this requirement is timing. One user stated "No way – the full review of the functional safety lifecycle will never be obtained in practice. How can a company go through all the preceding safety lifecycle steps while operations are breathing down the FSA assessor's neck?"[R2] Having a strategy already in place that plans for the FSA in advance and incrementally doing the earlier stages will save a project considerable headache resulting from damaging, last minute mandated inclusions.

Example of Flawed FSM Execution

In one large installation, FSM practices were not applied in a systematic manner. Specifically, the site failed to apply formal MOC procedures for quench water injection and furnace startup overrides. After the facility started up, several near-misses resulted and a FSA audit team was assembled to review gaps. Since there were numerous infractions (e.g. process safety time violations, inferior safeguard studies, deficient MOC studies), the facility start-up was delayed for a significant period and deferred productions cost was estimated in the millions USD.

B. Risk Assessment – Analysis Phase

Process Hazard Assessment (PHA) follows the process design step in the functional safety lifecycle. While both new process plant design and existing facilities have included risk, assessments associated with their unique processes for many years, it was not until 29 CFR 1910.119, Process Safety Management (PSM) for Highly Hazardous Chemicals [R4] that it became a formal requirement in the U.S for certain chemical manufacturers. Some equivalent requirements are usually present in other parts of the world, but not always.

Most process plants use HAZOP or some variation; the key item is to make sure that there is a systematic way to identify all potential hazards, so the risk team can determine how to manage each one

Quantitative Risk Assessment Utilizing Safety Integrity Level (SIL)

Use of qualitative risk ranking tools is relatively simple, but they can lead to inconsistencies between different PHA teams as well as the potential to under or over-estimate the risk. For lower level risks, this is not generally a significant concern, however, for higher risks there is a need for management to be able to make better informed decisions using a more consistent basis.

That requires a greater level of insight and is provided by more quantitative analysis techniques that determines if there is a risk or Safety Integrity Level (SIL) gap. In layman terms, SIL refers to "orders of risk reduction" as shown in the following table:

SIL	Risk Reduction
1	> 10 factor
2	> 100 factor
3	> 1,000 factor

Table 2. SIL & Risk Reduction

There are various graphical and numerical techniques to determine the required/targeted SIL value to achieve a tolerable risk. Examples include risk graph, qualitative and quantitative layers of protection analysis, and As Low as Reasonably Practicable. Each should give roughly the same SIL result if they're "calibrated" to the same tolerable risk. The real choice in technique depends more on what fits best with a company's existing risk management philosophy and procedures.

Layer of Protection Analysis (LOPA)

Layer of Protection Analysis (LOPA) is one of the more popular SIL methods utilized. LOPA is an engineering tool used to ensure that process risk is successfully mitigated to a tolerable level.

Layer of protection analysis (LOPA) has been utilized domestically now for over 20 years. As a result of its efficient and pragmatic approach to risk assessment/SIL determination, it has gained widespread popularity in the Gulf of Mexico process industries.

LOPA can be used at any point of time in the lifecycle of a project or process, but it is most effective when implemented during FEED when PFDs are complete, and P&IDs are under preparation. For existing processes, LOPA should be used during or after the HAZOP review or revalidation

Through this LOPA SIL methodology, it is hoped that strengthened PHAs and fewer systematic flaws will result.

Example where SIL Consistency was not Achieved

One client site that applied SIL determination had results for similar compressor applications that ranged from SIL0 to SIL3. Another had similar results for fired heater applications. It was found that the PHAs had considerable variance in consequence valuations and that each facilitator applied different independent protective layer (IPL) rule-sets.

To reconcile the differences, the PHA and SIL determination teams had to reconvene which resulted in unnecessary resource cost. Management is now wary of the SIL determination process and the team must redeem itself on future applications.

Safety Requirement Specification (SRS)

After it is determined that a process has SIL required Safety Instrumented Functions (SIFs), functional and integrity requirements will need to be documented in a Safety Requirements Specification (SRS).

The SRS is the primary reference for the remaining parts of the safety lifecycle. This document is especially important since it often marks the handoff of safety lifecycle responsibility from one company to another and is a key project communication document. Once these requirements are clearly laid out, they will significantly help the remaining design, installation, and operation phases of the safety systems lifecycle. The SRS addresses both functional and integrity specifications as stated below.

SRS Pitfalls

There are multiple faults that can creep into a good SRS. Several include:

- Issues could start with the input where the SIL determination team may have been given a poorly executed PHA as an input
- Problem with SRS documentation control which include when it:
    - Is not maintained
    - Had no revision control
    - Is missing important requirements.

Having a "living" SRS and competent implementers overcomes such issues.

Functional Safety Assessments (FSA) on Safety Requirement Specifications (SRS)

A functional safety assessment (FSA) is an independent assessment to verify that the SIS as specified will effectively manage plant risk. It should be conducted by a person or team that is inside or outside the company and is not biased to the project.

An assessment is different from an audit. While an audit basically checks that a company is following its procedures, the functional safety assessment makes the additional checks to verify that the procedures will effectively manage risk.

FSA team shall undertake functional safety assessment(s) prior to the hazards being present i.e. after validation and before operations. Modification activity shall not begin until FSA is completed. In cases where FSA is carried out on a modification the assessment shall consider the impact analysis carried out.

Generally, for higher SIL and more complex or novel systems, rigorous functional assessments must be done by people who are completely independent from the project, i.e. 3rd parties.

C. SIS Conceptual & Detail Design – Realization Phase

As shown previously in Figure 1, the SIS realization phase includes:

- Selecting technology, architecture, and testing intervals
- Detailed design, installation, commissioning, and validation

SIF Conceptual Designs

The SIS conceptual design step involves taking the preliminary SIF list based on the previous risk study where each SIF was identified as a protection layer. A team of process, operations, and instrument personnel will then verify each SIF's makeup of components, voting configuration, and testing plans before being used in the SIL verification calculations.

SIF Conceptual Design Issues

It is surprising how much time is required by a functional safety (FS) engineer to validate a SIF's conceptual design. Why? The problem is three-fold. First, the PHA hazard scenario is stated so vaguely that even the process engineer has a hard time specifying what is required to get to the safe state when reading the PHA post-facto. It is paramount to get the sensing and actions correct to meet stringent SIL requirements.

The second problematic issue with SIF conceptual design is accounting for all the components involved to meet the SIL target. In many cases, critical loops that have been verified using simplistic models are found deficient when all the mechanical and electrical diagrams have been reviewed.

Finally, is the proposed SIF fully independent of the initiating cause and other IPLs? Care should always be included in designs that involve control valves since they normally involve some level of common cause. Such issues are compounded if the facilitator and team for the SIL determination session do not fully understand IPL requirements or mis-apply IPL credits.

Example of Flawed SIF Conceptual Design

In a recent project, turbomachinery SIFs were being evaluated for a SIL2 application. All was thought to be fine since the system included SIL3 components for the sensors and logic solver. Also, the final element was partial stroke tested.

Unfortunately, late in the design cycle, an emergency trip device (ETD) was discovered in the mechanical drawings. This ETD was critical to the SIF since it acted as an interface component for dumping the hydraulic power fluid and had to be included in the SIL verification calculations.

Once the ETD's failure rate was included in the SIL calculations, the SIF was degraded to SIL1. Fortunately, the design team found an alternative solution to avoid adding an inline steam valve (estimated at $300k), but not every team is so fortunate.

SIL Verification Calculations ("Testing the Metal" of SIFs)

Each safety instrumented function (SIF) design must now be verified through probabilistic calculations. (See Clause 11.9.1) The key here is to do the probabilistic calculations for each SIF which will verify safety and spurious trip performance criteria as well as optimize design to economically meet the requirements for each different function. The spurious trip discovery could be quite significant for plant production where it has been stated that up to 18% of plant trips are associated with instrumentation.

If calculations show that the conceptual SIF design does not meet the SIL target, the choices are:

1. Shorten the testing interval, but not beyond the practical point for operations
2. Select better technology/equipment that has higher diagnostics &/or lower failure rates
3. Add redundancy or other IPLs

The conceptual design iterations will continue until the SIL or risk reduction target is met with the overall most economical system.

D. Start-up & Post-Installation – Operations Phase of SIS Safety Lifecycle

The longest phase that requires a joint PSM and SIS effort is the operations phase. In total, it includes the following activities:

- Startup
- Operation
- Maintenance
- Periodic Proof Tests
- Modifications
- Decommissioning

Proof Testing Practices

Since OSHA 1910.119 Process Safety Management covered many of the operations and maintenance requirements, most companies are performing some level of tests. For many, the test is a full functional test from the sensor to the final device done a periodic basis that aligns with their scheduled shutdowns.

So once installed, do proof test practices actually measure up? If the impact of letting the proof test time intervals slip and its impact upon safety is understood, companies would place it as a priority over standard operations. For example, in the following Figure 4, the probability of failure (PFD_{avg}) is rising at an alarming rate.

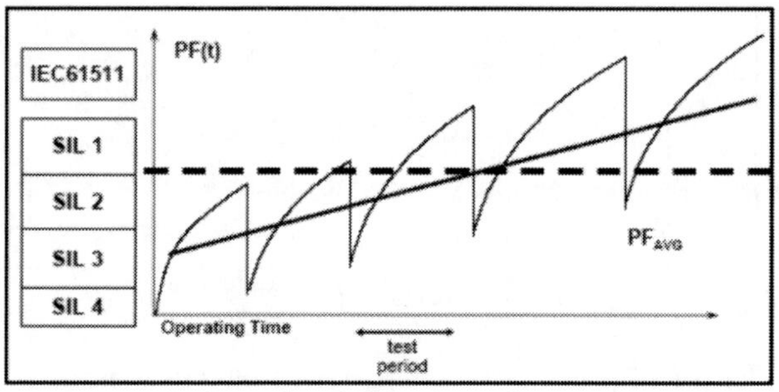

Figure 4. Graph of SIL Based Upon Testing

Consequently, the overall safety integrity level (SIL) is dropping as well. *Please remember, for each SIL level you lose, you have increased the probability of an event by 10 times!*

SIS Audits

Compliance audits are common in process safety management practices to meet the requirements of OSHA 1910, but how specific are they to SIS installation where the only focus on the system is when it causes a nuisance trip? The only requirement stated in ISA61511 is for a visual inspection of SIS equipment (Clause 16.3.2). Such practices should be improved for an optimum FSM program.

SIS Specific MOCs are Lacking

Based on discussions with many FS professionals, MOCs do not cover many of details that are SIS specific. Namely, they may not include items such as configuration history and adverse impact tests that are now spelled out in ISA61511. Such improvements need to be made to prevent disastrous software change mistakes that may not be discovered until an incident occurs.

Functional Safety Assessments during MOCs and Operations ("Gap Analysis")

Functional safety assessments (FSA) will provide a good diagnostic that a procedure has been thoroughly completed. With changes to a system, especially in regard to software, third party verification through an FSA will minimize error and should be performed on critical loops.

Concerning the operation and maintenance phase, the "Gap Analysis" format of the FSA will provide a direct measure on how close the SIS installation is to ISA61511 compliance. Since change in continual, the FSA will help point out any inadvertent deficiencies that have creeped in over time.

E. Finding a Better FSM Process Utilizing SharePoint®

As seen in the preceding sections, meeting the 300+ "shalls" in the functional safety lifecycle is overwhelming and it is no wonder that full ISA61511 compliance is still lacking. There are manual methods to reach the endpoint, but they are cumbersome at best and most are not implemented "real time". On previous projects, the complete kit to meet such a goal included 17 different procedures, checklists, and forms documents, which combined with reports, would result in 1000's of pages of documents.

To date, no company that the author has worked with in the last 15years have met all of these requirements. There has to be a better way!

Proprietary Alternatives not Accepted by IT

When pondering a pragmatic approach to compliance, several proprietary solutions were considered. The biggest problem with going with a pre-packaged software solution was getting it approved by the information technology (IT) group. Their corporate guidance was to limit any new applications unless fully vetted since they had such a plethora of legacy products, many of which had never met full expectations after millions of dollars spent. Such a software gatekeeper role prevented serious consideration for a project that had already started.

What About SharePoint®? We Already Use.

With IT blocking new software, the project team engineers sought a better way utilize existing platforms. When it was discovered that SharePoint®, already used for document sharing/management, could be coded for workflows, it was investigated. The problem now was timing: essentially, could the application be developed within the project timing constraints? When an independent programmer was asked, the answer was not positive in that 6 months to a year was estimated, well outside of the project constraints.

At that point, the team almost went back to manual FSM implementation methods, but fortunately, they found another option. A PSM business solutions provider had moved away from propriety platforms and had forged a SharePoint® solution. The project team looked at the documentation requirements and felt that 90% of the code was already developed. A contract was released shortly thereafter and within two months data was being loaded.

F. Case Study: Applying the SharePoint® Solution to a NH3 Terminal

The manual FSM process was quite laborious and only the SIS subject matter expert (SME) understood the process. Fortunately, the new SharePoint® now yields to better management and can be broken down to three (3) primary SharePoint® modules, namely:

- FSM Module1: Guides you through the functional safety management process
- PHA Module2: Support the PHA data import
- Safety Function Addition Module3: Utilized to complete the conceptual and detailed design for each SIF

Before reviewing the SharePoint solution, though, a review of the NH3 terminal process is in order. A simple process flow sketch follows to show the main equipment.

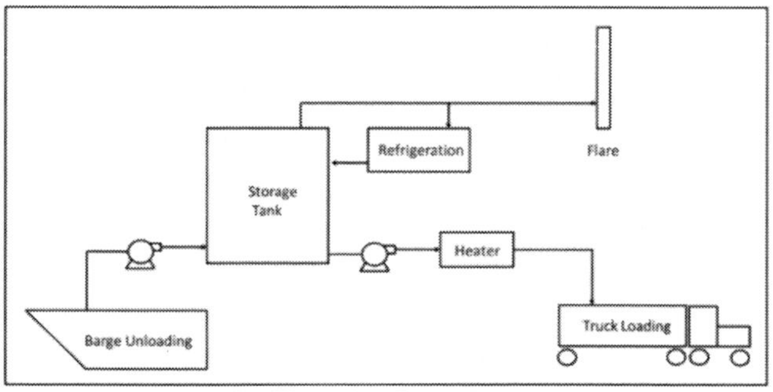

Figure 5. Sketch of NH3 Terminal

Liquid NH3 is unloaded from barges where it is sent to atmospheric tank storage. Since ambient heat vaporizes the liquid, refrigeration (compressors, condensers, and flash drums) is utilized to maintain storage temperature of -28° F (-33 °C). The liquid is then loaded when required to trucks after heating to ~40°F (4 °C).

In the following sections, each of the SharePoint® modules will be discussed as to how they supported the NH3 terminal case study.

SharePoint Module1: FSM (Functional Safety Management)

The FSM module was the key factor in gaining ISA61511 compliance at the NH3 terminal. To refresh, the reason to have a FSM process in place is to help minimize systematic issues which have contributed to major incidents in the past. In the past, to get through FSM, those responsible had to work through a mound of documentation and also "cattle prod" other contributors to do their part.

SharePoint now has economized the FSM process. The following screen print shows a sampling of the initial activities that were applied with the tool within the NH3 terminal FSM planning.

Figure 6. SharePoint® FSM Module Activities

The new SharePoint® application facilitated normal activities within FSM including:

1. Applying a pre-configured tool (e.g. Identifying resources for the project, competency logs, etc.)
2. Modifying or creating documents and storing them in a central repository (e.g. Procedures, templates, plans, etc.)
3. Verification that critical steps were completed satisfactorily

In projects prior to utilizing this tool, the best case is that the documentation was completed. (In almost all cases it was not!) But even for those rare situations where all the FSM documentation was completed, it was probably done outside of the best "ebb and flow" and inefficiencies resulted. SharePoint FSM helps force the logical progression of the safety lifecycle and once understood, PSM personnel will be able to take control.

SharePoint Module2: PHA

The PHA is basically the first step of the safety lifecycle, so it was necessary to find a bridge to import the work done expected proprietary platform. Such was the case for the NH3 terminal client.

To assist with the current NH3 terminal application, the SharePoint FSM module imports an Excel version of the PHA for further analysis by the functional safety (FS) engineer. The importer first validates if the fields are correct. For NH3 case, the format was perfect and the following "you can proceed with importing" message was received:

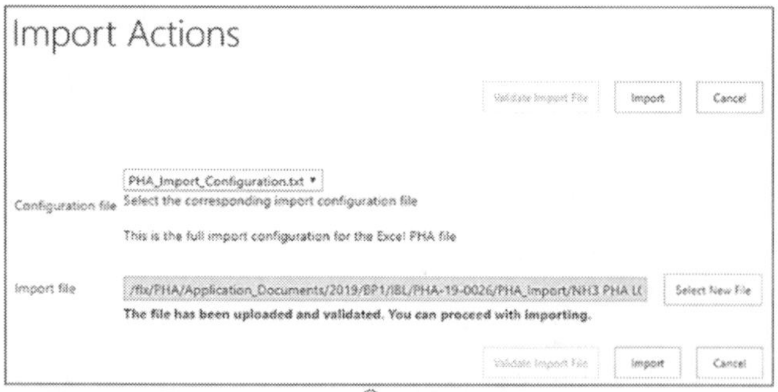

Figure 7. SharePoint® PHA Module Import

The imported file for review by a FS engineer is standard process. The real benefit is realized later when the PHA is modified. The SharePoint PHA Module automatically detects where new safeguards have been added and flags the FS with such changes! For a larger project that will be in state of flux where the PHA could be redone several times, this functionality is extraordinary and will safe a considerable amount of review time. Also, the safety factor of not overlooking new functions should be highly emphasized!

A second functionality that was recently added to the SharePoint PHA module is the ability for it to import a current safety instrumented function (SIF) list. Since most sites have some type of safeguard listing already in place, this feature was a well needed tool to overcome having to re-enter data. This action is shown in the following screenshot:

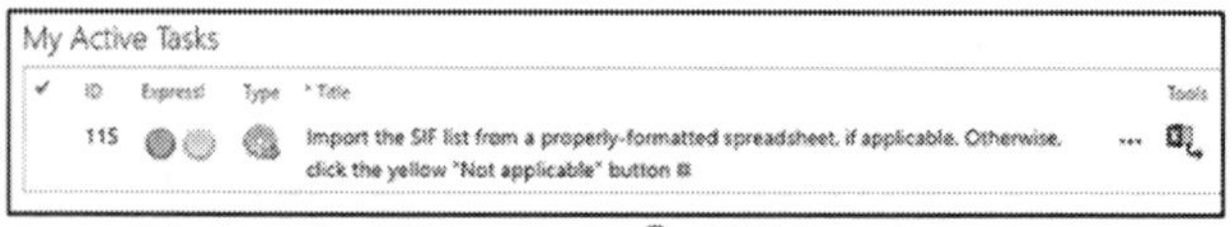

Figure 8. SharePoint® SIF List Import

The SIF list importer works similarly to the PHA importer in that it first validates the proper format of the data. In the first trial for the NH3 case study, it was not. See the following screenshot in Figure 9 a registered a "Failed to validate the imported data. Please select to review the log file" message:

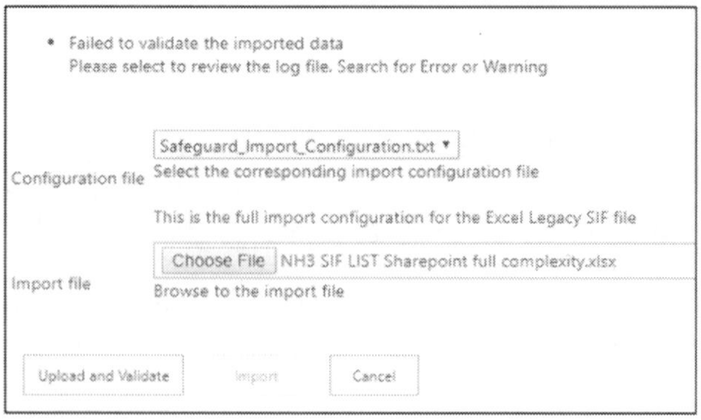

Figure 9. SharePoint® registered a "Failed to validate the imported data. Please select to review the log file"

FSA team shall undertake functional safety assessment(s) prior to the hazards being present i.e. after validation and before operations. Modification activity shall not begin until FSA is completed. In cases where FSA is carried out on a modification the assessment shall consider the impact analysis carried out.To assist in finding the error, the error "log file" was reviewed and the specific issue is shown in the next screen shot:

Figure 10. SharePoint® SIF List Import Log File

Naturally, this is an invaluable tool feature since not all sites have perfectly matched data sets. FS and PSM professionals will be able to get to the root of their import problems much more quickly.

Module3: Safety Function Addition (Updates)

After the FSM and PHA Modules, the SIF safeguards will have been identified and will require further review. For the NH3 study, ten (10) SIFs were identified as shown in the following Figure 11 and depicted by the green diamonds(◆):

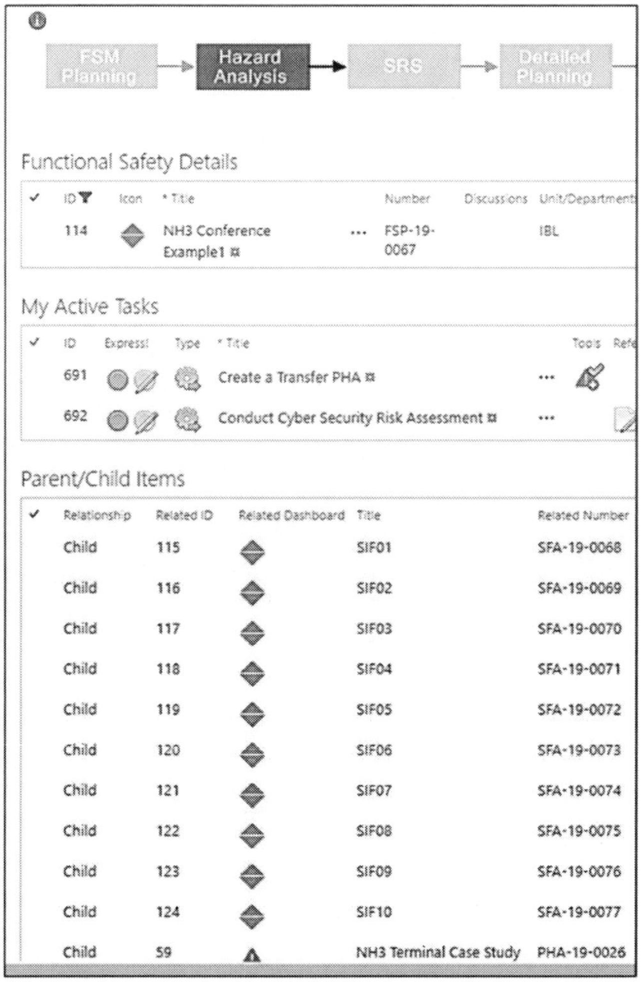

Figure 11. SharePoint® SIF List Design

Now each SIF will need to be fully designed. The two major stages for each SIF include:

1. Conceptual Design
2. Detailed Design

A partial listing of the actions required is shown in the following picture of the "Safeguard Detailed Requirements" dashboard which supports the workflow through both design states:

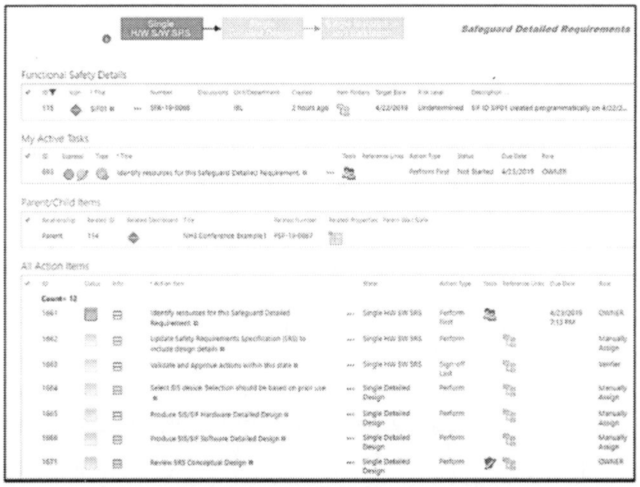

Figure 12. SharePoint® Safeguard Design

Once conceptual and detailed design actions were completed, the project and all SIFs were "locked up" as shown below.

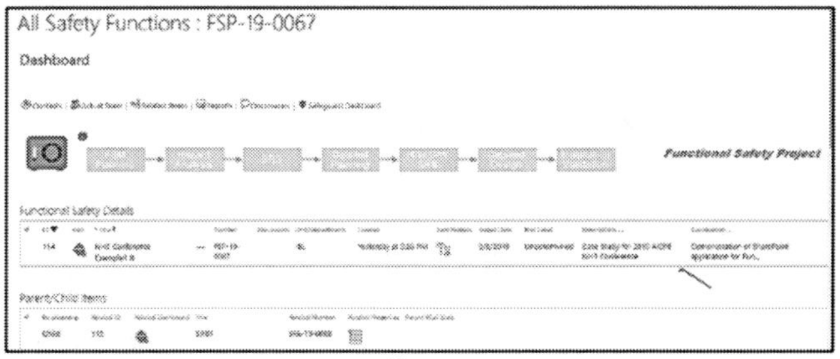

Figure13. SharePoint® Project/ SIF Locked

ISA61511 compliance has been met and any future changes will have an audit trail. Such a condition will be a paradigm shift for many facilities!

Conclusion - Benefits Utilizing SharePoint® for FSM Compliance

Executing an optimum FSM through a SharePoint automated workflows resulted in the following benefits:

1. Fully meeting the 300+ FS compliance requirements in a fast-paced project would have been practically impossible
2. PSM personnel are now working much more closely with SIS personnel

3. Proper front equipment selection was realized, and "over-specification" costs were minimized
4. It is expected that there was decrease in "specification errors" and that resulting accidents will never be realized
5. The company has now standardized on a universal process which offers substantial efficiencies and corporate KPI measures

Thus, each of these additive bottom line benefits, coupled with increased functional safety, can help justify the SharePoint® approach. Consider weighting them, putting in corporate estimates for both costs (implementation, spurious trips, and different accident scenarios) and gains (manpower efficiency and production), and see if they result in good news to present to the senior management team.

ABBREVIATIONS

API – American Petroleum Institute
ESD – Emergency Shutdown Device
FSA – Functional Safety Assessment
IPL – Independent Protection Layer
ISA – International Systems and Automation
FS – Functional Safety
FSM – Functional Safety Management
LOPA – Layer of Protection Analysis
PFD – Process Flow Diagram
PSM – Process Safety Management
SLC – Safety Lifecycle
SIF – Safety Instrumented Function
SIL – Safety Integrity Level
SIS – Safety Instrumented System
SME – Subject Matter Expert
SRS – Safety Requirements Specification

References:

[R1] ANSI/ISA61511, Application of Safety Instrumented Systems for the Process Industries, NC: Raleigh, ISA, 2018.

[R2] Miller, Curtis, Win/Win: A Manager's Guide to Functional Safety, 1st Edition, 2008

[R3] Out of Control: Why Control Systems go Wrong and How to Prevent Failure, U.K.: Sheffield, Health & Safety Executive, 1995

[R4] 29 CFR Part 1910.119, Process Safety Management of Highly Hazardous Chemicals, U.S. Federal Register, Feb. 24, 1992, http://www.osha.gov

Improving Energy Efficiency of IFFCO's Five Ammonia-Urea Plants

With the aim of improving the energy efficiency of Ammonia-Urea Complex of IFFCO's five Ammonia Urea Plants at three locations, IFFCO had undertaken a major Energy Saving Project (ESP). The target was to reduce energy consumption in all IFFCO's Ammonia-Urea Plants of different technologies and different vintages (commissioned in the span of 1974 to 1997).
Total project cost was around Indian Rs. 20 billion (approx. 300 million USD) to achieve average reduction in energy consumption of more than 0.40 GCal/t (1.67 GJ/t) of urea produced from IFFCO's Plants in India.
The paper covers the strategies adopted to identify the schemes and complete the project in less than 30 months with phase wise implementation to derive the benefits of energy savings at the earliest possible time frame and least downtime required for implementation of the schemes. The paper also covers various energy saving schemes implemented in each Plant and the problems faced during implementation of the project and remedial measures taken to achieve the targets successfully.

A. K. Singh and Piyush Misra
Indian Farmers Fertiliser Cooperative Limited, New Delhi, India

Introduction

Indian urea industry faces the challenge of providing urea to farmers at prices almost stagnant over decades in a scenario of continuously increasing energy cost. This is coupled with stringent government pressure to reduce energy consumption of urea plants to reduce subsidy burden. Further government policy of gas pooling has resulted in same gas price for all Indian urea plants, thus eliminating the benefit of lower gas price due to long term gas contracts. All these factors have made it essential to review feasibility of various energy saving schemes. Indian Farmers Fertiliser Cooperative Limited (IFFCO) undertook the daunting task of minimizing energy consumption in economically feasible manner without adversely affecting safety and reliability of plant operations.

IFFCO's Ammonia-Urea Plants at a Glance

IFFCO was established in 1967 and now it has five ammonia urea plants of different vintage and technology as indicated in Table 1.

Plant	Kalol	Phulpur-I (P-I)	Aonla-I (A-I)	Aonla-II (A-II)	Phulpur-II(P-II)
Commissioning	1974	1981	1989	1996	1997
Feed Stock	Natural Gas (NG)	Naphtha	NG + Naphtha	NG + Naphtha	Naphtha
Ammonia Capacity	900 tpd	900 tpd	1350 tpd	1350 tpd	1350 tpd
Ammonia Technology	MW Kellogg (Now KBR)		Haldor Topsoe (HTAS)		
Urea Capacity, tpd	1200	1500	2 X 1100	2 X 1100	2 X 1100
Urea Technology	Stamicarbon	Snamprogetti (Now Saipem)			
Power Source	State grid	Coal boiler/TG	(Gas Turbine (GT) / HRSG)		Oil Boiler/ TG

Table 1. Vintage, Feed Stock, Capacity, and Technology for IFFCO's Plants (As installed).

Over the years IFFCO had undertaken various modifications and major revamps to enhance the energy efficiency and plant capacity. Some of these modifications were purge gas recovery unit, primary reformer tubes with improved metallurgy, synthesis convertor modification, S-50 converter in ammonia plants. Pre-concentrator, replacement of CO_2 compressor and turbine for urea plant, conversion from Naphtha to NG.

Urea Production and Specific energy consumption for all the Urea produced from IFFCO's Plants since beginning is shown in Fig.-1.

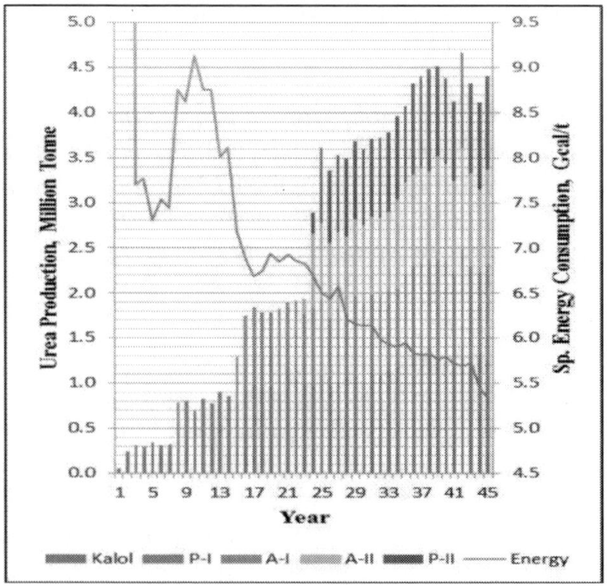

Figure 1. Annual Urea production and Specific Energy Consumption

ENERGY SAVING PROJECT (ESP) – Background.

As a first step towards ESP a group consisting of engineers from all the units and corporate office was formed on 16th February 2014 to explore the potential of energy savings. The group completed a preliminary study within two months and identified the major schemes to enhance the energy efficiency economically:

1. Synthesis Compressor Turbine of Kalol unit.
2. Air Compressor Turbine and Refrigeration Compressor Turbine of P-I unit.
3. Replacing Synthesis Gas Compressor rotor with 3D rotors for A-I, A-II and P-II units.
4. Turbo generator (TG) based power generation to be replaced by GT with heat recovery steam generator (HRSG) at Phulpur.
5. Hot potassium carbonate-based CO_2 removal system to be replaced by aMDEA based system for P-I, P-II, A-I and A-II units.
6. Energy efficient trays in Kalol and P-I.
7. Improved heat recovery in Primary Reformer furnace by installing new Process Air Coil in Kalol unit and new Mixed Feed Coil & Process Air Coil in P-I unit.
8. Replacement of CO_2 compressors with more efficient compressors for P-II, A-I & A-II.
9. Use of Vapor Absorption Machines (VAM) for cooling CO_2 at CO_2 Compressor inlet for Kalol, P-I & A-II units.

After preliminary analysis detailed discussions were held among group members and top management regarding following possible implementation methodologies:

a. Implementation of identified schemes in-house by each Unit.
b. Implementation of ESP for each unit by independently engaging consultants more suited to specific unit based on the plant technology and identified schemes for the unit.
c. Implementation of ESP for all the Units together engaging most suited consultant.
d. Implementation of the Project for all Ammonia Plants by engaging reputable ammonia technology supplier and for all urea plants by engaging reputable urea technology supplier.

Finally, option-c was chosen with M/s Casale SA, Switzerland as consultant for conducting feasibility study and basic engineering for identified schemes, along with identification of other possible schemes which could enhance energy efficiency in economically feasible manner.

Feasibility Study by M/s Casale

After finalization of scope of work for all the three stakeholders, i.e. M/s Casale, IFFCO and the detailed engineering contractor and after reaching the consensus on timeline of project, the job of carrying out feasibility study for ESP was awarded to M/s Casale on 28th May 2014.

Casale engineers visited all the sites to collect the required data and prepared the base case to explore the feasibility of various possible energy saving options and their combinations. A team of IFFCO engineers visited Casale's office for a week and discussed several combinations based on the findings of preliminary study along with the ones suggested by Casale. The most acceptable combinations were finalized for feasibility study.

Basic Engineering by M/s Casale

Detailed scope of work was finalized based on feasibility report and discussions held between IFFCO and Casale. Subsequently the Basic engineering contract was awarded to M/s Casale on 17th December 2014 (Project Zero Date).

To complete the Project at the earliest, it was decided that the enquiry of following major items shall be issued by IFFCO's in-house team based on the data sheet prepared by IFFCO's in-house team. However, at a later stage vendor will be provided the revised data sheets finalized by Casale before 30th January 2015 and Purchase Orders were to be placed based on revised data sheets.

i. GT-HRSG Package at Phulpur complex.
ii. Air Compressor Turbine and Refrigeration Compressor Turbine of P-I Unit.
iii. Revamping of Synthesis Gas Compressor for A-I, A-II and P-II Units.
iv. Installation of more efficient CO2 Compressor for P-II, A-I and A-II Units.

Other long lead items were identified, and enquiry of these long lead items were to be issued at the earliest. These long lead items included the Bulk Absorber, HP-LP Flash Vessel, and Lean and Semi Lean Pumps for the CO_2 Removal Section. Casale issued datasheets for above items by 2nd March 2015.

Draft Basic Engineering Packages for various units were issued in staggered manner between May to August 2015 for review by IFFCO. After many rounds of discussions, final issues of Basic Engineering Design packages were accomplished from Oct. 2015 to Apr. 2016.

It was decided to implement the schemes in phases. The schemes, which did not have long lead items, were targeted to be implemented in 2016. Other schemes which had long lead items were targeted to be implemented in 2017.

Casale also agreed to supply proprietary items viz., High efficiency trays for urea reactors, equipment and material for Split Flow Full Condenser System and Ammonia Wash System, Secondary Reformer Burner, and Ammonia Converter internals as per schedule for project completion.

Energy Saving Project Schemes

Major schemes implemented under ESP are as follows:

1. Additional Heat Recovery in Primary Reformer Convection Section:

Following modifications were carried out in different units to minimize the heat loss in primary reformer furnace in ammonia plant:

At Kalol Unit, feed gas desulphurization system was changed from activated carbon bed adsorption to Hot Catalytic Desulfurization System (HDS) based on experiences of same in P-I, P-II, A-I and A-II units,

The new feed pre-heat coil has been installed in convection section to supply required heat to increase feed gas temperature as required for HDS reactor.

Further, process air coil has been replaced with new coil having more heat transfer area to increase process air temperature to 525 ^{0}C (977 ^{0}F) which earlier used to be around 460 ^{0}C (860 ^{0}F). Higher temperature resulted in better conversion and more steam generation in waste heat boiler downstream.

To avoid flame impingement and catalyst bed sagging at higher air temperature, old air burner was replaced with a new Casale proprietary air burner in Secondary Reformer.

In P-I unit higher efficiency of primary reformer radiant section and plant operation at higher load resulted in lower outlet temperatures of mixed feed coil and process air coil. Therefore, the old mixed feed coil and process air coil were replaced by new coils with higher surface area in the same footprint.

In A-I unit exhaust flue gas temperature used to be about 165 to 185 ^{0}C (330 to 365 ^{0}F) depending upon season and Plant operating conditions. In order to recover more heat, air preheater has been replaced by new SS plate type air preheater with more surface area. This has reduced stack temperature by about 40 ^{0}C (72 ^{0}F).

2. Replacement of HTS Downstream Waste Heat Boiler by Steam Super heater:

At A-I unit, considering the lower steam requirement after ESP, HTS downstream waste heat boiler was replaced with new steam super heater to increase the temperature of steam generated from front end by 50^{0}C (90 ^{0}F) resulting in saving of NG in gas fired steam super heater.

3. Revamp of CO_2 Removal Section:

Hot potassium carbonate-based CO_2 removal system being used in P-I, P-II, A-I and A-II has been modified to more energy efficient aMDEA system which was in operation at Kalol. This reduced the regeneration energy from more than 800 kCal/Nm^3 (3350 kJ/Nm3) of CO_2 to 500-550 kCal/Nm3 (2100-2300 kJ/Nm3) of CO_2.

To get the full benefit of lower regeneration energy requirements, Steam to Carbon ratio (S/C) at Primary Reformer inlet was reduced from 3.3 to 3.0 in all the plants. Refractory liner of Primary waste heat boilers at Kalol and P-I Units have been replaced from INCOLOY 800-H to INCONEL 693 to avoid metal dusting.

4. Installation of Ammonia Wash Unit:

In order to remove moisture and oxides of carbon from make-up gas, it is normally mixed with converter effluent gas before entering into Ammonia Chiller and Separator. Removing moisture and oxides of carbon from the makeup gas before mixing it with recycle gas and changing the loop so as the makeup gas is diverted to convertor without going to chiller results in reduced compression power and chiller duty.

Based on experience of Kalol and P-I where ammonia wash unit was in use for more than a decade, it was also installed in P-II, A-I and A-II units. It's installation at the 3rd stage suction reduced synthesis compressor power requirement due to lower suction temperature of about 13 ^{0}C (55 ^{0}F) against earlier 35 ^{0}C (95 ^{0}F) and also due to lower chilling load as the makeup gas bypasses chillers due to loop re-piping.

5. Revamping of Synthesis Gas Compressor:

The less efficient synthesis gas compressors in A-I, A-II and P-II units have been replaced with new generation compressors. This has increased compressor efficiency from 67% to 74%.

6. Ammonia Convertor Internals Revamp:

In P-1 unit, since the ammonia convertor catalyst was due for replacement, the convertor has been revamped with more robust designed new internals and new catalyst loaded in it.

7. Installation of New Ammonia Pump:

In P-I, cold ammonia was not sent directly to urea plant and routed through ammonia storage.

A new Ammonia Booster Pump has been installed to supply required cold ammonia at -23 °C directly from Letdown Drum (107-F) to Urea Plant. This resulted in reduced vapor load on refrigeration compressor along with reduction in pumping power due to skipping of ammonia routing through storage.

8. Synthesis Loop Boiler Modifications:

At Kalol and P-I units, Ammonia synthesis converters have Feed-effluent exchanger inside the Converter shell. With Converter modifications carried out in the past, conversion has increased resulting in higher temperature of above 395 ^{0}C (743 ^{0}F).

To keep the conversion at highest level possible and utilize heat effectively, an additional waste heat boiler was installed generating medium pressure (MP) steam.

The additional MP steam thus produced is exported to urea plant, which requires MP steam at saturation for heating purposes. This reduces the extraction steam requirement thereby reducing HP steam consumption in CO_2 Compressor turbine.

At A-I and P-II units, vertical synthesis loop boilers were having leakage problem, due to which steam vent valve was partially kept open to maintain quality of steam. To avoid this steam loss, these boilers were replaced by horizontal boilers.

9. MP Process Condensate Stripper:

The process condensate separated in the front end of the Ammonia Plant is stripped by steam to remove impurities before using it as makeup water for steam generation. Condensate Strippers at Kalol and P-I Units were designed to operate at atmospheric pressure utilizing LP steam for stripping which was vented to atmosphere.

Based on experience at P-II, A-I and A-II units, the strippers at Kalol and P-I units were replaced by MP stripper. Stripped vapors along with process MP steam is now fed to primary reformer. This resulted in preventing steam loss.

10. Ammonia Recovery from Synthesis Loop LP Purge Gases:

In Kalol and P-I Units the low-pressure flash gas from Ammonia let-down drum (107-F) and refrigeration off-gas from flash gas chiller (126-C) were being used as fuel in the primary reformer.

Based on experience of ammonia recovery in P-II, A-I and A-II units, a new LP flash column, pumps and associated system have been installed to remove ammonia from LP purge gas and ammonical water is sent to ammonia recovery unit (already installed for HP purge gas) to recover ammonia in a distillation column.

11. Replacement of Inefficient Turbines:

At P-I Unit, old condensing type turbine drives for Process Air Compressor and Refrigeration Compressor were installed since the commissioning of the Plant. These have been replaced with more efficient new generation machines.

At Kalol Unit two steam turbines i.e. one back pressure (103-JAT) and one condensing (103-JBT) were in operation as Synthesis gas compressor drive since commissioning. Both these turbines were replaced by a single efficient steam turbine utilizing the existing civil foundation.

12. Replacement of CO2 Compressors:

CO_2 compressors of P-II, A-I and A-II units were replaced (2 in each Unit) with the efficient ones having 3-D designed impellers, in the manner similar to the replacement done in Kalol and P-I units in 2003 and 2008. As a result, overall compressor efficiency has increased from about 70% to more than 78%

13. Installation of VAM for CO_2 Cooling at CO_2 Compressor Suction:

At P-II and A-I units, VAM was used to cool CO_2 at suction of CO_2 compressor and was found to increase energy efficiency and capacity of compressors. The same has been replicated in Kalol, P-I and A-II units. In all the Units, CO_2 is now being cooled to a temperature between 15 to 25 ^{0}C (59 to 77 ^{0}F) depending upon loading on CO_2 Compressor.

At Kalol Unit, with the installation of VAM for CO_2 cooling, the old contact type cooler was removed. This has reduced the load of water treatment plant and also reduced the effluent generation, as there were frequent instances of leakage in contact type cooler

14. Installation of Ammonia Pre-heater in Urea Plants:

In all the units, new Ammonia Pre-Heaters have been installed to preheat high pressure ammonia going to the HP Section to utilize the waste heat in fruitful manner which otherwise was going to cooling water.

In Kalol Unit, heating of ammonia is achieved by recovering heat from hot condensate of Closed Circulation System before cooling it with cooling water. This additional heat recovered in Ammonia Preheater resulted in additional steam generation from HP System.

In P-I, P-II, A-I and A-II units required heat for Ammonia is provided by the stream going to LP Condenser. This has resulted in lower steam requirement in HP Stripper and also the reduced load in the LP Condenser.

15. High Efficiency Trays in Urea Reactor:

High Efficiency Trays procured from Casale on proprietary basis with guaranteed improvement in conversion by 1% were installed in P-I and Kalol in place of Sieve Trays used earlier.

16. High Pressure (HP) Split Flow Loop:

Owing to various modifications, Kalol urea plant is being operated at 145-150 % of installed capacity. To improve conversion at such higher load operation, the old conventional HP Loop is modified to Casale's HP Split Flow Loop.

In the original HP Loop, HP Stripper off gas was fed into HP Carbamate Condenser from the top along with feed ammonia. In the modified loop, HP Stripper off gas has been split into two parts; the major part goes to bottom of HP Carbamate Condenser and balance is fed into Urea Reactor. The HP Carbamate Condenser has been replaced with new one as per the requirement of Casale' Split Flow Loop.

The new HP Carbamate Ejector (P-1250) was also installed to transfer the carbamate from HP Scrubber to HP Carbamate Condenser at bottom using HP ammonia as motive fluid.

17. 2nd Stage Evaporator Heat Exchanger:

In Kalol unit, the old 2nd stage evaporator heat exchanger has been replaced by new one with more surface area suitable to use 0.4 MPa steam instead of 0.9 MPa steam used earlier.

18. Installation of Gas Turbine based Power Generation System at Phulpur:

A-I and A-II Units were having GT based power generation with steam generation from GT exhaust gases assisted by supplementary firing in HRSG unit.

Considering the high efficiency of GT/HRSG, existing low efficiency (less than 30%) turbo generators of 12 MW and 18 MW at P-I and P-II have been replaced with a single GT-HRSG of 30.2 MW (ISO) with 60 t/h of HP steam generation after taking into account the requirement of power and steam after ESP. The capacity and efficiency of the new GT is further increased by installation of VAM for suction cooling of air to about 12 ^{0}C (53 ^{0}F) utilizing hot condensate from both Urea Plants as heat source.

19. Optimization of Heat Recovery in BFW System:

At Kalol Unit, the Low Temperature Shift Converter (LTS) inlet temperature was maintained by the LP Boiler (112-C). Methanator feed inlet temperature was maintained in the methanator feed pre-heater (104-C) and the methanator exit process gas was cooled down with boiler feed water (BFW) in the methanator effluent BFW heater (114-C)

To optimize the heat recovery system, 104-C, 112-C and 114-C were removed and three new heat exchangers viz. HTS exit BFW heater, Methanator feed effluent exchanger, Methanator trim heater were installed.

At A-I unit, the synthesis loop BFW preheater at downstream of 2nd Synthesis Loop Boiler is replaced with a new exchanger having more surface area to reduce synthesis gas temperature at Preheater outlet to 244 ^{0}C (471 ^{0}F) from earlier design of 274 ^{0}C (525 ^{0}F). The heat carried by BFW resulted in higher steam generation.

20. Replacement of Water Coolers:

At Kalol unit, the Methanator exit cooler was replaced by new cooler with higher surface area and cooling water is shifted from shell side to tube side, to address the problem of frequent choking and leakage.

At A-I unit, the synthesis gas was cooled in synthesis loop water cooler, using DM water with corrosion inhibitor circulated by pump. Hot DM water was cooled in plate type heat exchangers with cooling water.

Direct cooling with cooling water eliminated the need to circulate DM water, thereby saving pumping power and synthesis gas is now cooler by 3 °C (5.4 °F). To implement this, a new exchanger with SS tubes was installed.

21. Reduction in Utility consumption leading to stoppages of Existing Systems:

At Phulpur Complex, the total utility steam requirement is reduced from about 330 t/h to about 100 t/h. This resulted in stopping of one gas fired boiler and one coal fired boiler. Now only one coal fired boiler remains in operation.

As the power requirement is met by Gas Turbine; both Turbo Generators are not in operation and one cooling tower with 3500 m3/h of cooling water flow is stopped.

At Aonla complex, the savings in steam requirements in the Ammonia and Urea Plants resulted in stoppage of the Service boiler.

Detailed Engineering by M/s PDIL

Based on the mutually agreed scope of work, IFFCO had awarded Contract for Detailed Engineering Design Package to M/s PDIL on the same day when the Contract for Basic Engineering Package was awarded to M/s Casale i.e., on 17th December 2014 (Project Zero Date).

To ensure the timely implementation of project, all enquiries and engineering activities viz. preparation of plot plan, rigging plan, required ground compaction and bill of quantities were started with the available preliminary information and were subsequently modified as and when more specific information is available preferably before placement of order.
In order to have better co-ordination with various vendors, it was decided that expediting and inspection shall be done by IFFCO. 3rd party inspection may be carried out as per the requirement. Further a team of engineers was deputed at PDIL office to provide required info and facilitate faster decision.

Procurement

PDIL assisted IFFCO in the procurement. Approved Vendor List (AVL) for ESP was prepared much before the procurement process started based on past experiences of PDIL and IFFCO with various vendors.

It was well understood that finalization of piping isometrics and hence finalization of BOQ for bulk items like piping, fittings, valves etc. cannot be done at an early stage of project. Therefore, it was decided to procure bulk items based on preliminary estimates on the basis of 1st issue of P&IDs and the remaining quantity to be procured at a later stage.

After order placement, a dedicated team of engineers from IFFCO were deputed in different areas of the country to have interaction with various vendors on regular basis for expediting, providing any clarifications required by vendor at the earliest and to arrange/ carryout inspection at vendor shop. Engineers deputed in various groups were rotated and also sent back to plant for some time to ensure that they are well aware of the progress of project at site.

Site Fabrication and Erection

To reduce the shut down time for implementation of project, various site fabrication and erection jobs, including laying of civil foundations and structural columns were completed in the running plant as far as possible. The equipment suppliers were pushed to supply load data at the earliest possible time frame and PDIL designed the foundations suiting to site layout and/ or modified the layout suited to required foundation. This proactive approach resulted in construction of the new equipment foundation in advance and optimized the turnaround period.

To make space for erection of new items, following actions were taken in running plant:

- Two Analyzer rooms in P-I Ammonia Plant were shifted to accommodate HP/LP Flash Vessels and MP Stripper.
- To accommodate the new Semi-Lean Pumps in P-I, GV Solution Tank was dismantled after installing a new tank at remote location.
- HP/LP Flash tower and flash gas compressor in P-II were installed in battery limits of urea plant crossing the road between Ammonia and Urea plants.

All the equipment, on arrival at site, were directly unloaded to their foundations. All the pipelines and fittings were erected and kept ready for hook up before taking the plant shut down. For all hookups to be made during Plant shutdown, piping spool segments were prefabricated, properly welded and tested much in advance considering ease of erection.

In order to have better co-ordination and to effectively utilize the experience of job execution at one unit in all other units, mechanical erection jobs at all the units was done by the same contractor similar thing was done for all civil erection jobs and electrical & instrumentation items erection.

For logistical reasons, the HP and LP Flash Vessels were transported separately and were to be welded at site. In P-I unit, due to constraints related to rigging of 60 meter (197 ft.) tall vessel, the placement of both towers for welding was to be done in middle of the Plant. This would have resulted in stoppage of movement of cranes resulting in additional three weeks shutdown. In order to reduce these additional shutdown days, it was decided to weld skirts of HP and LP Flash Vessel in vertical position against normal and conventional way of welding in horizontal position.

Theodolite was used for checking perfect verticality of both columns and then several guiding channels were welded to maintain LP column at its location on the top of HP column. Several welders were placed for welding at multiple locations, which was not feasible during welding in the horizontal position. One crane was positioned to hold LP flash vessel.

Pre-commissioning and Commissioning

Preparing punch-list to ensure completion of scheme was very challenging as the engineering jobs were not complete and final documentation was yet to be supplied. However, scheme wise information was made available to commissioning team based on various communications among all stakeholders.

Air blowing, cardboard blasting, and steam blowing was carried out as per the requirement and to ensure proper cleaning of all pipelines. All new items after erection were not chemically cleaned. Chemical cleaning was done only for the items where it was considered necessary.

Temporary pipelines, spool pieces and other fittings were made ready in advance by respective erection contractors as per the envisaged requirement.

Considering the past experiences of problems encountered in operation of CO_2 removal system in other plants, a detailed procedure for pre-commissioning and commissioning of CO_2 Removal System was finalized by the dedicated team of IFFCO engineers based on the input from other plants, Casale and BASF (who were the supplier of proprietary OASE System for CO_2 removal):

1. Draining of Old Solution:

The old solution was transferred to obsolete decommissioned fuel oil tanks and was concentrated using steam heating coils provided in the fuel tank reducing the transportation cost to authorized disposal agency for safe disposal.

2. Removal of Vanadium Passivation Layer:

Based on the feedback from other plants and BASF, removal of vanadium passivation layer from the system is essential for the satisfactory performance of aMDEA System. Considering that most of the pipelines were being replaced with SS, vanadium passivation layer in old equipment like Absorber, Stripper, Solution Preparation/ Storage Tanks etc. was of concern. The same was removed by grinding till the vanadium concentration came down to 0.1%. However, in cases where passivation layer was hard, copper grit blasting was also done.

3. Water/ Caustic Wash for Cleaning:

After completion of hookup job, the System was flushed with fire water for removing debris and leftover construction material, if any. After draining the fire water, DM water was filled, and System integrity was checked.

Considering that lot of construction were going on in the Plant, use of NG was not considered safe. Therefore, nitrogen gas available at 0.6 MPa was used. Further pressure was increased using Flash Gas Compressor to pressurize the absorber to 1.5 MPa and circulation of DM water was maintained for 8 hours at 80 ^{0}C (176 ^{0}F) and drained to tanks where the old potassium carbonate-based solvent was kept for disposal.

After cleaning with DM water, the system was to be cleaned with 3.5% (w/w) caustic solution. For this purpose, a small plunger pump was installed to pump the caustic from existing caustic tanks in offsite area through a new pipe line to a temporary manifold fabricated for spray of caustic in the middle of large diameter pipeline at pump suction.

DM water was again filled, and circulation was established with all pumps running. The caustic pump in offsite area was started to inject caustic with hourly lab analysis of caustic concentration in circulating DM water. After achieving the desired concentration in about 8 hours, caustic dosing was regulated to maintain the concentration. System temperature was maintained at 50 ^{0}C (122 ^{0}F) and caustic solution circulated at design flow rates with intermittent blow-down every 2 hours. After 32 hours, solution was drained.

The system was rinsed by refilling with DM water and circulation was established at design flow rates maintaining the temperature of 80 ^{0}C (176 ^{0}F) with intermittent blow-down every 2 hours to remove caustic from the system. After eight hours of circulation the system was drained. This step was repeated till the recommended water quality was achieved. After which 40% OASE solution, was charged and taken in towers up to normal level. Towers were pressurized with NG and all pumps were started one by one to achieve the solution circulation corresponding to 100% load.

Steam was injected in the reboiler to achieve the stripper bottom temperature to design value. Antifoam agent at 50 ppmw of was also added slowly. The OASE solution circulation was carried out for more than 72 hours maintaining PFD conditions for passivation after which process gas was introduced slowly into the absorber.

Guarantee Test Run (GTR)

GTR for individual equipment and machinery were carried out as per schedule. GTR for the Project were also conducted for each complex as per the pre-defined methodology agreed between IFFCO and Casale. Pre-shutdown data were collected before implementation of schemes and the operation was compared with base case to arrive at revised guarantees for energy savings.
After implementation of schemes post shutdown operating data was collected by IFFCO and sent to Casale for further analysis and submission of GTR reports for approval by IFFCO. Based on the final GTR report submitted by Casale all guarantees were considered to be met.

However, there were problems in performance of Bulk absorber in P-II, A-I and A-II units. On detailed review and inspection after opening the vessel during shutdown, this was found to be a problem of equipment internals. The equipment supplier has replaced the defective internals with improved design and the problem was rectified resulting in satisfactory performance.

Similarly, the performance of the Primary Reformer at Kalol Unit was not satisfactory. Process Air coil was modified as per the new design given by Casale, after which the performance is now satisfactory as envisioned.

Benefits Achieved

Various schemes implemented in all the five units of IFFCO have resulted in reduction in specific energy consumption by about 0.4 GCal/t (1.67 GJ/t) of urea in P-II, A-I, and A-II, 0.3 GCal/t (1.26 GJ/t) of Urea in Kalol, and 0.9 GCal/t (3.78 GJ/t) of urea in P-I unit.

Average energy reduction of urea produced from all IFFCO Plants in India was more than targeted reduction of 0.4 GCal/t (1.67 GJ/t) of urea.

Due to this reduction in consumption of NG and coal, Green House Gas (GHG) emissions have been reduced by about 1650 t of CO2 per day.

The Operational Excellence Journey to Prevent and Manage Accidental Release Incidents across the Ammonia Supply Chain

On the Operational Excellence journey, responsible collaboration between stakeholders in industry, government, and public safety promotes mutual understanding of risks and will reduce the threats associated with incident exposures. Community and regional based leadership and safety partnerships are fundamental to the success of proactive efforts such as geographically focused Chemical and Ammonia Safety Days. This paper describes some lessons learned in such collaborations and the elements of successful practice that lead to a "prevent them all or stop them small" success.

Kathryn Lawrence
U.S. Environmental Protection Agency

Gary Smith
Ammonia Safety Training Institute

Jerry L. Jones PE
Chemical Engineer and AIChE CCPS Staff Consultant

Introduction

The purpose of this paper is to provide examples of operational excellence practices being implemented to manage risks across the ammonia supply chain (storage, transportation, dispensing, and use of ammonia). Partnerships between (1) industry, (2) government regulators and (3) public safety agencies yields a combined perspective to prevent and respond to ammonia release incidents. Regardless of the point in the supply chain where an accidental release incident might occur, the connection between industry, government regulators, and public safety agencies is vital when implementing the 3-P's - Prevention, Protection, and Preparation. This shared responsibility can be visualized as a balanced and supportive tripod and it can be cultivated through local leadership and capacity building. The resulting community collaboration is centered on the Congressionally adopted national response system (NRS) framework that sets up local, state, and federal coordination of safety, chemical management, and emergency response practices. Hazard assessment, risk management, and process safety wisdom utilized by industry, first responders, public safety agencies, regulators and a myriad of professional safety consortiums provides the tools and shared best practices for success.

The "Tripod Approach" (see Figure 1) has a two-part goal: "Prevent Them All or Stop Them Small". Under the Tripod Approach ammonia producers, users, and transporters in partnership with public safety and regulatory agencies have adopted a variety of good operational excellence and risk management practices to:

- Prevent (minimize the likelihood of) accidental releases of anhydrous ammonia
- Prepare to respond to ammonia releases in coordination with stakeholders (e.g. government, regulatory, and emergency response agencies at all levels)

Training and collaboration are vital to ensure the necessary competencies, equipment, supplies and protocols exist to effectively and efficiently respond to releases and minimize the amounts and consequences.

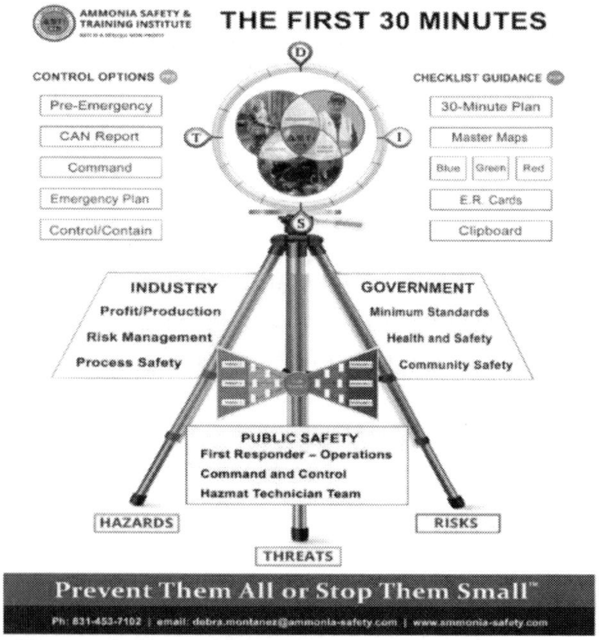

Figure 1. Tripod Approach to Prevention, Protection and Preparedness Partnerships

Point of Release

According to data from the National Response Center, hundreds of anhydrous ammonia releases are reported in the United States each year and over 25,000 since 1982. With respect to other toxic chemicals, anhydrous ammonia ranks first in the number of releases reported to the NRC. Many of these release reports identify related injuries and evacuations.

Due to its importance and value as a commodity in a number of industrial applications, ammonia is one of the most widely distributed toxic inhalation hazard (TIH) liquified gases used in the U.S. The broad distribution and usage frequency of ammonia has certainly influenced the number of reported ammonia emergency incidents most likely as a function of its ubiquity rather than its hazardous properties. Based on relative toxicity and approachability, other TIH liquified gases can present a more pronounced hazard than ammonia. Ammonia is simply in higher distribution. In consequence of that denominator, the number of incidents is greater. However, also greater is the opportunity to develop and adopt best practices in the pursuit of operational excellence regarding release prevention and response readiness.

Release volumes and situational impacts can, as would be expected, vary greatly. Historically, serious and large loss of containment incidents have occurred globally throughout the NH_3 supply chain:

- At NH_3 and fertilizer production facilities (also using NH_3 to produce nitric acid, ammonium nitrate (AN), urea (U), UAN solutions, and ammonium phosphate and sulfate fertilizers (MAP, DAP, AS)),
- During bulk transport by various overland or water methods (using vessels, pipelines, railcars, or tanker trucks),
- At distribution facilities of a wide range of capacities, and
- At end user sites for NH_3 e.g. during deliveries for on-farm uses, industrial uses as a chemical feedstock or reagent, and ammonia refrigerant, etc.)

A few examples of larger releases across the supply chain are shown in Table 1. It is important to note that there is not always a correlation of quantity released with severity of consequences (e.g. numbers of fatalities or serious injuries). Several of the release incidents listed in Table 1 resulted in no serious injuries or fatalities, but others involved multiple fatalities and injuries. Consequences are dependent not only on quantity released but also on the specific conditions at the location of the release. The life safety and environmental consequences of an indoor release are different than the threats associated with an outdoor release. Among other influences, the flammability and ability to escape an indoor release is much more of a concern than for an outdoor release. Alternatively, the release point, wind speed and atmospheric conditions associated with an outside release can invoke a higher threat to life and environment than an inside release that is confined and contained within a building.

Much smaller quantity releases than for the incidents listed in Table 1 have been deadly in numerous instances such as two accidents listed below in North America and Eastern Europe:

- 2001--Meat packing plant in USA: ~175 kg of NH_3 released during maintenance on wiped surface heat exchanger/cooler resulting in death of one maintenance worker and permanent blindness for second worker
- 2013—Large nitrogen fertilizer plant in, Ukraine: ~600 kg of NH_3 released into compressor room from pipe failure during a maintenance operation resulting in five fatalities and numerous injuries onsite.

The large number and broad distribution of facilities that store and use anhydrous ammonia provide a compelling argument for operational improvements related to prevention, protection, and preparation for response readiness.

The following are examples of escalating factors that are recognized as high risk and threat to life safety, environment, and property.

- High pressure aerosol release headed downwind to a populated area or near a body of water.
- A release in low wind, high humidity, and cool/moist environment.
- Ammonia vapor contained within a room, building or enclosure with sources of ignition.

Year	Location	NH_3 Source	Loss
1970	Blair, NE USA	Refrigerated atm. pressure NH_3 storage tank--overfilled at storage terminal-	~150 tons
1971	Floral, AR USA	Refrigerated atm. pressure NH_3 tank-- rapid over pressure rise & vessel failure	~600 tons
1973	Potchefstroom South Africa	NH_3 bullet tank failure at fertilizer mfg. site	38 tons
1976	Port of Landskrona, Sweden	Ocean vessel loaded with NH_3 at port—transfer hose failure during offloading	~180 mt
1976	Houston, TX USA	NH_3 tank truck crash and tank BLEVE in transit accident	19 tons
1989	Jonova, Lithuania USSR	Refrigerated atm pressure NH_3 storage tank failure at mfg. site	~7,000 mt
1994	Mazingarbe, France	railway car of NH_3 at fertilizer mfg. site--disconnection of transfer line during unloading	27 mt
1994	Terra, Port Neal, IA USA	Refrigerated atm. pressure NH_3 storage tank--puncture of tank from projectile after reactor explosion	~5,700 tons
2002	Minot, ND USA	Railway tank cars of NH_3 derailed and damaged	~600 tons
2003	Calamus, IA	"Nurse tank" for transport-catastrophic failure of tank weld after filling at coop	~3 tons
2004	Kingman, KS USA	NH_3 pipeline failure	~500 tons
2010	Theodore, AL USA	NH_3 refrigeration system piping failure from hydraulic shock	16 tons
2016	Tekamah, NE USA	NH_3 pipeline failure	~800 tons

Table 1. Large Quantity Releases of Anhydrous Ammonia in Transport and Use, at Production Sites, and at Storage/Distribution Facilities

- Ammonia vapor greater than 300 ppm for more than 30 minutes of exposure.

The emergency planner, regulator, and responder must prevent, provide mitigative protection measures, and prepare to confine, contain, and control (3-C's) the loss of hazard containment.

Operational Excellence Overview and Key Questions to Answer

Operational Excellence (OE) should not be defined only in terms of management system business process improvement tools or tactics such Six Sigma, Lean Mfg., Kaizen, 5S, etc.

OE is better described as an organizational mindset or culture dedicated to excellence and the relentless pursuit of continual improvement in performance, with goals often set in 3 categories:
- Operational efficiency/effectiveness in managing a range of risks as shown in Figure 2
- Workforce capabilities, employee engagement, job satisfaction
- Financial results showing that great safety programs yield a higher net corporate profit.

Three key questions for producers, transporters, and users that are on the OE journey include:

1. Are we doing the "right" things? Organizations often use consensus standards or guidelines to help answer this question.

2. How well are we doing them? What "good" looks like? Need to know what you don't know.

3. Are we cultivating a spirit of interdependence of teams, integrating business processes/tools/ assets, and allocating resources appropriately based on risks?

Source: Adapted from "Oliver Wight ABCD Checklist for Operational Excellence", Fifth Edition (2000); Latest edition now known as "The Oliver Wight Class A Standard for Business Excellence", Seventh Edition (2017)

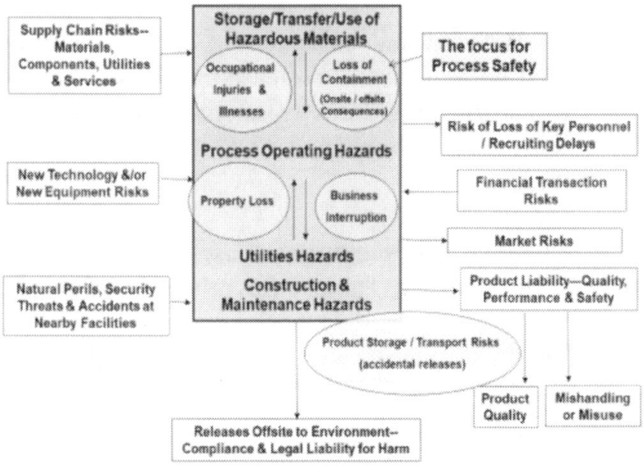

Figure 2. Manufacturing and Supply Chain Risks

Answering Key OE Questions

Good practices for preventing loss of containment incidents in the chemical process industries is offered by a multitude of technical consortiums. For example, the AIChE Center for Chemical Process Safety (CCPS) Risk Based Process Safety Guidelines (Wiley, 2007), and other related guidelines and concepts books are widely viewed as excellent sources for what "good" looks like to manage process risks and prevent loss of hazard containment. Such guidance documents describe "how to" examples for implementing effective risk-based management system practices that are proactive and frequently go well beyond regulatory requirements.

AICHE's CCPS Vision 2020 strategy for continuous improvement in risk-based process safety provides further specific guidance on topics directly related to the Tripod Approach in terms of:

3 of the core tenets of the strategy –
- o Disciplined Adherence to Standards
- o Intentional Competency Development
- o Enhanced Application & Sharing of Lessons Learned

3 of the societal themes of the strategy –
- o Responsible Collaboration

- Enhanced Stakeholder Knowledge
- Harmonization of Standards

Process Safety Management Systems	
Process Safety Mgt. Systems	AIChE CCPS *Guidelines for Risk Based Process Safety* (2007)
Process Safety-Asset Integrity and Reliability	AIChE CCPS *Guidelines for Asset Integrity Management* (2016)
Process Safety-Mgt. of Change	AIChE CCPS *Guidelines for the Management of Change for Process Safety* (2008)
Process Safety-Safe Work Practices	AIChE CCPS website with examples of good practices for different types of work permits: https://www.aiche.org/ccps/resources/tools/safe-work-practices
Process Safety-Conduct of Operations	CCPS Concept Book on *Conduct of Operations and Operational Discipline: For Improving Process Safety in Industry* (2011)

Table 2. Examples of Good Practices in Process Risk Management

Operational Excellence activities should address all six of these elements operationally integrated between industry, government, and public safety. The partnership model offered in the Tripod Approach promotes mutual understanding of risks, encourages results-oriented process safety ownership, enhances emergency response readiness and can reduces chemical accident risk in small and large communities. Annual commodity-specific Safety Day events provide a local area forum for sharing learnings, prevention and emergency readiness exercises, and, most importantly, Tripod understanding of the value of process safety and community partnership and engagement.

Experience and Continuing Efforts Using the Tripod Approach

For decades industry, regulatory agencies, and emergency response agencies from across the US have collaborated on training and capacity building programs. These efforts have been led and contributed to by organizations such as AIChE, the Ammonia Safety Training Institute (ASTI), International Institute of Ammonia Refrigeration (IIAR), American Society of Heating, Refrigerating and Air-conditions Engineers (ASHRAE), Compressed Gas Association (CGA), Refrigerating Engineers & Technicians Association (RETA), Global Cold Chain Alliance (GCCA), U.S. Chemical Safety Board (CSB), and The Fertilizer Institute (TFI), and the International Association of Fire Chiefs - Hazmat Center.

These programs are intended to ensure that first responders from industry and emergency response agencies understand the hazards and risks of ammonia release incidents and are prepared to respond. For example, ASTI performs an annual "live" release drill at Ft. Ord in California to collaborate around training activities for managing and responding to ammonia releases. Hundreds of stakeholders from industry, government, and public safety have been trained under realistic field conditions. Numerous response techniques, field equipment, instruments and personal protective equipment have been tested and evaluated as part of these drills. Leading companies involved in the production and distribution of anhydrous ammonia have sponsored attendance by firefighters from many smaller communities across the US.

Building on these operational excellence successes, a series of nationally distributed and geographically focused Chemical and Ammonia Safety Days were developed to amplify best practices. The Safety Day events cultivate and showcase chemical emergency prevention and planning efforts between Industry, and non-governmental organizations, and a myriad of public agencies including fire departments, health departments, Local Emergency Planning Committees (LEPCs), State Emergency Response Commissions (SERCs), local regulators, and state and federal agencies such as the EPA, CSB, DHS and OSHA. Within the context of Chemical Accidents and Releases, the focus of these efforts has been to "Prevent Them All or Stop Them Small".

Safety Day programs are designed for and with local community stakeholders from industry, governmental leaders and local first response agencies with a goal of seeding a geographic collaboration for chemical accident prevention and response readiness and partnership. Agenda and curriculum elements include technical training focused on local chemical threats, first responder case studies, accident prevention program prioritization, personnel accountability, bowtie analyses/descriptions of event propagation and management (Figure 3), and emergency response planning centered on the first 30 minutes and one-plan approach. The programs often include field demonstrations and exercises for fire departments and facility responders.

The Right Side of the Bowtie

Operational Excellence extends to the right side of the bowtie analysis when emergency planners address preparedness measures that safely confine, contain, and control (3-C's) a loss of containment.

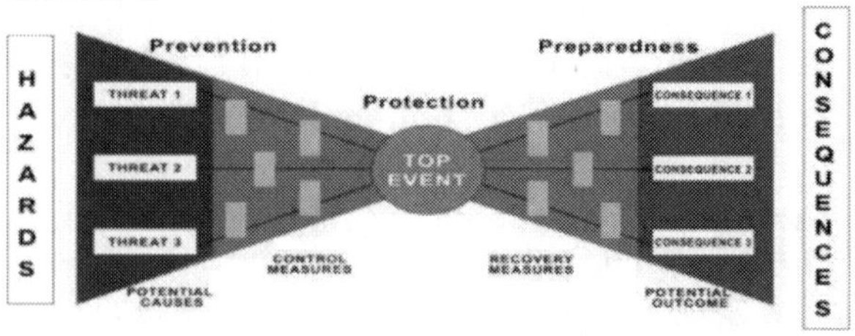

Figure 3. Bowtie Analysis of Preparedness Measures that control a Top Event

The emergency responder must be prepared to predict and engage incident action plans that safely mitigate escalating factors. The following points identify critical tasks and crisis management concerns associated with a loss of containment:

- System shutdown to reduce the potential consequences onsite and offsite of loss of containment from toxic effects or fire and explosion
- Response procedures to safely escape an ammonia release and engage the response team.
- Personal protective equipment to reduce the immediate and on-going threats for those who work near and respond to ammonia emergencies.
- Incident Command System capability of emergency first responders and hazmat technician leaders.
- Teaming and joint-planning of industrial and public safety responders that comply with governmental safety requirements.
- Clarity of expectation and capability of facility and community responders as depicted in emergency plans.
- Synchronized command readiness as supported by checklists and playbooks that clarify 3-C key points for each hazard zone within the industrial plant.

The Right Tools: The Right Plan

Plans are both process and tool for response readiness and for response. They must provide structure but also be situationally adaptable and flexible. To instruct and inform, ASTI has developed and recommends a playbook approach around the functional phases of an incident: Discovery, Initial Response, Sustained Response (as needed) and Termination as illustrated in Figure 4.

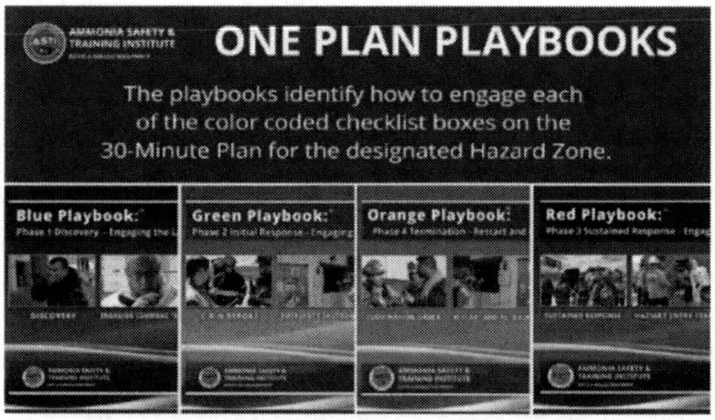

Figure 4. Example of a One Plan Playbook

(1) **Blue Playbook:** Life safety, alert the command team, notify 911 and other mandatory responders, establish command, and escape/evacuate the isolation zone to rally points and then evacuation points (which would include shelter-in-place locations.

(2) **Green Playbook:** begins with the need to provide a Conditions-Actions-Needs report to command team members and transitions to a SIMPLE approach for systems shutdown procedure that addresses flammability, isolation of leak source, management of pressure, pressurized ventilation, life safety assessment within the isolation zone, and the need for an emergency response plan.

(3) **Red Playbook:** Used by hazmat technicians who enter an isolation zone to engage 3-C's based upon an incident action plan and safety plan approved by the Incident Commander. The playbook shows the location of key valves and ammonia system hazards.

(4) **Orange Playbook:** Used by those in charge of recovery, restart, and crisis management issues that follow an emergency event. The need for an effective "hot wash" incident review and the need for emergency plan improvements are accomplished within the scope of the Orange Playbook recommendations.

Whether one plan, four, or a dozen, a thoughtful, communications and risk-based approach to emergency preparedness plans can serve response activation both as a training tool and as an incident management guide. The One Plan Playbook example of Operational Excellence on the right side of the bowtie has already produced demonstrated results at a number of facilities who have embraced and embodied the model in their response preparedness activities. Such tools can be and have been developed in the One Plan framework to serve a multiplicity of scenarios besides hazmat response to include natural disasters, personnel injuries and other emergency events that require swift, competent action and coordinated communication.

The Tripod as a National Approach

Beyond the Safety Day construct, the Tripod Approach employs methods that are consistent with best practices findings developed by others. Most recently at a February 5-6, 2019 meeting in Chantilly, VA, DOT Pipeline and Hazardous Materials Safety Administration, FEMA, U.S. Fire Administration, and the International Association of Fire Chiefs met and developed a roundtable report focused on hazmat response readiness practices.

Several key themes identified during the Hazardous Materials Roundtable relate to the Tripod Approach presented in this paper:

- Leadership at the local and state level is critical in developing both the systems and partnerships necessary for a safe and effective response to the wide variety of and ever-changing hazardous materials risks and threats in the modern age.
- Building an integrated planning and response capability to better protect local communities requires the support and direct involvement of the "whole community," including responders, planners, regulators, government, and industry with expertise in all areas of hazardous materials development, manufacture, transportation, storage, and use.
- The application and use of risk-based planning and response processes based upon current national consensus standards provides a safe and effective foundation for the emergency preparedness activities. Nonetheless, there are still many issues and challenges to address to strengthen the nation's preparedness capability.
- The role of Local Emergency Planning Committees (LEPCs) is a critical and

foundational element in providing the planning and risk assessment services needed by the response community and requires sustained emphasis and support at both the local and state levels.

- Driven by homeland security efforts, the last decade has seen substantial improvements in the gathering, coordination, and sharing of critical information throughout the emergency preparedness community. However, there remains a need for systems and processes within the hazardous materials preparedness community that can gather, analyze, package, and distribute critical information in a timely and "user-friendly" manner.

Safety Days Grow and go Global

Tripod Safety Days now command a growing national and international audience and community of practice. They derived their origins from many prior communities of practice. With respect to the model that is the subject of this paper, the origins were modest and locally focused. In 1997 a group of leaders from industry, government, and public safety organized the first ammonia safety day in Central California. To their amazement over 150 participants came to the Santa Cruz County Fairgrounds from all over Central California. That experience spawned a Safety Day movement in California that moved to Salinas, Modesto, Turlock, Bakersfield, Los Angeles, and then to Phoenix, Salt Lake City, Portland, and Seattle. It has now grown throughout the U.S., Canada, South America and Australia.

Chemical Safety Days sponsored by ASTI, RETA, IIAR and other non-profit organizations annually draw over 6,000 local area participants from industry, government, and public safety. The experience has created a stronger link between the Tripod leadership and the local area end-users and emergency responders. This movement has created networks and working relationships that reflect the wisdom of the high-powered collective impact of those who have attained successful outcomes when dealing with chemical emergencies.

As they have grown in number and scope, chemical safety days now often address other hazardous chemicals/materials including chlorine, sulfur dioxide, higher volatility crude oil, LPG, etc. A community-based hazard awareness inventory can inform the scope and best value of a given Safety Day focus. First Responders, Local Emergency Planners, and Industry representatives can contribute to that inventory as well as to the best framework for the Safety Day engagement.

Conclusions

Responsible Collaboration between industry, regulatory and investigatory authorities, labor organizations, communities, public safety agencies, and technical organizations promotes mutual understanding of risks and can reduce NH_3 incident risk exposures in both small and large communities. Chemical Safety Days provide opportunity for chemical emergency preparedness, first responder, and technician responder readiness.

Community based safety partnerships and leadership are fundamental to the success of such efforts. The Tripod Approach to Accident Prevention and Response Readiness provides a model of operational excellence and is a demonstrated formula for successful community of practice engagement.

Prevent Them All or Stop Them Small.

Lessons Learnt in Emergency Shipments of Catalysts

Fulfilling urgent customer requirements after an unplanned catalyst demand requires an effective operation of emergency logistics and the ability for the operating plant to swiftly respond to quickly changing dynamics.

This paper will highlight lessons learnt while arranging emergency shipments and describe some of the issues that can prevent replacement goods from moving as fast as may be desired. Examples include INCO terms, local regional product registration, custom clearance, preflight checks, payment for aircraft bookings, availability of critical logistics (drivers with the right permits, plane space), re-labeling of re-directed goods to comply with destination country regulations – all being critical matters, which take finite time and must be done correctly.

John Brightling, Dave Hawkes, Charlie Peters
Johnson Matthey

Introduction

Over a number of decades ammonia catalysts and production technologies have been improved through new innovations and technologies, operating efficiencies, reliability and monitoring.

A significant innovation and trend within production technology has been the increasing use of economies of scale, with the largest operating plant having evolved from 600-1000 mtpd 50 years ago to plants operating up to 3300-3700 mtpd today. In respect of the impact on catalyst volumes required, Figure 1 illustrates how plant scale has led to increased volumes of both HTS and LTS catalyst in the reactors of ammonia plants. Clearly with the investment in a large plant the opportunity cost losses increase for any time the asset is not running – such as if catalyst beds have been damaged. Given the increase in plant scale, emergency replacement of catalysts

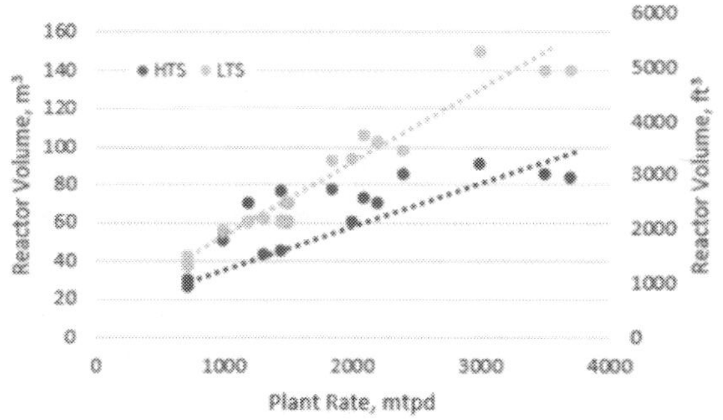

Figure 1. HTS/LTS reactor vol. v plant rate

may now weigh in the range of 200 - 500 mt. Shipments of an urgent order of this size of catalyst can prove challenging and expensive and require careful management.

Currently the industry is focused on using digitalization to progress towards goals of reliability and predictability. While digitalization does improve the reliability of the operation, it cannot fully safeguard against unplanned events that may result in emergencies for plants whereby catalysts have become unexpectedly damaged.

Catalysts for the most part do their work in a quiet, sustained way and perform their tasks as we would like them to, running in service life's measured in years or even decades. However, that can beguile operators into a false sense of security, forgetting about the unexpected failure modes and how critical they are and how difficult and expensive it can be if catalyst is unexpectedly needed in a matter of days, not weeks or months.

Plants, equipment, and catalysts deliver incredible reliability today and therefore it may be a long time between serious problems. Operators who were present for challenges and upsets from years prior may have moved on, taking with them corporate knowledge and experience gained managing those emergency situations.

Experience gained

In earlier papers a series of Catalyst Catastrophes (References 1-5) have been presented by Johnson Matthey at the AIChE Ammonia Safety Conference. These have detailed real events, structured by reactor duty, and included generic considerations such as loading/discharge and transient operations.

In most of the Catalyst Catastrophes the catalyst system had been damaged by a plant incident and there was often an emergency need for catalyst delivery. Throughout our years developing and producing ammonia catalysts and process technologies, we have encountered many hurdles and learned valuable lessons managing emergency shipments. We hope to pass along our lessons learnt.

While suppliers of catalysts carry unallocated emergency stocks, these are not limitless in terms of volume and may not be sufficient for large unplanned demands. In these cases, suppliers may need to work with their supply chains and customers to utilize stock previously earmarked for other 'in-works' orders and dispatches.

Transportation issues and hurdles

Some of the critical issues and hurdles that can prevent replacement goods moving as fast as may be desired are INCO terms, local and regional product registration, custom clearance, preflight checks, payment for aircraft bookings, availability of critical logistics (drivers with the right permits, plane space), relabeling of re-directed goods to comply with destination country regulations and repackaging according to IATA regulations.

INCO terms

The choice of INCO term, Table 1, for an emergency delivery will largely depend on mode of transport and who has responsibility for it.

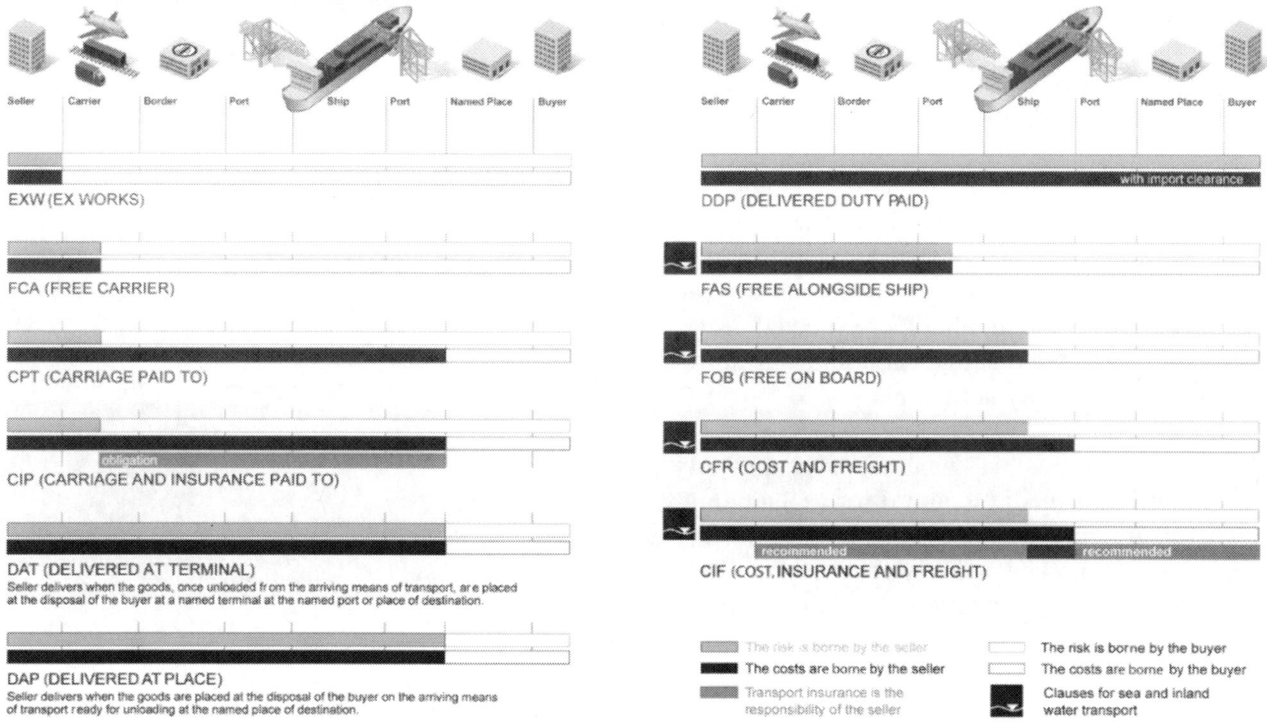

Table 1. INCO Terms

Customer collection:
FCA - customer freight forwarder handles logistics (seller arranges customs export clearance)
EXW - customer freight forwarder handles logistics (and buyer arranges customs export clearance)

Seller arranges delivery:
CIF/CFR - supplier arranges sea freight to customer nominated seaport
CIP/CPT - supplier arranges airfreight to customer nominated airport
DAP - supplier arranges delivery to named place, buyer imports into destination country

DDP - supplier arranges delivery to named place, seller imports into destination country.

Note that DDP cannot usually be arranged. The seller's legal entity is unlikely to have a presence in all destination countries hence they will not be able to arrange the import.

Local country requirements

A lack of understanding of the local requirements for importing goods can cause delays. Reasons for the delays differ from country to country, following may be some of the causes:

- Country product registration

In order to protect the interests and safety of their population, many countries maintain strict regulations for all products that enter their borders, especially controlled substances. In these cases, a product needs to be registered with various government agencies so that the importer can get the clearance to import and sell such products within their borders e.g. import permits required in Saudi Arabia

- Country packaging standards

The primary focus when transporting hazardous chemicals is the hazardous nature of the product.

However, there are other factors to be considered, for example labelling and packaging. Different country legislations have different regulatory and specification needs regarding labelling with UN standards for drums.

What can be less obvious and has been known to cause delays and issues for deliveries is pallet quality. For some circumstances, especially if commoditization of supply choice has driven buyers to lower tier suppliers, unexpected issues can arise due to the fact that economies mean required standards may not be met.

An example for this, was where an urgent supply of inert support media required for a plant shutdown emergency was impounded at importing country customs. The pallets were non-compliant with local country regulations - lacking fumigation certification and having bark still attached to the wood, which was deemed very serious by the Customs and Border Protection Service, resulting in the goods being quarantined and inaccessible to the customer.

Considering the sophisticated nature and technical complexity of an ammonia plant with its standards and controls in place, it is hard to imagine that the simple natural technology of wooden pallets can be a potential show stopper if not considered upfront, see Figure 2.

Figure 2. Wooden pallet approval mark

Forestry Commission

United Kingdom Wood Packaging
Material Marking Programme

for Wood Packaging Material Manufactured in Great Britain

This is to certify that

T.O. Mounter & Sons Ltd
Sawmill, Park Street, Willington, Crook
Co Durham, DL15 0ER

(manufacturer name and address)

Heat Treatment (HT)
Approved measure

FC0144
Registration number

18 April 2019
Certificate valid until

Forestry Commission
timcon

Figure 3. Wooden pallet certification

Without the appropriate certification, for simple components such as a pallet, Figure 3, a delivery can be found to be non-compliant and delayed or blocked completely by government authorities and control of the urgent delivery lost.

A quarantine officer doesn't feel the same economic pressures that motivate an operating company to prioritize getting the plant back on-line as soon as possible.

- Country MSDS/SDS

Almost every country has regulations that require the MSDS/SDS to be in native language(s) of that country, so an MSDS/SDS in the U.S. or U.K. must be in English, one in Norway must be in Norwegian, an SDS in Canada must be available in both English and French etc.

Most countries have already adopted the Globally Harmonized system of classification and labelling of chemicals. Those that have not, are working towards, this, and good practice dictates that where there is no standard, adoption of GHS is appropriate, although the building block concept and varying, but mandatory, substance hazard classification in certain countries, means that GHS is more of a concept than a reality. Safety Data Sheet requirements also vary across the world, in terms of both content, and language. Most countries have now adopted the standard 16-section SDS format, whilst not all enforce it. GHS, EU-REACH and global equivalents mean that the hazard classification of substances and associated products is now changing more than it ever did.

One of the activities that is critical in the event of redirection of catalyst goods during an emergency shipment is the relabeling of drums with the correct hazard labels.

GHS defines and classifies the hazards of chemical products, communicating health and safety information on labels and safety data sheets. It aims to provide the same set of rules for classifying hazards, and the same format and content for labels and safety data sheets (SDS) will be adopted and used around the world.

The target audiences for GHS include workers, transport workers, emergency responders and consumers. The two major elements of GHS are:

1. Classification of the hazards of chemicals according to the GHS rules:
GHS provides guidance on classifying pure chemicals and mixtures according to its criteria or rules.

2. Communication of the hazards and precautionary information using Safety Data Sheets and labels:

A high-level summary of the classification for ammonia plant catalysts is:

Duty	**Hazard Class, UN numbers**
HDS	9, UN3077 & 4.2, UN3190
ZnO	9, UN3077
Reforming	Non haz for transport
HTS	9, UN3077
LTS	9, UN3077
Methanation	9, UN3077
Ammonia Syn	9, UN3077 & 4.2, UN2881

4.2 = Flammable solid
Class 9 = Miscellaneous
UN3077 = Environmentally Hazardous Substance, Solid
UN2881 = Metal Catalyst Dry
UN3190 = Self-heating Solid

Airfreight

Normally for planned catalyst deliveries sea freight is used to move catalyst goods in 20' or 40' ISO-containers.

In a plant emergency where every day of lost production potentially costs $1M's, faster transport is needed, leading to airfreight.

The costs of air freight are significantly higher than those associated with sea freight and are usually calculated based on weight rather than volume (which is used for sea cargo).

Air-freight can be summarized as - "Not as fast as hoped and more expensive than everything else!"

Compared to peoples' expectation and experience of being a passenger on a plane, shipping goods by air is a much more complex matter. As an air-passenger with modern transport infrastructure and flight routings, it is probably possible to get yourself from any major city to any ammonia plant in the world in <48 hours, often less - providing that you are prepared to pay a high airfare and have the necessary visa.

By comparison the fast movement of a consignment of catalyst goods by air freight takes significantly more than 24 hours due to safety, trade documentation and/or other factors.

There are restrictions on what can be safely and legally transported using air freight. From a Security perspective, it is the responsibility of airlines to comply with in-country and international regulations. Security measures are risk-based and applied upstream in the supply chain with cargos requiring special packaging and physically delivering to airlines well before a scheduled flight for pre-flight inspection and security screens.

Many catalyst products are classified Dangerous Goods for air transport. These can either be restricted to carriage on cargo aircraft or can be permitted on passenger aircraft subject to compliance with IATA requirements.

Around 80% of air cargo is carried on passenger aircraft so it is important that all air cargo is robustly screened to ensure that prohibited articles are not part of the flown cargo. Goods transported by airfreight must be 'SPX' (secure) cargo that can be transported on any aircraft. Goods can become 'SPX' by either passing screening at the airport, or by originating from a 'Known Consignor' and being transported to the airport by a 'Regulated Agent'. If screening is required, the primary method is x-ray, secondary methods include Explosive Trace Detection or hand searches. Drums of catalyst are dense, and most x-ray machines will not be able to see through the package so Johnson Matthey's warehouse has been authorized as a known consignor. This provides a cost savings in screening costs but even more notably, without known consignor status, there is a higher risk of damage in transit due to invasive screening methods.

Dangerous goods being sent by air must be in perfect condition otherwise they will be rejected by the airline. The slightest dent in a metal drum is enough to stop it from flying. Time must be spent meticulously inspecting every drum and repacking any drums with imperfections. It takes roughly a day to repack 80 drums.

Figure 4. Special packing for air freight

Thus, airfreight of a large quantity of catalyst takes a finite amount of time, with Figure 5 providing a typical schedule to move 150 tons of goods.

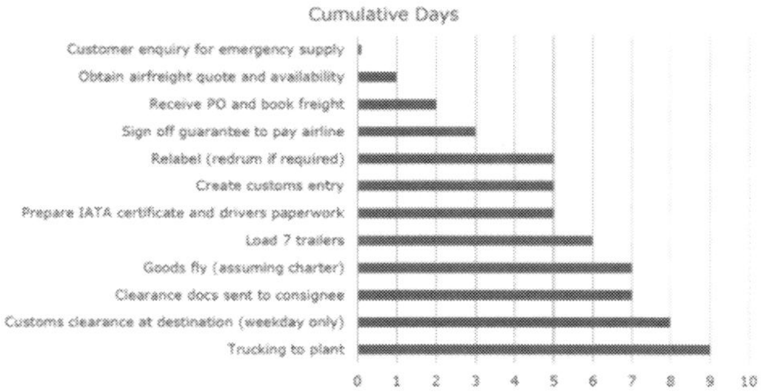

Figure 5. Typical air freight to move 150tes of catalyst

Types of airfreight

A common feature of emergencies it that they can call for airfreight, there are several different choices when it comes to arranging an air freight delivery:

Consolidated freight – where consignment will be split over several different flights. These shipments might be placed in the hold area of passenger planes or in dedicated freight planes, which can accommodate up to 20 ton of cargo on one flight, subject to plane size. Frequency of flights to specific destinations will vary dramatically and must be factored into planning if your goods will travel over multiple flights.

Charter services – occasionally, and at a very high cost – entire freight planes can be chartered for a single air freight delivery.

Costs for charter flights are typically 2-3 times higher than for consolidated cargo. To move a 200 ton load of catalyst via a dedicated Charter from one continent to another could cost between \$500,000 to \$1,000,000.

A final negative impact from use of airfreight in moving goods internationally, is an additional carbon footprint with x100 greater CO_2 emissions from airfreight compared to sea transport.

Advance payment guarantee

Despite the differences between travelling routinely as an air passenger or less routinely as an emergency cargo - some aspects of the process are identical. No airlines will hold time-critical space for you as a booking until a formal commitment has been made with guaranteed payment. Thus, fast confirmation for acceptance of airfreight costs is important.

In most cases, airfreight shipments are sent on a CIP (supplier arranged flight) Destination Airport basis and the customer will need to arrange clearance and onward delivery upon arrival.

Transportation regulations

Catalysts are by design very active materials, for some products especially where they are pretreated this brings to bear additional transport restrictions, which can vary regionally, e.g. within Europe the ADR (Accord Dangereux Routier) regulations apply to the transport of any hazardous goods. (Ref 6)

Figure 6 provides an example of the transport classification for **KATALCO** 35-8A Ammonia Synthesis Catalyst on a North American format SDS as follows, the key hazard being it is a pre-reduced catalyst which is hazard class 4.2

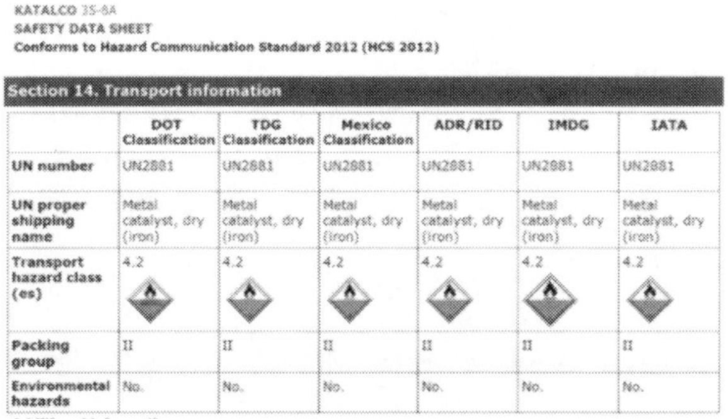

KATALCO 35-8A
SAFETY DATA SHEET
Conforms to Hazard Communication Standard 2012 (HCS 2012)

Section 14. Transport information

	DOT Classification	TDG Classification	Mexico Classification	ADR/RID	IMDG	IATA
UN number	UN2881	UN2881	UN2881	UN2881	UN2881	UN2881
UN proper shipping name	Metal catalyst, dry (iron)	Metal catalyst, dry (iron)	Metal catalyst, dry (iron)	Metal catalyst, dry (iron)	Metal catalyst, dry (iron)	Metal catalyst, dry (iron)
Transport hazard class (es)	4.2	4.2	4.2	4.2	4.2	4.2
Packing group	II	II	II	II	II	II
Environmental hazards	No.	No.	No.	No.	No.	No.

Figure 6. ***KATALCO*** *35-8A transport classifications*

This catalyst falls under transport hazard class 4.2 which is described as a "Spontaneously combustible substance". This catalyst can self-heat (without exposure to a heat source) when exposed to oxygen.

Figure 7 shows how the hazard classifications differ for **KATALCO** 83-3X, a copper oxide/zinc oxide based, low methanol LTS catalyst, which is classified as environmentally hazardous to aquatic life.

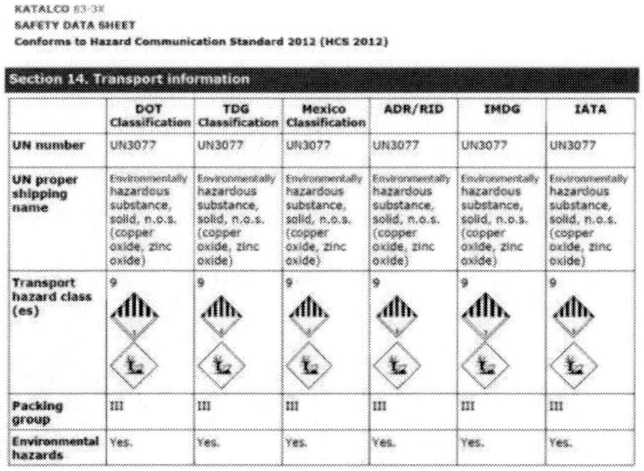

KATALCO 83-3X
SAFETY DATA SHEET
Conforms to Hazard Communication Standard 2012 (HCS 2012)

Section 14. Transport information

	DOT Classification	TDG Classification	Mexico Classification	ADR/RID	IMDG	IATA
UN number	UN3077	UN3077	UN3077	UN3077	UN3077	UN3077
UN proper shipping name	Environmentally hazardous substance, solid, n.o.s. (copper oxide, zinc oxide)	Environmentally hazardous substance, solid, n.o.s. (copper oxide, zinc oxide)	Environmentally hazardous substance, solid, n.o.s. (copper oxide, zinc oxide)	Environmentally hazardous substance, solid, n.o.s. (copper oxide, zinc oxide)	Environmentally hazardous substance, solid, n.o.s. (copper oxide, zinc oxide)	Environmentally hazardous substance, solid, n.o.s. (copper oxide, zinc oxide)
Transport hazard class (es)	9	9	9	9	9	9
Packing group	III	III	III	III	III	III
Environmental hazards	Yes.	Yes.	Yes.	Yes.	Yes.	Yes.

Figure 7. ***KATALCO*** *83-3X transport classifications*

This relevant transportation information is provided on the MSDA/SDS. Additionally, there is a requirement to label each package (e.g. drum or bulk KATPAC, Octobox) with a HAZ label.

Note that **KATALCO** 35-8A is packing group II and **KATALCO** 83-3X is packing group III. Packing groups identify levels of risk, with I being highest and III being lowest. The standards of packaging are shown by a UN drum mark such as:

un 1A2/X368/S/**/GB/****

This code is interpreted as follows:

1A" steel removable head drum
X for pPG, I, II, III
368 maximum gross weight in kgs
S approved for solids
** last two digits of year of manufacture
GB country of certification
**** represents the number of the manufacturers certificate

Johnson Matthey drums are rated to hold packing group I, even though our products are a maximum of packing group II, Figure 8.

Figure 8. JM drum marking of UN standard

HAZ Labels - With GHS, certain information will appear on the label, for example the chemical identity may be required or standardized hazard statements. Signal words and symbols will appear on the label according to the classification of that chemical or mixture. Precautionary statements may also be necessary, if required by the local regulatory authority. The requirements for precautionary statements on a HAZ label differ regionally and naturally there are different language groupings required to communicate hazards for different regions/countries.

Figures 9 and 10, show how the HAZ labels differ for the same product Low Methanol LTS, **KATALCO** 83-3X being shipped in the same 240kg drum, from same GB origin to different goods destinations in US or Russia. Note that between Figures 9 and 10 there are different GHS hazard diamonds due to different classification thresholds within the destination countries.

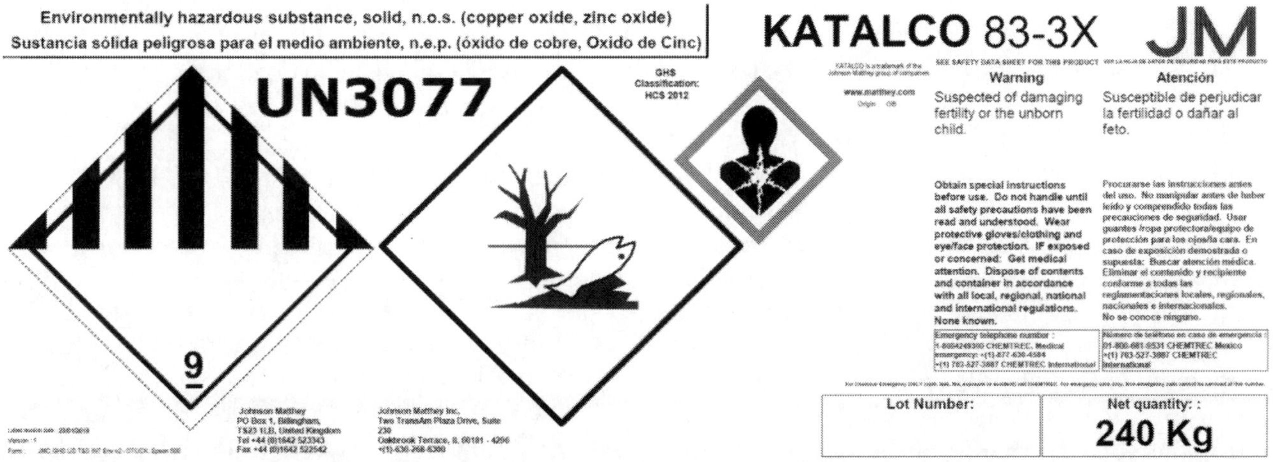

Figure 9. North American market HAZ chem in English and Spanish to the US form standards.

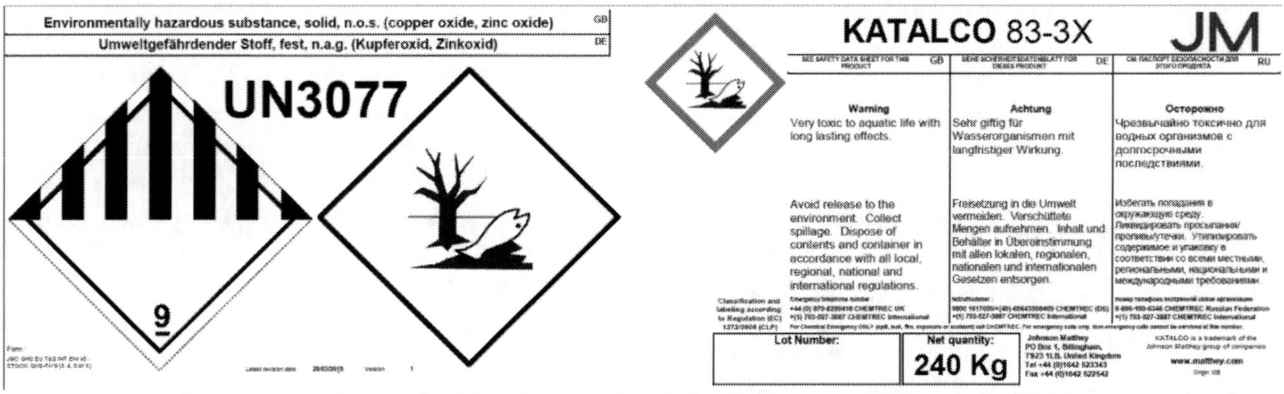

Figure 10. Russian market with HAZ chem in English, German and Russian in EU form standard.

For a recent unforecasted emergency catalyst order, JM was asked to urgently supply 150m^3 (5,300 ft^3) of **KATALCO** 83-3X, to a customer in Russia.

To fulfil such a large demand of catalyst, it was necessary to re-allocate material that had been originally destined for the North American market. Some of these goods had not yet been dispatched from JM's supply chain from the UK to US and equally a smaller quantity needed to be airfreighted back from the US to the UK as consolidated freight. Such rerouting of goods involves significant work logistically including removing the original HAZ label and ensuring it was replaced with the correct one for the Russian market on all 651 packages, Figure 11.

As with many regions cross border multi-language documentation requirements can be demanding. If there are errors or inconsistencies between the goods labels and documentation clearance delays or impounding of goods are possible at a border.

Figure 11. 651 Drums after relabeling

Case study incidents and transportation learning

The following examples provide case studies of plant emergencies that resulted in emergency catalyst demands with challenging logistics and support requirements

Case study 1 - Emergency LTS to Asia

Replacement due to high pressure drop after re-used support media failed under a fresh charge of **KATALCO** 83-3X. An Ammonia plant in Asia had loaded a full charge of 62m^3 (2190 ft^3) of **KATALCO** 83-3X catalyst, which was reduced without any problems. However, when operating at full rate the Start Of Run (SOR) pressure drop was measured at 2.1 bar (30.5 psi) rising where previous charge had only been 1 bar (15 psi) End Of Run (EOR).

The problem was investigated including adding an additional pressure tapping using the catalyst drop out pipe to the vessel, to help characterize the pressure drop contribution. As a result of this measurement, the pressure drop was quantified as 0.3 bar (4.4 psi) from catalyst bed as expected and 1.8 bar (26.1 psi) below the catalyst bed which was not expected.

The vessel was discharged, and a detailed investigation completed. This showed that a number of different types of support ball had been used and that one of the types installed had failed catastrophically, Figure 12.

Figure 12. Failure of reused support balls

The cause of the failure was due to the rapid drying of the balls which had become wet during long term storage. In the reactor, conditions of high heat transfer and/or depressurisation leading to vaporization of trapped moisture within the 25mm (1") ball causing them to shatter/explode.

Whilst the original catalyst charge that had been installed before the incident was shipped on a lead time of 12 weeks, once the incident had developed, an immediate need for a full charge of 62m^3 (2190 ft^3) **KATALCO** 83-3X LTS catalyst and 20m^3 (706 ft^3) **KATALCO** 92-2 new alumina-silicate ceramic balls totalling 121tes of product was required urgently.

To do this, a dedicated Boeing 737 cargo plane, was contracted to fly a replacement charge, at air freight cost of around $500,000.

Case study 2 - Emergency primary reformer catalyst to Europe

Plant suffered steam loss in aux boiler trip, resulting in carbon formation and high pressure drop, investigation revealed catalyst needed full replacement of a charge compromising of 10m^3 (353 ft^3) **KATALCO** 25-4Q and 20m^3 (706 ft^3) **KATALCO** 57-4GQ.

The incident happened over the weekend and Johnson Matthey was able to check stocks and respond to enquiry over the weekend. Goods were loaded out in 2 x 20te 'curtain sided' trucks, Figure 13, and delivered within 48 hours.

Figure 13. 20-ton Road truck loading

The goods were not hazardous for transport, so ADR license is not required. Goods could travel via a number of ferry routes, for example Dover to Calais, Dover to Zeebrugge or Immingham to Rotterdam, or via Channel Tunnel.

Case study 3 – Emergency HDS/ ZnO to MEA region

An ammonia plant had an emergency requirement for HDS catalyst and zinc oxide absorbent. They first noticed an issue when the primary reformer methane slip increased significantly, and hot spots appeared on the tubes. The plant technical staff immediately began a process of trouble shooting in close collaboration with the Johnson Matthey regional team and UK HQ.

It had run for many years on natural gas with very low sulphur (probably circa 0.1 ppm), so monitoring the purification system had not been a high priority. They changed to a different gas supply with significantly higher sulphur but had not initiated monitoring or even routine analysis. It was diagnosed that the CoMo HDS had become deactivated by continuous low-sulphur operation leading to over-reduction. Organic sulphur compounds were therefore not fully converted to H_2S and the zinc oxide was also close to saturation.

Emergency supply was arranged by Johnson Matthey, 70m^3 (2472 ft^3) of zinc oxide was flown from the UK and 25 m3 of HDS had to be flown from North America. Both required

Johnson Matthey to renegotiate deliveries with other customers, for whom supply was fortunately less urgent. The zinc oxide was repackaged into Octoboxes but there was not time to do so for the HDS, so that had to be supplied in drums. The process of customer procurement and emergency shipping took approximately 2 weeks.

Fortunately, the plant was able to operate at almost full load in the meantime, by increasing S/C ratio and periodically suspending production and steaming the primary reformer catalyst to remove the build-up of sulphur and resultant carbon. Tube appearance and pressure drop was monitored closely to make sure catastrophic carbon build-up was not occurring. Each steaming episode took around 24 – 36 hours, following JM's procedure. The reforming catalyst proved robust and operated to the next scheduled turnaround.

Case study 4 – Emergency multiple catalyst supply to North America

A refinery had a large site-wide set of trips due to external events during extreme weather conditions in a country that can experience temperatures as low as -40°C (-40°F).

Attempted restarts in the prevailing conditions resulted in wetting and damage to the catalysts, which led to a full load of replacement catalyst needed. This included primary reforming, HTS, LTS and methanation, totals volumes as Table 2.

Duty	Gross Weight '000, lbs	Gross weight, '000, kgs	Vol, ft^3	Vol, m3
Primary	232	105	3712	105
HTS	548	248	6800	192
LTS	552	251	6,600	187
Methana tion	56	26	795	23
TOTAL	1,388	630	17,907	507

Table 2. Catalyst emergency delivery totals

Two separate logistical routes were required to fulfill this emergency requirement in the shortest possible time:

a) Over land by truck from JM warehouse in Houston USA

b) By air freight from JM warehouse in UK to closest international airport which had a long enough runway to land a Boeing 747-8F wide-body plane which was still some 740 km from the plant and required trucking. Although the 747 used had a maximum volume capacity of 400 m3, the maximum allowable payload weight was 110,000 kg (242,506 lbs).

510T went by road from the USA, 120T by airfreight from the UK, of which 110T went on the chartered 747, with a further 10T sent on a passenger flight as 'consolidated freight'.

Timeline summary for scoping requirements and movement of this large quantity of goods

Day 1 – Notification 1st official request
Day 3 – Quote issued
Day 6 – Final clarification of quote
Day 7 - Customer PO
Day 8 – First truck dispatch
Day 10 to 13 - On-land delivery complete
Day 14 - Air freight delivery complete

The organisation of the transport of this quantity of material is a significant undertaking with the availability of around 40 trucks being an issue. The challenges of cross border shipment and the need to have truck drivers with the appropriate hazmat license and number of available drivers prepared to drive such a long distance being equally onerous due to Easter weekend timing.

Communications.

Fulfilling urgent customer requirements after an unplanned catalyst demand, requires an effective operation of emergency logistics with a quick response critical for the operating plant.

Simple matters can cause problems, as has been highlighted in the above case studies. Documentation requirements and controls have become more rigorous over time and failure to follow approved procedures can cause logistics to grind to a halt at times when pressure for delivery is high and the focus is for speed.

Summary / Conclusions

Fulfilling urgent customer requirements after an unplanned catalyst demand requires an effective operation of emergency logistics and a quick response from the operating plant, as the lost opportunity cost for world scale ammonia plants can be easily $1M's per day off-line.

A summary of some of the key factors that are essential for fast response to an urgent need are, a full understanding of requirements so that clear customer instructions (ideally outlined in a Purchase Order) can be established to enable allocation of material, appropriate labelling and agreed transport booked.

Good relationships with suppliers and freight forwarders are essential to a quick response time. All parties should have a solid understanding of INCO terms and where responsibilities transfer. Knowledge and experience of cross border movements is vital as is regulations around the movement of hazardous goods.

While not always possible, given the limits and costs of air-freight in respect of large catalyst volumes that can be 100s m^3 (1,000s ft^3), to reduce the risk of emergencies for large plants that have the highest opportunity cost losses, a spare catalyst charge or purchase of safety stock in advance of planned changes may help mitigate the risk of unexpected emergencies.

References

1. "Catalyst Catastrophes", M P Roberts, Proceedings of 48th AIChE Ammonia Safety Symposium, Orlando 2003 (AIChE Technical Manual 2003, Vol 44).

2. "Catalyst Catastrophes II", J R Brightling and M P Roberts, Proceedings of 50th AIChE Ammonia Safety Symposium, Toronto 2005 (AIChE Technical Manual 2005, Vol 46).

3. "What Not to Do on Your Plant (Catalyst Catastrophes III)", P V Broadhurst and S D Ransome, Proceedings of 56th AIChE Ammonia Safety Symposium, Montreal 2011 (AIChE Technical Manual 2011, Vol 52).

4. "Catalyst Catastrophes IV: Yet More Tales of the Unexpected", P V Broadhurst, Nitrogen and Syngas Conference, Berlin 2013.

5. "The Importance of Catalyst Design in Managing the Impact of Transient Operating Conditions in Ammonia Plants", P Farnell, M Carlsson, Proceedings of 59th AIChE Ammonia Safety Symposium, Vancouver 2014 (AIChE Technical Manual 2014, Vol 55).

6. HSE, Carriage of dangerous goods

ADR

http://www.hse.gov.uk/cdg/manual/classification.htm#ADR

Packaging
http://www.hse.gov.uk/cdg/manual/packaging.htm

Gap in pagination due to unavailable paper.

Pages 235-244

Reformer Incident Caused by Carryover of Dust from Desulphurization Unit

On March 13, 2013, the Ammonia Plant-1 primary reformer tubes at Maaden Phosphate Company suffered overheating due to the passing of zinc-oxide (ZnO) catalyst from the desulphurizer reactor. One of the reformer tubes ruptured and many others developed cracks. The incident happened when one of the desulphurizers was taken in service after catalyst replacement. The natural gas preheat coils and primary reformer tubes were found filled with ZnO and bulk of catalyst in one desulphurizer was found missing. This paper presents the findings from the investigation into the cause of the failure, subsequent modifications, change in operating philosophy and lessons learned from incident.

Abrar Rafiq
Maaden Phosphate Company

Abdulrahman Abdullah Al-Nassar
Maaden Phosphate Company

Introduction

Maaden Phosphate Company operates two of the world largest ammonia plants at Ras-al-Khair, 60 km north of Jubail on the eastern coast of Saudi Arabia. The first ammonia plant is licensed by TKIS and was commissioned in February 2011 with a nameplate capacity of 3300 MTPD. The second Ammonia plant is also a TKIS licensed plant, commissioned in Aug-2016 having name-plate capacity of 3300 MTPD. Ammonia plant-1 has seen some improvements over the years and currently operating at 3500 MTPD. Maaden Phosphate also produces sulphuric acid and phosphoric acid at site and manufactures DAP and MAP.

Background

Ammonia plant-1 is typical TKIS dual pressure plant producing ammonia at two different pressures. The natural gas purification section has two hydrotreaters in series (a second was added in 2014) for conversion of all organic sulphur to inorganic compounds. These are followed by two desulphurizers in a lead-lag arrangement. The primary reformer is top fired TKIS design having 408 catalyst tubes and 243 Callidus design burners. The natural gas is preheated to 370°C (698°F) in the convection section and then passed through the hydrotreater and desulphurizers. It is mixed with process steam, further heated up in convection section to 540°C (1004°F) and then enters the primary reformer.

History

Every 4-6 months one of the two desulphurizers requires catalyst replacement. This is based on the level of sulphur in the natural gas.

The reactor with fresh catalyst is taken in service as lag reactor to protect the primary reformer catalyst.

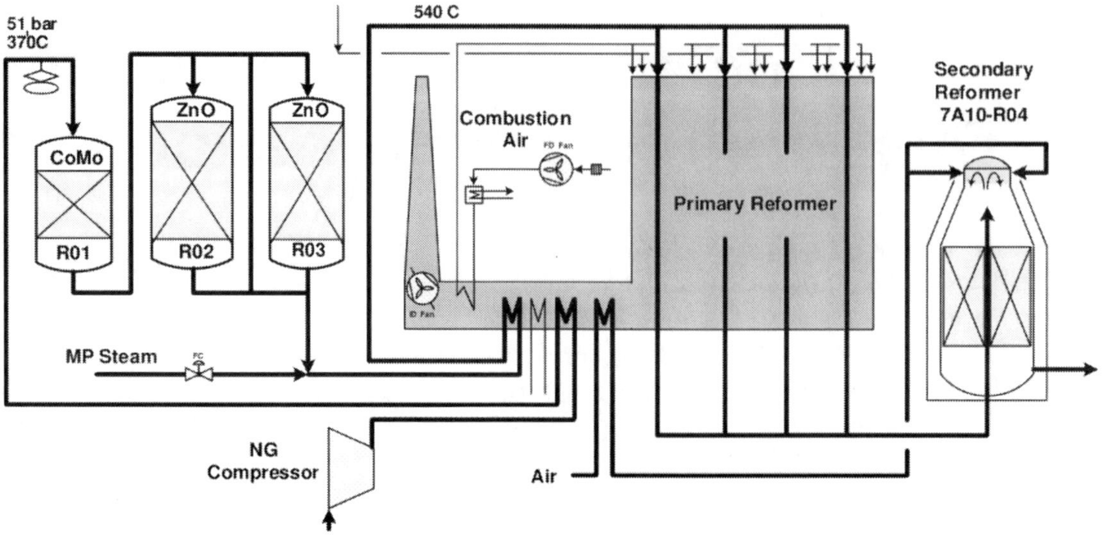

Figure 1. Ammonia plant purification and reforming section

Incident

On 30th March 2013, the ammonia plant was running normally at a load of 99 %. Catalyst in one of desulphurizers R02 was just replaced. Heating of R02 catalyst completed and as per procedure it was carried out in reverse direction. At 15:00 hrs., a planned work activity started to line up the R02 as lag reactor.

As soon as R02 became on line, various abnormal conditions were observed at the primary reformer:

- Increase in the ΔP across the Reformer catalyst Tubes from 3.18 bar at 1500 hrs. to 3.58 bar at 1533 hrs.
- Increase of CH_4 slip outlet of primary reformer. Methane slip increased from 11.86% at 1500 hrs. to 16.06 % in a short time, which indicated a sudden change of primary reformer performance.
- Increase of the primary reformer outlet temperature from 814 °C (1497°F) at 1459 to 830 °C (1526°F) at 1533 hrs.
- Significant increase of the tube metal temperatures, the upper five meters of the tubes color changed to red hot, seen by visual inspection. Measured tube metal temperature found approaching 1000C (1832°F) while the normal values are up to 917 C(1682°F).

The reformer load was reduced by decreasing the natural gas feed flow. At the same time, the Steam / Carbon Ratio was increased. This helped to improve Reformer catalyst performance and reduced CH4 slippage which came down to 12.58% at 19:40 hrs. However, the improvement of CH4 slip didn't result in improving the ΔP across the tubes or in significant improvements of the tubes skin temperatures.

During the plant upset and while trying to adjust the operating parameters, the secondary reformer tripped (the air flow was interrupted) which resulted in high fluctuation of steam pressure and flow, leading to the partial trip and later a total trip of the primary reformer.

As a result of excessive overheating, one tube ruptured, and few tubes reported with bulging. Tubes are micro alloy 25/35% Ni/Cr.

Incident Investigation

R02 reactor was opened and the catalyst bed was found at lower elevation. The loss of bed height indicated the catalyst has passed through the bottom mesh, through pre-heat coils and to the primary reformer tubes.

Figure 2. R02 reactor top

As seen in the figure, the bed is lower on one side where the bulk quantity had escaped. The top mesh is also inclined with alumina balls sliding downwards.

Figure 3. R03 bottom mesh visible as folded after removing catalyst

The reactor R02 catalyst was unloaded and the bottom mesh found displaced/folded from one side as seen in figure-3. Below the mesh is punch plate which act as a support for the catalyst. Punch plate has few mm gap b/w the vessel wall. The displacement of mesh kept the gap wide open and catalyst escaped to the downstream piping and equipment.

Figure 4. Unloaded catalyst from reformer tubes

The process gas/steam mix feed enter the reformer tubes through four headers at the top. On opening of the header manholes, ZnO was found all the way forming a thick layer.

All the primary reformer tubes were opened and the zinc-oxide (ZnO) catalyst was found in every tube. It coated the catalyst, effectively reducing its activity.

These initial findings suggested that soon after the R02 was lined up, a significant amount of ZnO catalyst passed through with the gas and settled on the catalyst of primary reformer.

The ZnO catalyst in the reformer tubes started blocking the catalyst surface, reducing the effective reaction.

Figure 5. ZnO in reformer top gas distribution header

The pressure drop across the catalyst tubes increased dramatically. The methane slippage increased in a short time due to the blockage of the holes of primary reformer catalyst pellets, with the outer surface coated with the ZnO catalyst dust.

The high tube metal temperatures were due to low catalyst performance and reduced flow through the tubes with the same firing going on. As a result of continuous operation, one tube ruptured and four other tubes overheated and bulged.

ZnO catalyst is in an extruded form but after exposure to high temperature in the convection coils, it was converted into a powder. This powder required removal by cleaning from the R02 outlet up through the primary reformer outlet. The Reformed Gas (RG) Boiler after the primary reformer had only little amount of dust, indicating all ZnO in section upstream of that point.

The displaced mesh was further investigated. This review found that prior to placing a new bed of ZnO catalyst in service, the reactor is heated through reverse flow of natural gas. The gas for heating comes from outlet of the in-service reactor (R03) and passed through R02 from bottom to top where it was sent to flare by opening of the R02 inlet PSV bypass. The reverse heating had a potential to lift / change position of bottom punch plate. This could affect the overlap of mesh and the bottom punch plate. Calculation has shown that a differential pressure of 0.75 bar across the catalyst bed was enough to lift punch plate and catalyst.

Damage Assessment and Recovery:

NDT testing of tube was performed and out of 408 tubes total, 69 tubes were found with crack indications: these were replaced at a later date.

ZnO powder was accumulated in a huge circuit starting from R02 outlet up to reformer tubes. To clear all piping and equipment, specific cleaning method was required. Pipe cutting was done at different locations to form various cleaning loops. Initially air blowing, and steaming was carried out but it was not affective as ZnO was sticking to the piping. It was decided to do chemical cleaning for major loops. Citric acid (5%) was used for this purpose as it has good solubility for ZnO.

The most difficult part of the recovery was the cleaning of feed gas/steam preheat convection coils E-11/12. The problem was faced that Citric acid was going selectively to part of the coil, leaving major portion with very less flow. Thermal imaging confirmed non-uniform distribution of flow.

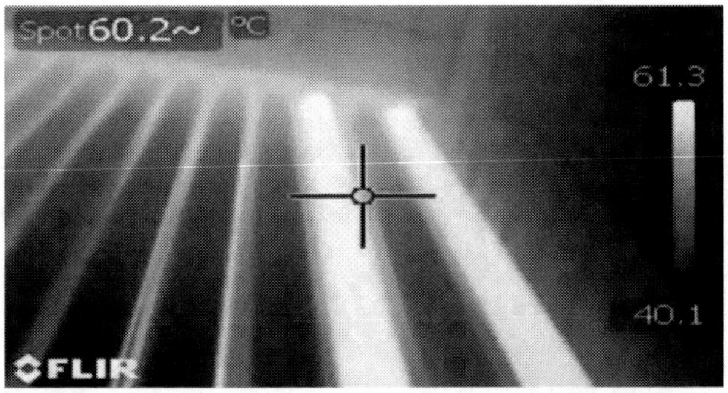

Figure 6. Except two coils, others have less flow

The main header for the coils was cut and a person went inside the header to clean/rinse each part of the coil separately.

Hydro milling carried out for the E-12 coil outlet vertical piping to the primary reformer. The process of "Hydro Milling" uses water for the purpose of cleaning in a circulatory motion along the length of the pipe using high pressure. The hydro milling tool moves forward inch by inch. The cleaning of the pipe is carried out with high pressure jets of water as it travels the required length. All the catalyst of reformer was unloaded, the tubes were cleaned thoroughly, and fresh catalyst loaded.

Resolution:

To avoid recurrence, the following actions were taken:

- All the front-end reactors have been provided with a floating ring at the bottom. This ring sits over the mesh and ensures there is no movement of mesh away from reactor wall.

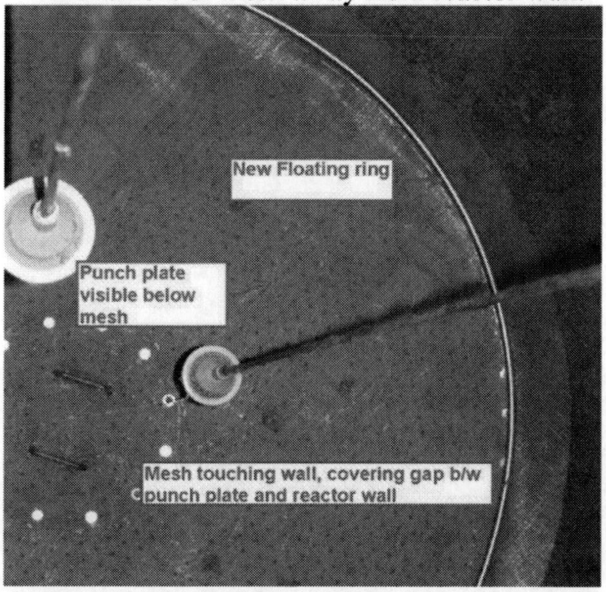

Figure 7. New floating ring provided above bottom mesh

- Reverse heating of the catalyst has been discontinued. Downstream vents are now in place to allow the heat up of the catalyst in the forward direction only.
- Also, the procedure for fresh catalyst lead/lag positions has changed. Now, whenever a desulphurizer catalyst is replaced, it is always aligned as lead reactor for 3-4 days before switching it to the lag reactor position. This ensures that any loss of catalyst will go to lag reactor. In that case, the lag reactor ΔP will increase and the potential for a major failure or upset in the plant is mitigated.

Methanator vessel overheating Why still a "Hot Topic"?

Several Yara ammonia plants have recently experienced an overheating of their methanator vessels leading to complete catalyst unloading for internal inspection. The purposes of this paper are to review the facts and root causes of these incidents, review the inspection activities and the associated recommendations, evaluate the probability and consequences of failure and finally describe the correction measures to avoid the re-occurrence of such incidents

Rodney Babwah
Yara Trinidad Ltd

Michel Warzee, Giuseppe Franceschini
Yara International

Introduction

The reaction occurring in the methanator is the final stage of the synthesis gas purification: the traces of CO and CO_2 are removed from the process gas, as they are strong poisons for the ammonia synthesis catalyst.
The nickel-based catalyst reactions are:

$$CO + 3H_2 \rightarrow CH_4 + H_2O$$
$$\Delta H = -206 \text{ KJ/mol}$$

$$CO_2 + 4H_2O \rightarrow CH_4 + 2H_2O$$
$$\Delta H = -165 \text{ KJ/mol}$$

These reactions are highly exothermic with a temperature rise of approximately 165 °F (74 °C) for 1 % CO or 140 °F (60 °C) for 1 % CO_2 at the inlet. With high enough carbon oxides in the feed and all layers of protection defeated, the temperature inside the vessel can easily increase beyond its integrity operating window.

Risk assessment for overheating of methanator vessel

Risk assessment for methanator overheating is traditionally executed using a Risk Matrix including the Probability of Failure (POF) and the Consequences of failures (COF).
For the methanator overheating risk, without any layer of protection:

The COF is classified as "Major", involving, in the worst case, the complete rupture of the equipment with a major release of flammable gasses leading to fire and eventual explosion and potentially causing:

- Extensive people injuries or fatalities
- Destruction of the surrounding structures and equipment
- Huge economic consequences including loss of production, liabilities, penalties and induced costs between 5 to 20 million USD

The POF is a Medium frequency, between 0,1 to 1 per year for the overheating. As no rupture of a methanator being reported in more than 50 years of operation, we reduce the POF of a full vessel rupture to Low Frequency, between 0,01 to 0,1 per year.

According to the Yara Risk Matrix shown in Figure 1), a **Safety Integrity Level 2** (SIL2) protection is therefore required.

Frequency POF / Consequence COF	F	E	D	C	B	A
	Frequent	Medium Frequency	Low frequency	Very rare	Extremely rare	Nearly impossible
	..... down to 1/yr	>10-1/yr - 1/yr	>10-2/yr to 10-1/yr	>10-3/yr to 10-2/yr	>10-4/y to 10-3/yr	<10-4/yr to 10-4/yr
1 Severe	Redesign	4	3	2	1	ALARP
2 Major	Redesign	3	2	1	ALARP	ALARP
3 Moderate	Redesign	2	1	ALARP	ALARP	OK
4 Minor	Redesign	1	ALARP	ALARP	OK	OK
5 Minimal		ALARP	ALARP	OK	OK	OK

Figure 1. Yara Risk Matrix. Numbers in red cells refer to SIL level. (ALARP = as low as reasonably . practical)

In order to be compliant with a SIL2 and therefore reduce the risk to a tolerable level, several instrumented layers of protection have to be properly designed and installed [1]:

- Two automatic "fail close" block valves to isolate the methanator in case of high CO and CO_2 content in the feed gas. These valves must have quick closure time (ideally less than 10 seconds), high tightness rate (class 5) and high reliability
- An automatic process valve upstream the isolation valves to automatically vent the feed gas keeping the plant front-end stable at a defined pressure
- Automatic methanator trip closing the two inlet block valves in case of high temperature in the catalyst bed. Catalyst bed temperature measurements should be redundant and always within the active catalyst.
- Automatic methanator trip closing the two inlet block valves in case of low flow of the CO_2 removal solution. The flow measurements should also be redundant (ideally 2 out of 3)
- An Emergency Shutdown System (ESD)
- Additionally, to achieve long maintenance intervals and to avoid hidden failures, partial stroking or automatic certified intermediate stop recording should be available

Process incidents

Over the last years, there have been four overheating incidents involving methanator reactors in different Yara ammonia production sites.

Incident 1

An overheating incident of a methanator vessel occurred in 2013. The incident was caused by reduced flow in the semi-lean solution stream to the CO_2 absorber thus increasing the CO_2 slippage and ultimately raising the methanator bed temperature. The temperature of the catalyst bed went up to 1113 °F (600 °C) for a period of 3 minutes at 320 psig (22 barg). The overall temperature excursion lasted 3 hours with an average temperature of 1000 °F (535 °C), well beyond the vessel design temperature of 800 °F (425 °C).

Analysis

A Fitness for Service (FFS) assessment was carried out after the event and it determined the plant could continue to run using the reactor till the planned upcoming turnaround in nine months, at which time the internal inspection of the vessel could be carried out. The FFS assessment also indicated that the vessel could have experienced (because of this overheating event) some creep damage with a reduction of approximately 5% of its creep rupture life.

Root cause

The root cause of this incident was that the low flow trip for the lean solution bulk stream was not set high enough to prevent the CO_2 content increase. When the main lean solution valve to the absorber failed closed, the bypass around it was still partially open as it always was during normal operation in order to reduce the excessive vibrations on the main control valve. The flow rate through this bypass was greater than the minimum flow trip set point.

If the load to the CO_2 absorber had been reduced at the time of the control valve failure, the severity of the excursion would have been less.

Consequences

- External and internal inspections
- Minor repairs of internal superficial linear indications on two welds
- Catalyst removal and replacement

Incident 2

An overheating incident of a high pressure methanator vessel occurred in 2013, after a loss of plant power and because of a non-operational solenoid trip system on the single inlet trip valve. The temperature excursion lasted for 16.5 hours reaching 1456 °F (790°C) for 30 minutes at 1741 psig (120 barg). The normal operating condition was 625 °F (330 °C) at 1800 psig (125 barg). Overall, the vessel took almost 24 hours to cool below the vessel design temperature of 900 °F (480°C).

Analysis

The preliminary FFS assessment, carried out immediately after the event, indicated that the reactor consumed more than 50% of its creep rupture life in this single event.

Based on this initial assessment, the plant was not allowed to restart without having first performed a detailed internal inspection of the methanator and then executed the eventual and consequent recommendations.

Root cause

The temperature excursion was caused by the inlet shut off valve failing to close as per trip schedule.

A mechanical failure occurred on the polyurethane seating area (on the rocker assembly) of the 3-Ways acting trip relay valve (Figure 2).

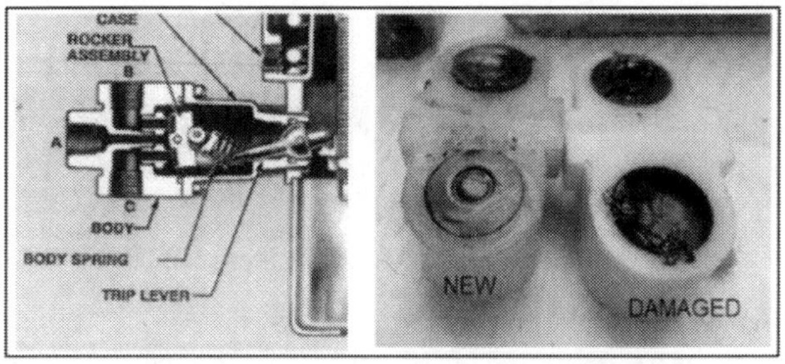

Figure 2. 3-Way trip relay valve

Evidence of water, due to the poor quality of the instrument air, was found in the trip relay valve. Polyurethane seats were most likely hydrolyzed and softened enough to be removed by the constant instrument air flow on the open way.
This failure allowed the process trip valve to stay open and permitted the bad quality gas to go through the catalyst for about 25 minutes accelerating the reaction before being stopped.

Operators attempted to manage the temperature increase in the vessel (40 seconds after the first high temperature alarm), but they didn't check the real closure of the trip valve.

Consequences

- Catalyst removal and replacement
- External and internal inspections
- Removal of all internal crack-like indications. An assessment of these grinded areas was also done resulting in four zones to be repaired by welding. These welding repairs required dehydrogenization, preheating, post weld heat treatment, inspection and hydrotest
- Overall the plant lost 22 days
- Cost for catalyst replacement
- Inspection and repairs cost
- Methanator replacement at the next plant turnaround

Incident 3

In May 2018 a level upset in the Benfield system caused the actuation of the methanator trip valve due to high temperature increase rate on the methanator.

This trip logic has the safety function to protect the methanator from exceeding its design temperature of 840 °F (450 °C) and it may be initiated in any of the following cases:

- When the run status of both the Benfield solution pumps is off for more than 10 seconds "AND" there is a low flow of Benfield solution
- When 2 out of 3 bed temperature measurements are at 790 °F (420°C) or above
- When 2 out of 3 temperature rate measurements are at 0.6 °C/sec (or more) for 5 consecutive seconds and the bed temperature is above 680 °F (360 °C)
- When high level (2 out of 3) is measured in the separator down stream of CO_2 absorber

Analysis

At the time of the trip actuation, the hot spot in the catalyst bed was at 694 °F (368 °C), still well below the design temperature of 840 °F (450 °C). In less than fifteen minutes since the actuation of the methanator trip, a low level in the front-end waste heat boiler steam drum caused a trip of the complete plant.

The catalyst bed temperatures continued to rise for about 12 hours and all the measurements exceeded the vessel design temperature with a maximum recorded value of more than 1200 °F (650 °C).

Root Cause

The automatic methanator trip should have been initiated by the low flow in the Benfield solution stream and the Benfield pumps both offline: at this point in time the methanator bed temperatures had not even started rising.

However, the low flow calculation (based on the addition of semi-lean and lean Benfield streams) was not correct because the lean Benfield solution flow measurement was not in operation due to a faulty flow meter: this prevented the trip to be actuated.

Risk assessment as part of force log process could have prevented this from happening. No force/defeat log was filled in for the faulty flow meter despite that it was part of a trip sequence: the unavailability of a document that lists all critical instruments, the defeat of whom should have been required this process to initiate, contributed to the event.

The overwhelming number of alarms in a tense situation didn't help the panel operator to correctly identify the underlying cause of the on-going event: after this event it was therefore established that the alarm management philosophy needed to be reviewed.

Finally, the methanator was not depressurized from the CO_2 contaminated gas until the next morning and the specific procedure to cool down the methanator after a high temperature trip was not followed

This event raised important questions about the effectiveness of the current refresher training mechanisms and on the full control of the emergency handling procedures by the operational personnel.

Consequences

- Catalyst removal and replacement
- External and internal inspections (no indications found)
- Cost for production loss, inspection activities and catalyst replacement

In addition, the following actions were taken:

- Modify the methanator trip logic to remove the "AND" gate between the low flow of the Benfield solution stream and the run status of the solution pumps

- Develop a list of critical instruments, the defeat of whom, shall require a force log risk assessment to initiate
- Review the alarm management philosophy during trips and implement corrective actions to channel the right information to the panel operators
- Review the existing periodic refresher training program about emergency handling procedures and implement actions to verify the effectiveness of this training program
- Develop a plan to do periodic stroke tests of control valves that stay in one position for extended duration

Incident 4

In 2019, due to instability of the aMDEA CO_2 removal stripper column level, caused by foaming, the suction pressure of the semi-lean pump was significantly varying. The operator by-passed the trip logic to avoid the auto-start of the back-up pump; doing this, he didn't realize that, in the same trip logic, he was also bypassing the solution low flow trip and the high temperature trip of the methanator.

When the running pump tripped, because of low suction pressure, gas with high CO_2 content was sent to the methanator. Seeing the methanator temperature rising, the operator activated manually the trip valves and vented the gas upfront the vessel depressurizing it.

The catalyst temperatures inside the vessel increased to approximately 1160 °F (625°C)

Analysis

The vessel, built in 1966 in 13CrMo44 material, has a design temperature of 950 °F (510 °C) and operating pressure of 425 psig (29.3 barg).

During this event, the temperature inside the vessel was above the design for more than 8 hours but with reduced internal pressure.

The expected damage mechanisms for this overheating event were identified and their probability of failure was estimated being low. It was recommended to externally inspect the vessel before re-starting and looking for: possible deformations (visual inspection) and cracking (Time of Flight Diffraction, ToFD) on the longitudinal and circumferential welds of the vessel.

Root Cause

The CO_2 removal section trip logic from the ESD system was modified in the previous plant turnaround wrongly mixing different safety functions like the methanator high temperature trip protection. This was not even effectively communicated to the operators.

Consequences

No indications were detected with the external inspection of the vessel.

The safety logic was corrected (dividing the safety functions wrongly mixed together during the previous plant turnaround) and tested before the plant start-up.

Catalyst behavior in case of overheating

Total carbon oxides of about 4% will lead to approximately 930 °F (500 °C) at the exit of the methanator. However, an operating failure occurring in the upstream equipment (such that an high concentration of carbon oxides is fed to the catalyst) can quickly put the methanator at risk, potentially increasing the temperature over 1290 °F (700 °C). When this kind of event occur, the process gas must be immediately shut off and vented upstream to prevent excessive catalyst bed temperature.

Acceptable cooling media that can be used in the event of catalyst overheating are:

- Hydrogen
- An inert gas, such as nitrogen
- Cold process gas, once the CO_2 removal system is under control
- Ammonia synthesis gas

There is a potential for coke formation in case methane is present in the cooling gas, therefore natural gas is not recommended as cooling medium.

Long periods of operations above 968 °F (520 °C) or higher will cause catalyst sintering and permanent loss of activity.

In case of temperature upsets for a short time, even above 650°C (1200 °F), experiences have shown that the catalyst will still be operational and that it can be kept in place for some years.

Damage mechanisms associated with overheating

The damage mechanisms expected for the methanator vessels are creep, short-term overheating, thermal cracking (indicated by API 571 [2] as a form of "thermal shock") and High Temperature Hydrogen Attack (HTHA).

All of these damage mechanisms are dependent on materials of construction and operating conditions. HTHA is not a damage mechanism associated with overheating but very relevant to methanators.

Methanators are mostly constructed of low alloy steels, mainly 1.25Cr - 0.5Mo and 2.25Cr - 1Mo. There are also some older plants using carbon steel and C – 0.5Mo steel.

These materials all have specified limits in terms of temperature, creep life and susceptibility to HTHA and have been in consistent operation for many years based on the temperature and pressure conditions of the respective plants, when threshold limits are exceeded the material can be damaged.

Creep

Creep is the slow and continuous deformation of metal components at high temperature and under constant load below the yield stress of the material. It is a continuous process, involving development of voids at grain boundaries, which coalesce to form micro-cracks, which in turn interconnect to form macro-cracks. The rate of creep is a function of the material, load, and temperature. An increase of about 18 °F to 25 °F or an increase of about 15% on stress can cut the creep rupture life in half or more, depending on the alloy.

As a general rule, creep becomes significant if the operating metal temperature exceeds one third of the alloy melting temperature.

API 571 [2] states specific threshold temperature limits for creep (Table 1), which, when exceeded, can result in creep damage.

Material	Threshold temperature
Carbon Steel	700 °F (370 °C)
C - 1/2Mo	750 °F (400 °C)
1 1/4Cr - 1/2Mo	800 °F (425 °C)
2 1/4Cr - 1Mo	800 °F (425 °C)

Table 1. API 571 [2] creep temperature limits

In most cases the vessel welds will typically experience more creep damage than the parent material because of higher stress concentrations and eventual pre-existing construction flaws in the welded joints.

Methanator materials, especially 1Cr - 0.5Mo and 1.25Cr - 0.5Mo steels, suffer from low creep ductility. Therefore, quite often, there is very little deformation on these specific materials when they exceed threshold temperature limit. If visible deformation or cracking is found during inspection, it means that creep has already passed the initial stages and can rapidly progress.

The early stages of creep can only be detected by metallography, however a typical inspection plan for creep (including the other expected damage mechanisms previously mentioned: short-term overheating, thermal cracking, HTHA) would include:

- Internal visual inspection
- Geometrical measurements (diameter, thickness, local deformations, out of roundness)
- Metallographic replica
- Hardness testing
- Destructive sampling (if practical)
- Wet Fluorescent Magnetic Particle Testing (WFMT) on the welds and Heat Affected Zones (HAZ)
- Advanced Ultrasound Techniques (UT) such as Time Of Flight Diffraction (TOFD) and Phased Array (PAUT) can be applied when catalyst is not being replaced during a plant turnaround and in cases where excursions would have occurred but not to the extent warranting a full internal inspection

Short-term overheating

Short term overheating is the deformation occurring at relatively low stress levels because of localized overheating above the metal design temperature. This usually results in bulging and eventually failure by stress rupture.

Quite often in methanator overheating incidents, temperatures would exceed the design limits at low pressures which can result in short-term overheating.

If there is damage from short-term overheating, there would be localized deformation or bulging on the order of 3% to 10% or more, depending on the alloy, temperature and stress level.
Inspection for short-term overheating is primarily visual inspection, hardness testing and dimensional measurements (out of roundness of the reactor) focusing on the thinner parts of the reactor.

In order to have an indication about the severity of the overheating event, a preliminary assessment can also be done using API 579 Part 11 [5]. This part of API 579 [5] provides FFS procedures for pressurized components that are potentially damaged by exposure to a fire. However, these procedures are just as applicable for process upsets due to a chemical reaction within process vessels, which can cause, in case of methanators:

- Mechanical distortion and structural damage
- Degradation of mechanical properties,
- Degradation of metallurgical microstructure,
- Presence of crack-like flaws in the pressure boundary

Another qualitative indication about the severity of the overheating event can be obtained using the procedure outlined in the first edition of API 581 Appendix J [9]. With this procedure, the short-term overheating probability of failure is estimated considering the severity of the temperature upset (the difference between the metal temperature during the over-heating event and the design metal temperature) and the duration of the event. Result may be expressed in the form of graph (Figure 3), which gives useful indication about the probability that the event has caused some damage on the equipment.

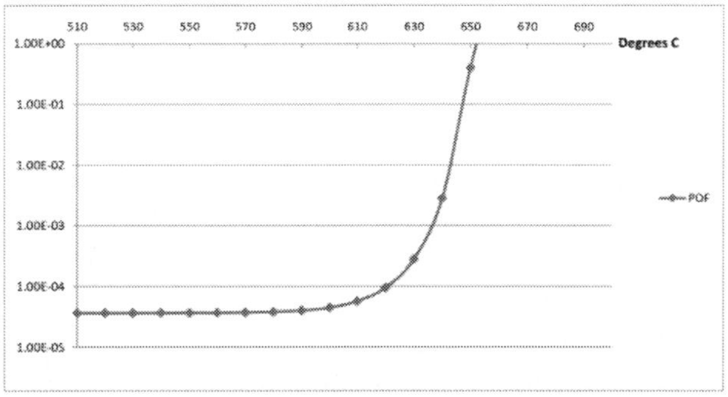

Figure 3. Short-term overheating POF [9] as function of metal temperature, design temperature of the vessel = 510 °C

Thermal cracking

A form of thermal cracking can occur when high thermal stresses develop over a relatively short time in a piece of equipment due to thermal expansion (or contraction). If the thermal expansion (or contraction) is restrained, stresses even above the yield strength of the material can result [2].

Materials that have lost ductility, such as Cr-Mo steels equipment in hydrogen service (temper and hydrogen embrittlement) can be particularly susceptible to thermal cracking depending on the severity of the temperature upset.

Therefore, it is always recommended, after a severe overheating event in a methanator (typically an event for which the metal temperature exceeded the design metal temperature by more than 200 °F) to execute an inspection for cracking on the vessel welds: externally by UT TOFD, or internally by WFMP.

High Temperature Hydrogen Attack (HTHA)

HTHA results from the exposure to hydrogen at elevated temperatures and pressures. The atomic hydrogen dissolves into the steel and reacts with carbides in the steel to form methane (CH_4) which cannot diffuse through the steel: methane pressure builds up, forming bubbles or cavities, micro fissures and fissures that may combine to form cracks.

For a specific material, HTHA is dependent on temperature, hydrogen partial pressure, time and stress in relation to the cumulative service exposure time.

As per API 571 [2] materials most affected by HTHA in terms of lowest resistance to higher resistance are carbon steel, C - 0.5Mo, 1Cr - 0.5Mo, 1.25Cr - 0.5Mo, 2.25Cr - 1Mo, 3Cr - 1Mo, 5Cr - 0.5Mo.
For methanators, based on hydrogen partial pressure, operating history, materials of construction and post weld heat treatment applied during construction, the levels of susceptibility to HTHA are typically medium or high. The last revision of API 581 (Figure 4, [3]) assigns high susceptibility to carbon steels and C - 0.5Mo steels equipment operating above 350 °F (177 °C) and a hydrogen partial pressure exceeding 50 psia (0.345 MPa).

Yara has reviewed all its equipment in hydrogen service, including methanators, since the last revisions of API 581 [3] and API 941 [4] and redefined the levels of susceptibility to HTHA included in the risk assessment and a management plan for these items.

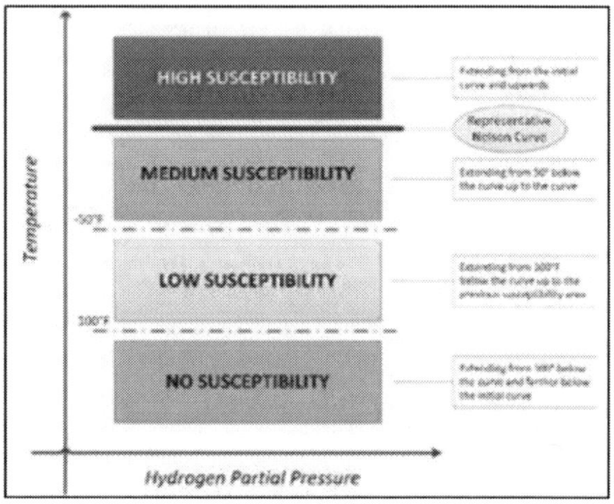

Figure 4. Example of HTHA Susceptibility for Cr-Mo Low Alloy Steels, API 581 [3]

The likelihood of HTHA occurring has proven to be difficult to predict and also the detection of HTHA by inspection is itself problematic. Yara has adopted a standard approach of using the ultrasonic method Time of Flight Diffraction (ToFD) for HTHA inspection in combination with Phased Array UT (PAUT) and other conventional methods such as manual shear wave UT, visual inspection and metallography.

Initial FFS assessment, inspection, repair and remaining life after an overheating event

In determining if and to what extent an inspection is required after an overheating event, it is important to have all the information available in a very short time. This will ensure that the right recommendations and decisions will be made in assessing the integrity of the reactor to continue the operation.

The following information should be recovered:

- Temperature and pressure conditions during the event (review of DCS trends and Integrity Operating Windows, IOW, alarms)
- Duration of the event down to the last minute
- Any previous excursion events from its operating history
- Previous inspection and repair history
- Already active damage mechanisms
- Eventual local thinned areas
- Materials of construction
- Design and operating process conditions
- Type of catalyst
- Design of thermocouples and length embedded into catalyst as well as information on any thermocouples directly attached to reactor shell
- Any piping or equipment after the reactor that would have also experienced overheating

Initial FFS assessment

The initial assessment, immediately after the overheating event, may be executed as below:

1. Define the extent of exposure in terms of how much the temperature exceeded its design limits, duration of the event and pressure during the event
2. A Level 1 FFS assessment as per API 579 Part 10 [5] can be done using the screening curves relevant for the material and based on the overheating conditions (temperature, pressure, duration of the event). When it is known that there already is some previously accumulated creep damage, this should be added to any potential damage caused by the overheating incident (Figure 5)
3. Based on the results of this initial Level 1 FFS assessment, some recommendations, in relation to inspection, may be given. In particular, if the creep damage is still tolerable (25% or less), an immediate internal inspection of the vessel will not strictly be required and an external advanced UT inspection (looking for eventual thermal cracks) of the welds will be executed
4. If the Level 1 criterion above defined is not satisfied (creep damage exceeds 25%), then a Level 2 FFS assessment may eventually be executed as per API 579 Part 10 [5]. When the creep damage exceeds, or is close to, the maximum allowable creep damage (80%) immediate external and internal inspections should be executed

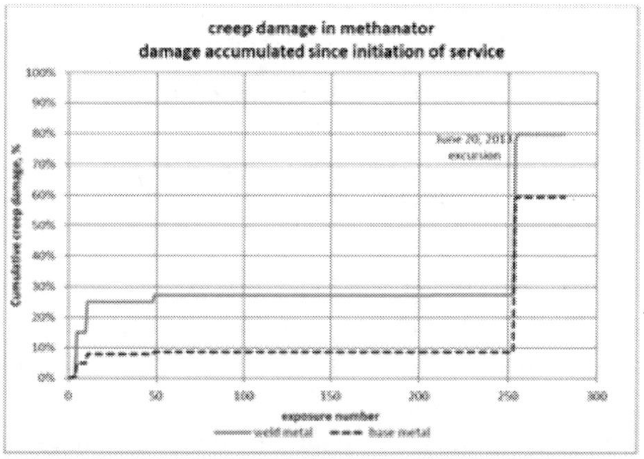

Figure 5. One of the methanators in Yara showing the extent of increase in the creep damage that occurred when a single overheating incident occurred

External Inspection

When an external inspection is executed to assess the integrity of the reactor after an overheating incident, this involves the inspection of the vessel welds (other than visual inspection and eventual geometrical measurements of the vessel).

The recommended technique for this weld inspection is ToFD (Figure 6) supported with conventional UT Shear Wave (UTSW).

The preparation for this technique requires cooling the reactor down to ambient conditions, removal of insulation and cleaning of HAZ areas on either side of welds. It should here be noted that ToFD can also be done at high temperatures (but a specific procedure should be qualified for this).
Also to be noted that in any decision to do ToFD, or any form of advanced UT externally, one must take into consideration what can be detected by external advanced UT (minimum detectable size versus the maximum allowable defect) and the skills of the NDT technician.

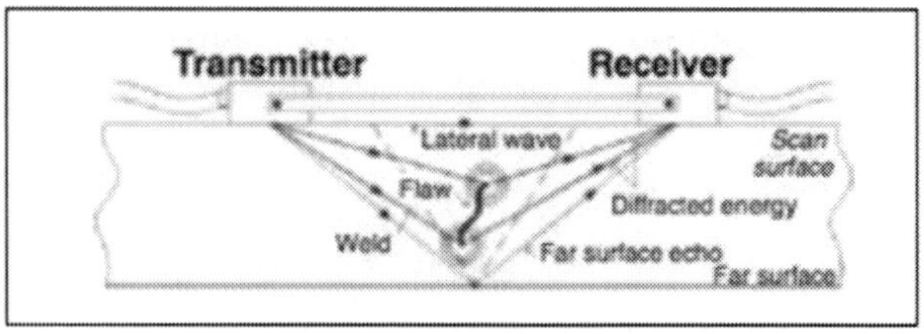

Figure 6. ToFD configuration

There are many benefits in choosing to perform ToFD inspection on the reactor welds:

- It offers greater accuracy for measuring crack-like flaws. An accuracy greater than ±1 mm can be obtained in a wide range of material thickness
- High speed of detecting discontinuities and high probability of detection
- Accurate positioning and sizing of discontinuities
- Permanent digital records
- Detection is independent of type and orientation of discontinuities
- Compliant to ASME Code Case 2235
- Defined acceptance criteria as per EN 15617

The results from ToFD can subsequently be used in the FFS assessment. For an FFS assessment of a detected flaw from the inspection, information about flaw length and depth is required as well as the orientation in relation to the principle stress.

From the results of the external inspection and depending on what acceptance criteria is used (ASME or EN), if flaws are detected and characterized and they are within the acceptance criteria, then a decision can be made to return reactor to service and define a time period for another inspection using the same methods to monitor for any flaw growth.

If the detected flaws exceed the ASME or EN acceptance criteria, assessment should be done as per API 579 Part 9 [5] in order to determine the maximum allowable flaws.

At this point of the external inspection, when the flaws exceed the defined acceptance criteria, a detailed internal inspection is mandatory.

Internal Inspection

The internal inspection will be focused on damage mechanisms associated with an overheating event (creep, short-term overheating, thermal cracking, HTHA) and on the verification, on the root surface of the welds, of any indications eventually detected with the external ToFD inspection.

The inspection plan to inspect for these damage mechanisms includes:

- Detailed visual internal
- Dimensional verification
- WFMT (or Dye Penetrant Testing)
- Field Metallography
- Hardness measurements
- UTSW

A specific Inspection and Test Plan (ITP) should be developed and used to track all the inspection being done and any subsequent repairs.
In carrying out the internal inspection, all indications should be documented including any original fabrication defects.

At this time acceptance criteria should have also been defined in terms of permissible size and type of flaws, out of tolerance dimensions, and hardness tolerances.

Once all flaws have been identified and mapped (Figure 7), a repair plan can be developed for those that have exceeded the acceptance criteria.

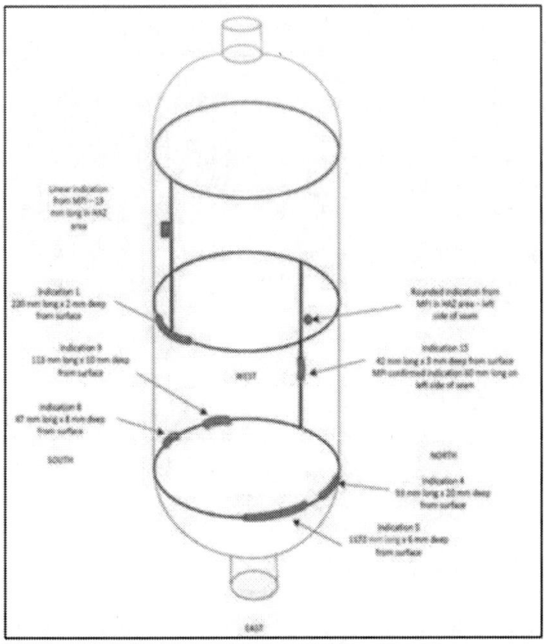

Figure 7. Example of one of the Yara Methanators showing the extent of the indications from inspection

Repairs

All the required repairs should be executed according to the pressure vessel code of construction and any applicable required authority regulations (e.g. ASME, National Board, etc.). In addition, the repair organization should also be competent and certified when required.

Removal of crack-like flaws

The repair plan should clearly indicate the steps required for the removal of detected flaws and the inspection hold points for validation. It should also indicate the allowable depth of material that can be removed during excavation of flaws (e.g. by grinding) without the need for re-welding. For example, in one of the Yara Methanator overheating incidents, the excavated flaws were initially assessed as local thinned areas per API 579 Part 5 [5] which would have resulted in derating the reactor: repairing by welding these local thinned areas resulted in maintaining the original design intent.

When practical, excavation of a "boat sample" of a detected indication would be useful for a more in-depth material failure analysis in order to understand which damage mechanism caused the flaw.

Weld repairs

Repair by welding of removed flaws requires qualified and approved procedures as well as a qualified repair organization.

Six key elements that can ensure a successful repair are:

1. Dehydrogenization (heat soak) to remove atomic hydrogen trapped in the material prior to any welding
2. Controlled preheat and welding using calculated soak bands for thermal gradients
3. Inspection after welding
4. Post Weld Heat Treatment (PWHT) using calculated soak bands for thermal gradients with soak temperatures based on material and thickness off reactor (Figure 8)
5. Final inspection after PWHT
6. Hydro-test at temperature above the ductile to brittle transition temperature (to be noted that not in all cases a hydro-test is required. However, it can also be mandated by local Authority requirements)

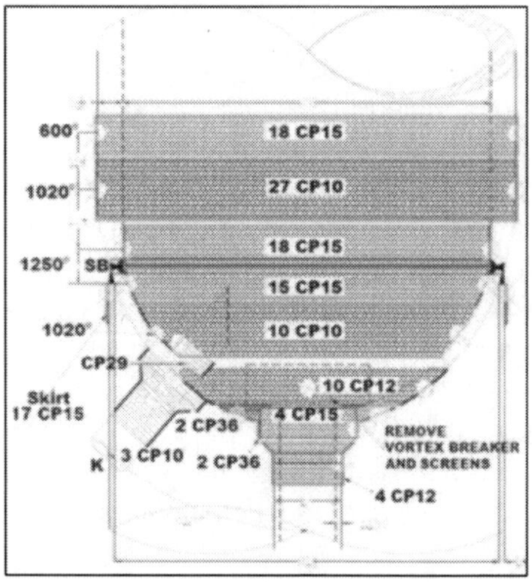

Figure 8. Example of thermal gradient band for heating elements used in PWHT

Remaining Life Assessment

Based on the operating history of the reactor, extent of overheating incident and repairs carried out, a remaining life assessment can be done for future operation. This may be carried out considering all of these below:

- Applying the same principles of Fitness for Service used in assessing the initial overheating incident
- Stress analysis of the locally thinned areas that were not repaired by welding
- Evaluation of the inspection results from the field metallography, hardness measurements and dimensional data
- Material failure analysis of any excavated samples from the detected flaws
- Further evaluation of reactor shell and welds for creep damage combining all cycles and operating history
- Future risks of HTHA based on overheating incident and past operating history

By combining the results of these assessments, the remaining life of the reactor can be estimated and recommendations for its future operation may be given such as the review of the existing protections and further limitations on operating conditions.

For example, in the case of one of the Yara Methanator overheating incidents, the FFS assessment indicated that the creep damage reached the maximum allowed (approximately 80%, Figure 9): the future operation was therefore limited to re-defined maximum working pressure and temperature and a recommendation to replace the reactor at the next planned plant turnaround was given.

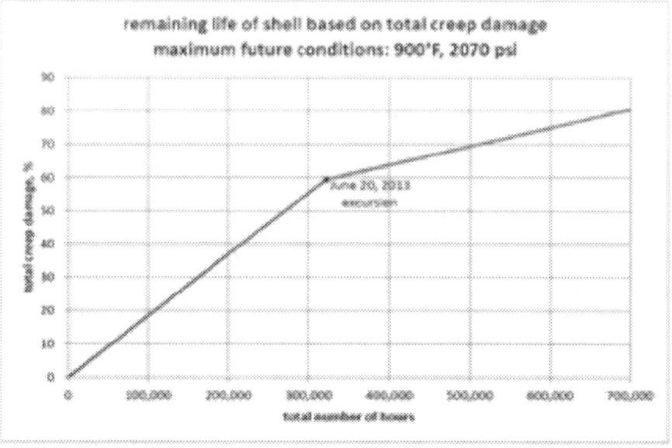

Figure 9. Example of remaining life assessment based on creep damage for one of the Yara's Methanators

What went wrong?

The four described incidents showed how overheating can occur very quickly when key operational, control and safety measures fail.

The causes of each described incident were different; however, the human factor was predominant in all of the four, in particular in respect with the override of the safety trip logics and the management of change of the ESD logics.

Recommendations

- Ensure that the CO_2 removal and methanator trip systems are properly working
- Verify that the trip thermocouples are correctly located
- Verify that proper maintenance and inspection routines are correctly executed on trip

systems, trip valves, solenoids, relays, thermocouples
- Verify that the loop testing of trip circuits are properly carried out during turnarounds
- The override of the safety functions shall not be allowed unless a complete risk assessment is done, and all the required mitigating actions are in place
- In almost all the cases, we could question if the metal wall temperature of the vessel was really raised at the same temperature of the bulk catalyst. The only way to answer to this interrogation would be to install metal temperature measurements on the vessel which will record the real metal temperature of it. This cannot be considered as an active prevention for the methanator overheating event, but it could eventually reduce the economic impact of it, immediately indicating if the integrity of the vessel was threatened or not by the event itself
- In case of overheating methanator vessel event, do not restart the plant without having executed a proper integrity assessment of the vessel

Conclusion

The recommendations were subsequently shared with all plants with similar design and operation. We have all the technology and knowledge to prevent overheating events in methanator vessels: we know what to do and how to do it.

In most cases, the proper layers of protection have been put in place, requiring significant but necessary capital expenditures.

We also have proper techniques and procedures to assess if the equipment may continue to operate after an overheating event or if it needs to be repaired or replaced.
However, mistakes, both in design and operation, still lead to bad experiences. There is unfortunately a certain normalization or acceptance about these events, showing a lack of risk awareness. Statistically, these events represent potential major incidents.

We still have to work hard to increase the awareness about this kind of risk in order to avoid that the methanator overheating event will remain a "hot topic" in our ammonia plants.

Reference

1. Safety Improvements in a methanator vessel, M.Walton, P. Sharp,T. Southerton (AIChE Ammonia Technical manual Vol x, 2007)
2. API RP 571 SECOND EDITION, APRIL 2011
3. API RP 581 THIRD EDITION, APRIL 2016
4. API RP 941 EIGHTH EDITION, FEBRUARY 2016
5. API 579-1/ASME FFS-1, June 2016
6. Risk Reduction Projects Combating HTHA, John Brightling, Stephen Shapcott (AIChE Ammonia Technical manual Vol 59, 2018)
7. Fitness-for-Service Case Studies in HTHA Equipment, Michael Nugent, Trace P. Silfies, Paul J. Kowalski, Phillip E. Prueter (AIChE Ammonia Technical manual Vol 59, 2018)
8. Operating manual METH 130 Clariant
9. API PUBLICATION 581 FIRST EDITION, MAY 2000

Cracked Dissimilar Welds in Outlet Headers of a Primary Reformer After 34 Years of Operation

This paper describes cracking of the 14 dissimilar field welds in the outlet manifolds of a 1983 Kellogg design primary reformer of OCI Nitrogen's AFA3 Ammonia plant in Geleen (NL). First, the primary reformer design and operation is described. Thereafter, the sequence of events, root cause analysis, learnings and actions taken to avoid this kind of failure in this type of primary reformer in ammonia plants will be highlighted.

M. Grooten, R. Dubois, J. Stoffels, S. Schepers
Sitech Services

F. Raeymaekers
OCI Nitrogen

M. Keltjens
Sintra Engineers

R. Gommans
Materials & Corrosion Know-How

Introduction

Located at Geleen in The Netherlands, OCI Nitrogen operates two ammonia plants, each with a nameplate capacity of 1360 MTPD of ammonia. One of these plants, Ammonia 3 (AFA3), was designed and constructed by M.W. Kellogg based on Kellogg's reduced energy ammonia technology for steam reforming of natural gas. The plant was commissioned in 1984. The current capacity is 1550 MTPD.

Description of the reformer furnace and radiant harps

The primary reformer furnace is a Kellogg design with 364 radiant tubes, seven radiant tube rows, with 52 tubes per row. Figure 1 shows the reformer furnace layout. There are eight arch burner rows with 22 conventional burners per row, a total of 176 burners. In addition, there are eight conventional tunnel burners and 15 low-NO_x steam superheat burners. A mixture of gaseous ammonia and steam is injected at the inlet of the convection section to reduce the NO_x concentration in flue gas from the radiant section (SNCR).

The convection section consists of two coils in the "hot leg", the air-steam coil and the mixed feed preheat coil. In the "cold leg", there are four coils, the steam superheat coil, the boiler feed water preheat coil, the feedgas coil and the fuel gas preheat coil. A cast-iron combustion air preheater further cools down the flue gasses before they are vented to the atmosphere.

Figure 2 shows the radiant harp layout. Each row of catalyst tubes has one inlet manifold, which feeds each tube with the preheated mixed feed through a flexible pigtail, providing for thermal expansion. Each row of tubes has an externally insulated outlet manifold at the bottom to which the catalyst tubes are connected by means of weldolets. The reactants pass through holes drilled in the outlet manifold and flow upwards through a riser, which is welded to the middle of the outlet manifold with a weldolet. The radiant harps are each assembled from three prefabricated sections, connected with dissimilar field welds in the outlet manifold. The riser from each row is connected to a common overhead transfer line. The transfer line is outside of the firebox and is internally insulated which permits the use of a carbon steel pressure shell. For additional safety, a level-controlled water jacket surrounds the entire transfer line. The reformer design incorporates a carefully designed spring support system for the radiant harps, inlet manifolds and transfer line, so that controlled movement of the whole assembly can take place during all phases of operation without any overstressing of components due to thermal expansion. Table 1 shows the materials and sizing of the radiant harps.

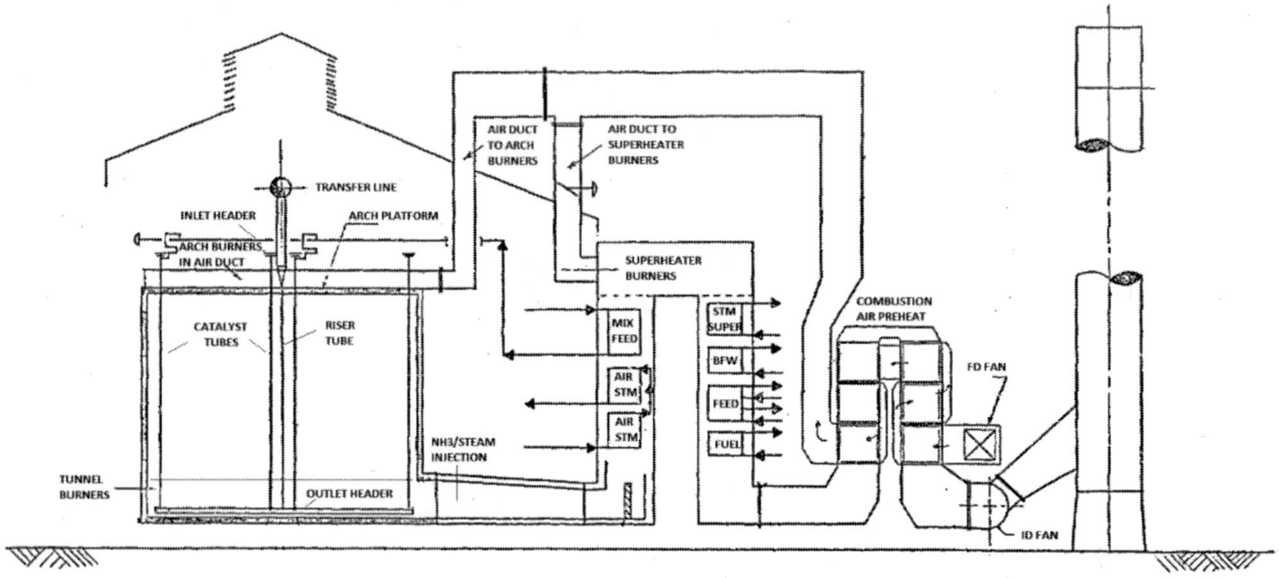

Figure 1. Reformer furnace layout

Item	Material	Size (ID x MSW mm)
Inlet manifold and pigtail	A213 gr TP304H	
Catalyst tube	35Ni-25Cr-Nb (0.4%C), centrifugally cast	88.9 x 12.2
Outlet manifold	32Ni-20Cr-Nb (0.1%C), centrifugally cast	152.4 x 18.4
Riser tube	35Ni-25Cr-15W-5Co (0.5%C), centrifugally cast	124.5 x 18.0
Weldolet, riser extension and transition piece	Alloy 800H	

Table 1. Materials and sizing of the radiant harps

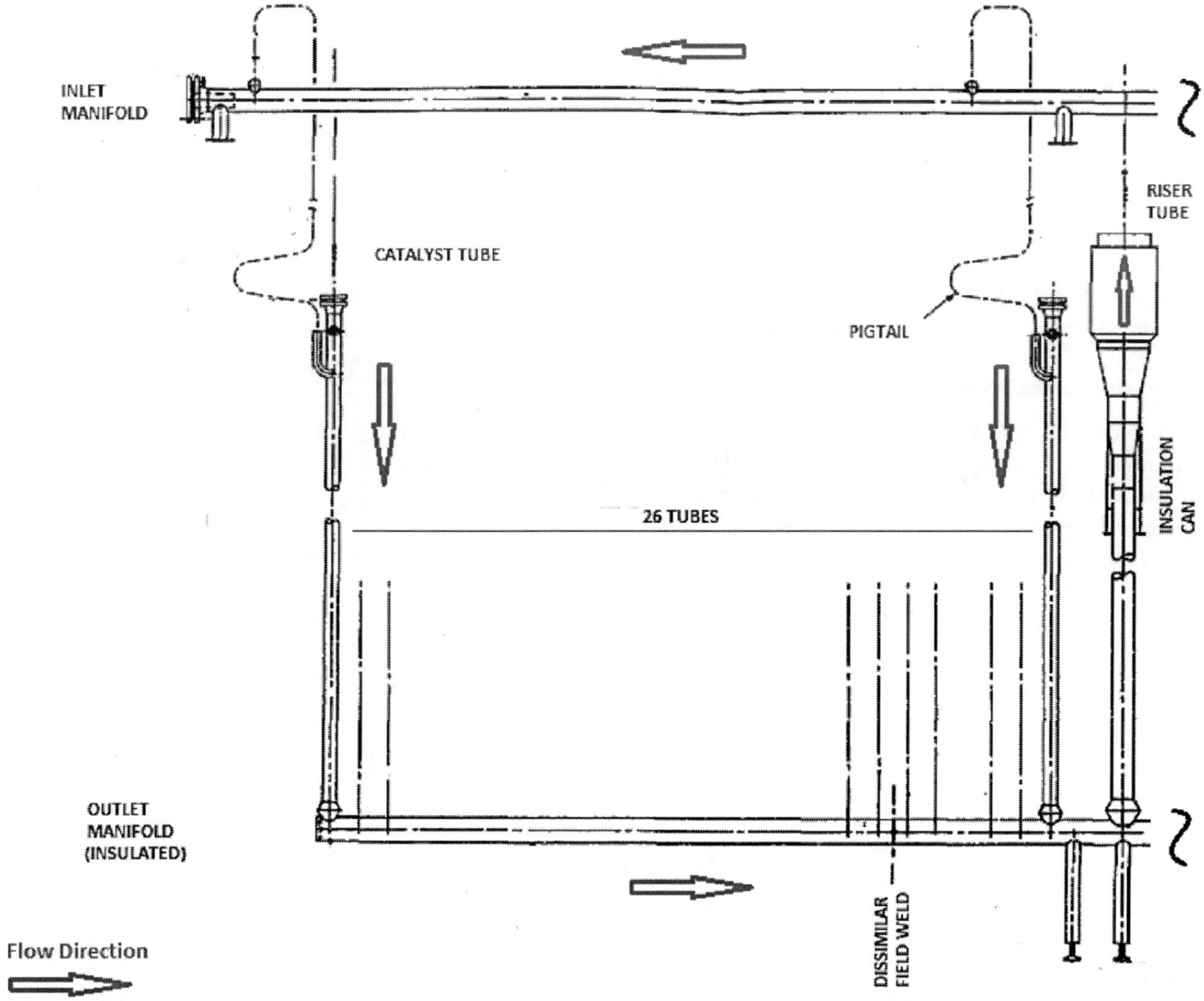

Figure 2. Radiant harp layout, schematic of a left half row, including riser

Operation of the reformer furnace and radiant harps

The reformer furnace and radiant harps have operated without major problems over a period of 34 years. Table 2 shows the average process- and design conditions concerning the radiant harps. It follows that the furnace has been operated relatively mild compared to design. The tunnel burners are not in operation. This would increase the air-steam- and mix feed temperature, but it would also add additional mechanical stresses on the coils. Tube wall temperatures in the radiant section are monitored regularly with a pyrometer and are kept below 900°C / 1652°F and 914°C / 1677°F for the catalyst tubes and riser tubes respectively. Apart from incidental performance issues of catalyst, burners or riser insulation cans, the temperatures limits were respected.

Item	Process Condition	Design Condition
Inlet manifold and pigtail	38 bara / 551 psia 530°C / 986°F	44 bara / 638 psia 635°C / 1175°F
Catalyst tube	38 bara / 551 psia 780°C / 1436°F	40.8 bara / 592 psia 895°C / 1643°F
Outlet manifold	36.5 bara / 529 psia 780°C / 1436°F	40.0 bara / 580 psia 801°C / 1474°F
Riser tube	36.5 bara / 529 psia 800°C / 1472 °F	40.0 bara / 580 psia 942°C / 1728°F
Hot air-steam coil	750°C / 1382°F	830°C / 1526°F
Mixed feed coil	530°C / 986°F	635°C / 1175°F

Table 2. Process and design conditions in reformer furnace.

OCI Nitrogen has a shutdown interval of four years for its ammonia plants. Additional to a thorough visual inspection, several nondestructive testing (NDT) methods are performed on the radiant harps during the shutdown. These reflect the main damage mechanisms and corresponding inspection strategies for reformer tubes and outlet components. A detailed description is given in *Reference 1.*

- Diameter measurement of outlet manifold
- Replica investigation of outlet manifold microstructure
- Automated OD tube diameter measurement and eddy current testing of catalyst and riser tubes
- Automated ID Laser Profilometry measurement of catalyst tube as part of catalyst exchange procedure
- Dye penetrant testing of critical welds.

Based on the inspections and fitness for service assessments, in the past 34 years only three catalyst tubes had to be replaced. One in 2012 and two in 2016. Furthermore, in 2012, cracks were found in the welds between the weldolets and riser tubes and between the weldolets and catalyst tubes adjacent to the risers. The cracks originate from low cycle fatigue; a detailed description is given in *Reference 2*. Several repairs had to be performed. Nevertheless, reliable and safe operation of the furnace harps was expected until the next maintenance shutdown in 2020.

Failure of welds in outlet manifolds: Sequence of Events

On Friday October 20th, 2017, the ammonia plant was shut down for a short-planned maintenance stop. During startup on Thursday October 26th, 15 knm^3/h process gas was introduced at 22h at a primary reformer exit temperature of 737°C. According to Standard Operating Procedures, the primary reformer is monitored during startup by the operator via the peep holes. At two positions on the outlet manifolds, flames were observed. The insulation of the outlet manifolds (headers) between the catalyst tubes 32 and 33 was observed to be burned and damaged on top at row 4 and row 5. See the photograph in Figure 3. At 23.30h, process gas was shut off due to a leakage of process gas in the primary reformer. The plant was further stopped and cooled immediately.

The cracks were located in the outlet manifolds in the middle harp piece, at the side of the riser, at the location of the dissimilar field weld. See for example the photograph in Figure 4. On the basis of the observed damage, the spring hangers of row 4 and 5 were inspected. No deviations were found, no blocked hangers and all spring indicators were correct. It is estimated that the failure, after 34 years of operation, occurred after about 170 start/stop-cycles of the furnace.

Figure 3. Damaged external insulation.

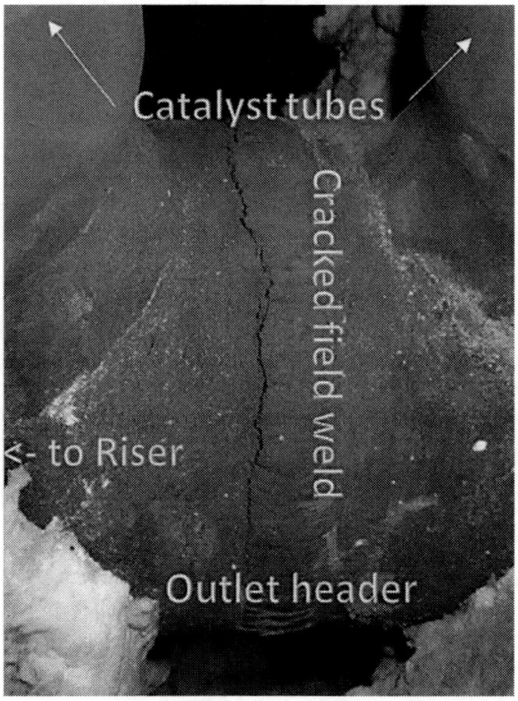

Figure 4. Cracked field weld no. 10 row 5.

Material research

Further inspections revealed that leakage occurred at two field welds in the outlet manifold in row 4 and 5. A third field weld in row 2 between catalyst tube 32 and 33 showed visually observable cracks of 5 cm length at clock position 12h. Dye penetrant tests (DPT) showed that all comparable field welds of the outlet manifold, 14 in total, have unacceptable linear indications all in the range of clock position 9h to 3h. To verify the internal structure and integrity, a radiographic examination (RT) was performed at two field welds 9 and 10, both row 5. An overview of the field welds is shown in Figure 5. The selection was based on one weld being cracked / with leak, another weld being cracked / no leak. Both radiographic films showed internal longitudinal cracks. Considering the results of the NDT examinations it was concluded that all 14 field welds needed to be replaced. All these 14 field welds originate from the initial plant construction in 1983. Repair led to a shutdown period of multiple weeks.

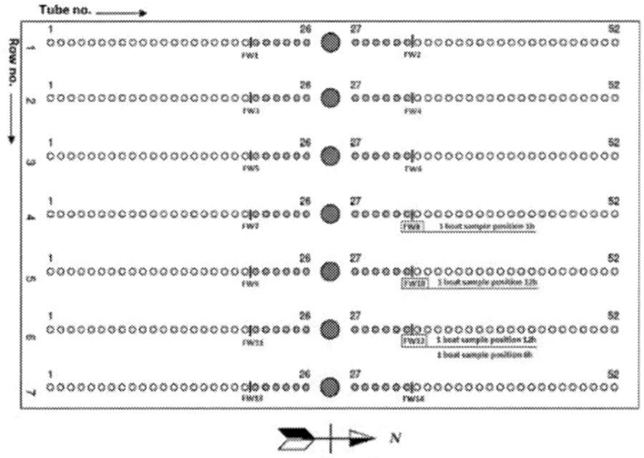

Figure 5. Top view of radiant harps. Field welds (blue) and boat sample positions (red).

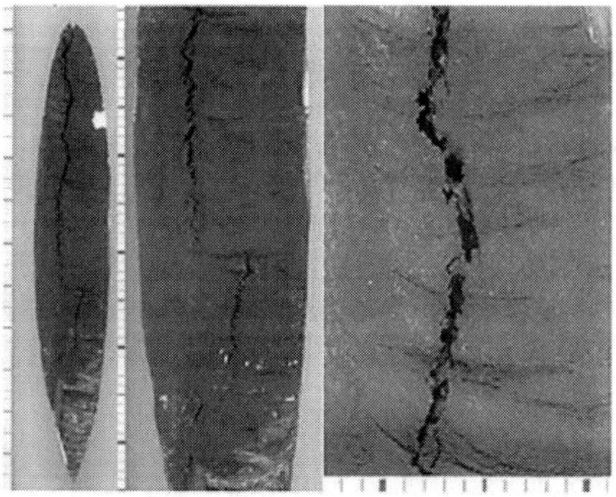

Figure 6. Boat samples.

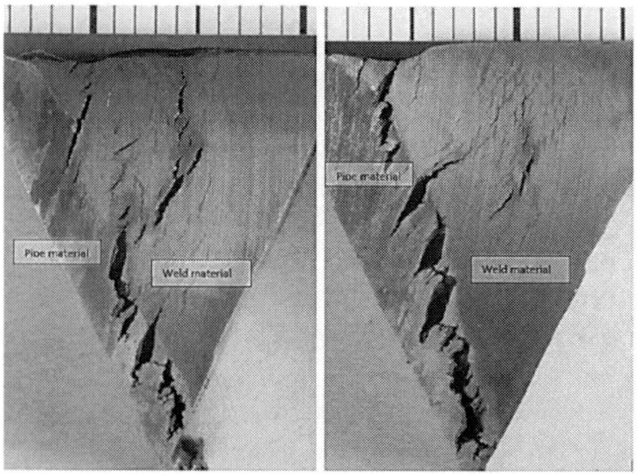

Figure 7. Photography of Field Weld #8.

Before cutting the field welds, four boat samples were taken out at different locations:

- Row 4 field weld 8, clock position 1h,
- Row 5 field weld 10, clock position 12h,
- Row 6 field weld 12, clock position 12h,
- Row 6 field weld 12, clock position 6h.

Typical boat samples are shown in Figure 6. On these boat samples a failure investigation has been performed together with a third party, Element Materials Technology. The conclusions of this investigation were:

- The failure investigation of the circular welds in the outlet manifold shows that creep damage has been developed in the heat affected zone, ranging from class 2B to 3B (primary failure mechanism). This creep damage developed primarily due to difference in thermal expansion coefficient between the weld metal (type 625) and the base material (32Ni-20Cr-Nb).
- Within the wall thickness, from clock position 9h – 12h - 3h, wide open cracks (creep damage class 4 and 5) were present. At clock position 6h much less creep damage was present. This indicates a relatively large deformation across the weld as a result of mechanical stresses (secondary failure mechanism). See Figure 7 for an example of the wide-open cracks developed both radially inward and outward.

From the material research, the following lessons have been learned:

- The expected primary failure mechanism is creep, which usually starts from the outside. That's why all historical examinations have always been executed on the outer surface of the radiant harps. The failure investigation showed that in this case the creep damage starts from within the wall thickness. Creep damage would not have been visible with the applied NDT.
- Further study showed that API 579 - Fitness for Service standard edition 2016 - describes that the initial inter-granular voiding or cracking develops within the wall of the component. Failure occurs by crack link-up and propagation to the surfaces. Cracks do not normally initiate either at the inside or outside surfaces.
- Remarkably, API RP-571 - Damage Mechanisms Affecting Fixed Equipment in the Refining Industry - does not warn for this type of dissimilar weld cracking, where Ni-base weld metal is joining Fe-base components.
- The base metal 32Ni-20Cr-Nb of the outlet manifold was in a good condition, even after 34 years at elevated, but normal, reformer operation temperatures! Creep damage such as void formation and diameter increase (swelling) has not been observed.

- At the time of construction of AFA3 the choice of the type 625 filler metal was standard industry practice. However, after 34 years of service the creep damage in the heat affected zone has weakened the dissimilar field welds. The normal design lifetime of radiant harps is much lower. The filler metal applied for repair was type 625 again, for practical reasons.

Finite Element Modeling

In order to explain the reason for both crack initiation on the inside of the manifold wall and crack initiation at clock position 12h, different simulation steps are performed. 3DS ABAQUS 2018 is used for all finite element simulations. The first step is an axisymmetric model of a dissimilar groove weld between two header parts. This step assesses the effect of the dissimilar field weld on the location of failure over the wall thickness at uniform temperature and the effect of temperature difference over the wall during startup and shutdown.

Step by step, the model is increased in complexity (by introducing non-linearity). A 3D model of one half of the outlet header is simulated, with a bending moment applied on one side. Aim of this step is to assess whether a bending moment has an influence on the clock position of failure (clock position 12h or 6h). Next, the 3D model is expanded with one weldolet, see Figure 8. Aim of this step is to assess the effect of the presence of the weldolet and its dissimilar weld to the outlet manifold. The last step includes weldolets at either side of the field weld in a 3D model. Using this final step, the effect of the presence of weldolets on the location of failure is confirmed.

The model features 2 material models: 32Ni-20Cr-Nb for the base material and Alloy 625 for the welds. 32Ni-20Cr-Nb has a higher thermal expansion ($1.87 \cdot 10^{-5}$ K^{-1} versus $1.6 \cdot 10^{-5}$ K^{-1} at 800 °C). Plasticity is included through an elastic-ideal plastic material model.

The models are subjected to the following load cycle. At first, internal pressure is applied. This

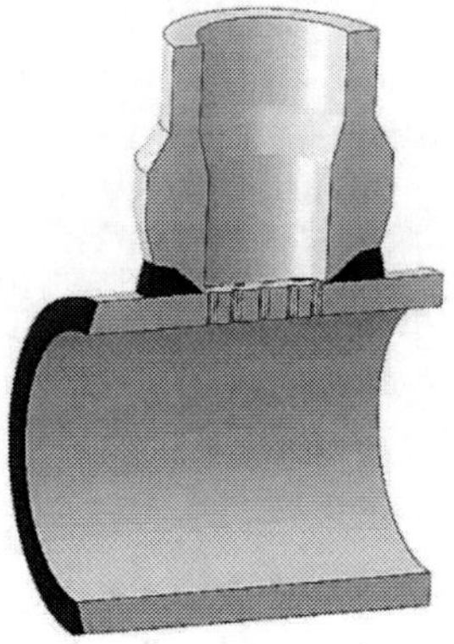

Figure 8. Dissimilar welds in simulation model. Field weld in outlet manifold (left) and catalyst tube weldolet (top)

is followed by a temperature increase to 800°C. After this, a viscoelastic step of 3000 hours is used to simulate creep behavior in the header. After 3000 hours the temperature is lowered to 20°C. This cycle is then repeated. Figure 9 presents the axial stress after startup, creep and shutdown at the dissimilar field weld in the outlet manifold. After startup, a compressive stress above the yield limit is seen at the root of the weld (-190 MPa). This is due to the difference in thermal expansion of the 32Ni-20Cr-Nb and Alloy 625 of the weld. As a result of creep relaxation, the compressive stress is reduced to -40 MPa. This difference will also lead to a deformed shape since the weld will expand less (due to thermal expansion difference), preventing the header from expanding freely.

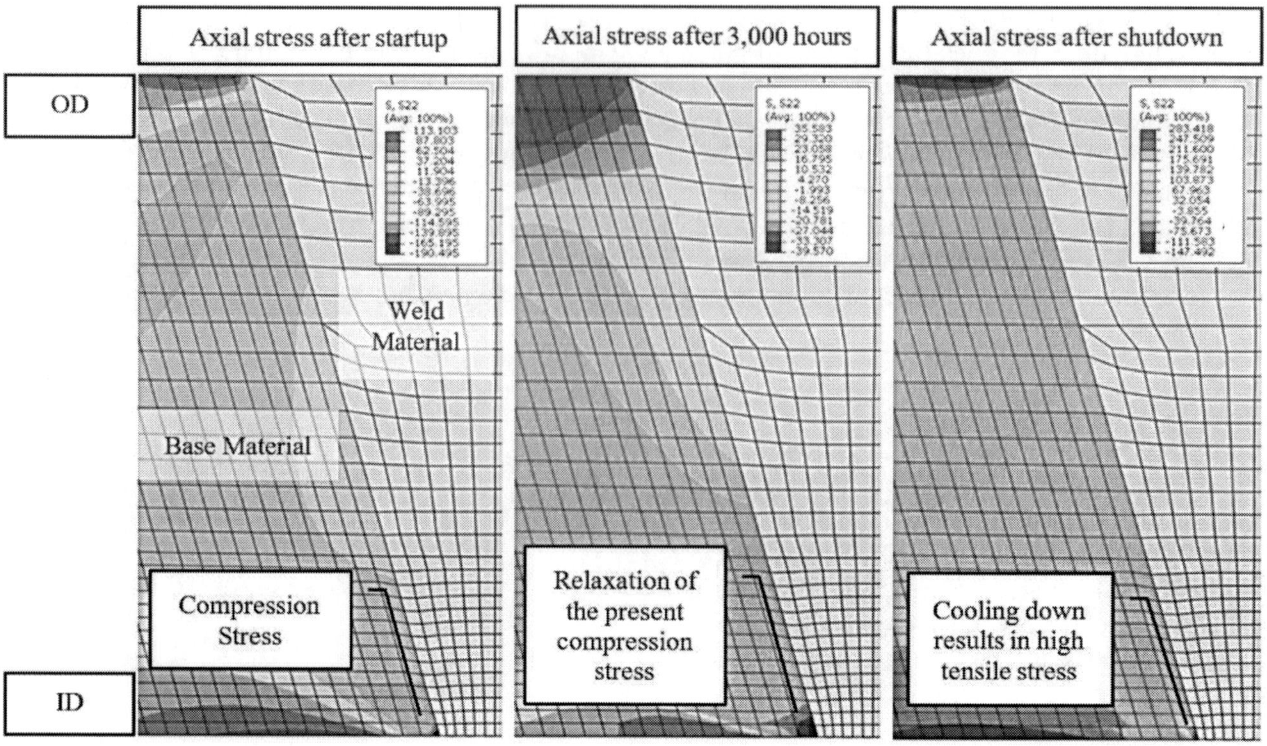

Figure 9. Stress plots of axial stress after start-up, creep and shutdown. Stress in [MPa]

After shutdown the header will revert to its original form. This results in high tensile stresses at the root (inside surface) and toe (outside surface) of the weld.

All stresses are secondary in nature and will shake down (decrease in plastic strain range with each incremental cycle) to some extent. However, the amount of shake down in the root of the weld is less than in the cyclic equivalent strain range on the toe of the weld. Figure 10 reveals that after 2 cycles the location with the highest plastic equivalent strain range switches from outside to inside.

This explains why the location of failure does not initiate at the outer wall.

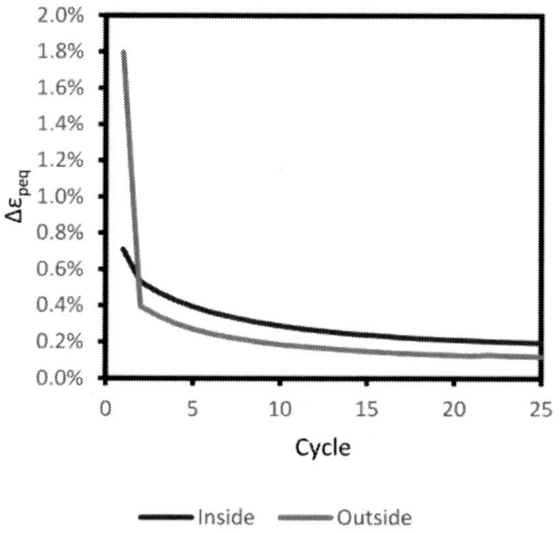

Figure 10. Equivalent plastic strain range at the in and outside of the manifold wall

The final step of the assessment is to include 2 weldolets, one on each side of the outlet manifolds dissimilar field weld.

Like the field weld, the weldolets are welded using Alloy 625. The difference in thermal expansion of the weld material and the header prevents the topside of the header from expanding freely. This constraint results in higher cyclic plasticity at the top side when compared to the bottom. This is seen in Figure 11.

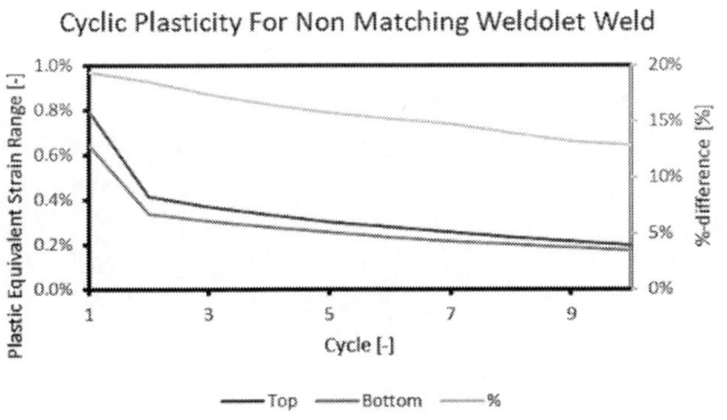

Figure 11. Cyclic plasticity for a non-matching weldolet weld

With an average difference of 15% in plasticity between the top and bottom, the presence of weldolets can be designated as the cause for the failure at clock position 12h. This combined with an external bending moment due to the use of spring hangers and temperature difference over the wall during startup and shutdown is the cause of failure on the inside at the top location.

It is therefore advised to use a matching weld consumable for assembly of new radiant harps, to ensure that thermal expansion of the weld and base material are near-equal. No Finite Element simulations with a matching filler have been performed (yet). In addition, it is advised to minimize stresses on the outlet manifold by use of constant load hangers as opposed to spring hangers in an installation that is operated within the creep range. Difference in thermal expansion of the catalyst tubes and riser pipe causes an additional bending moment, which is better accommodated using constant load hangers.

Conclusions

The 1983 Kellogg design primary reformer furnace has been operated without major problems for 34 years at relatively mild conditions. During a startup in October 2017, cracks in the dissimilar field welds in the outlet manifolds were observed. Repair led to multiple weeks of shutdown. The primary failure mechanism of the dissimilar field welds in the outlet manifolds of the primary reformer is creep in the heat affected zone. Crack growth starts below the material surface and was not detectable with the inspection methods used. This failure mode led to cracks after 34 years of operation and has not been observed at this position in this design in the ammonia industry so far, by our knowledge. The difference in thermal expansion of the base and welding material causes highest strain and stresses near the inner diameter under cyclic loads in the creep range of the material. The nearby dissimilar welds of the weldolets between the catalyst tube ends and the outlet manifolds increase the stresses at clock position 12h at the dissimilar welds in the outlet manifolds. The difference in expansion of the risers to the catalyst tubes with the spring hangers further increase these stresses.

Learnings

The risk of cracking in the outlet manifolds of the primary reformer could be minimized by applying matching welding material, a modified design for the connection between catalyst tubes and outlet headers and less or a better controlled difference in expansion between risers and catalyst tubes. Given the presence of more dissimilar welds at other critical positions in the primary reformer and the uncertain condition of these welds, a reharping project for the AFA3 primary reformer is under development.

References

Reference 1: Gommans, R., Jakobi, D. and Jimenez, J.L., "Inspection Strategies of Reformer Tubes and Outlet Components", Safety in Ammonia Plants & Related Facilities Symposium (2002)

Reference 2: Stoffels, J., "Failures in Aging Ammonia Plants", Safety in Ammonia Plants & Related Facilities Symposium (2016)

Improving Ammonia Plant Reliability and Safety with Island Mode Operation

Yara Pilbara Fertilisers is a 2200MTPD KBR purifier plant that operates in island mode and is powered by 2 X 22MW steam turbine generators. Steam is generated from process heat and two package boilers (150T and 50T). A new Technical Ammonium Nitrate Plant obtains its steam and electricity form the ammonia plant, the additional utility demand, increases the risk of ammonia plant trips and complete power loss (blackout). With the help of the plant licensor, KBR, a study was initiated to identify control system and logic changes needed to increase the reliability of the electrical and steam system to prevent blackout situations during plant upset conditions. The study also identified risks and helped to improve the safety and reliability of the ammonia plant, nitrate plant and the shared utility plant.

Alan Smith, Neel Lalchan
Yara Pilbara

Introduction

The Yara Pilbara ammonia plant located in Western Australia has no external grid power supply. Power for the site is generated from two captive power plants (CPPs) fed by 46 bar (667 psi) steam generated from two package boilers (50 t/h and 150 t/h) and process heat. Island mode operation can expose the site to dangerous blackouts during steam system upsets or any electrical system faults.

Bibliography

CPP- Combined power plants, steam turbine generators with capacity of 22MW each
EDG- Emergency diesel generator, capacity 5MW
aMDEA – Activated methyl diethanolamine used to absorb carbon dioxide from the process gas after reforming.

Steam System

The three levels of steam are available on site are as follows:

High pressure (HP) steam at 120 bar (1740 psi) and 510°c (950°F). This steam is used to power the turbines which drive the syngas compressor (103-MJT) and ammonia refrigeration compressor (105-MJT).

Medium pressure steam at 46 bar (667 psi) and 388 °C (730°F) is fed to the process gas before the primary reformer for the reforming reaction. The remainder is used to power offsite CPPs and the various ammonia unit turbines (boiler feed water, air compressor, semi lean aMDEA, reformer ID/FD fans and boiler fans).

Low Pressure (LP) steam is generated from MP to LP extraction turbines at 3.5bar (50 psi) and 175°C (347°F). it is used for sealing steam to turbines, supplementary steam to the process air compressor turbine and to make desalinated water.

The 50T and 150T package boilers provide steam to the MP header and can ramp up automatically when the demand is required.
A general network schematic is provided in Figure 1 for reference.

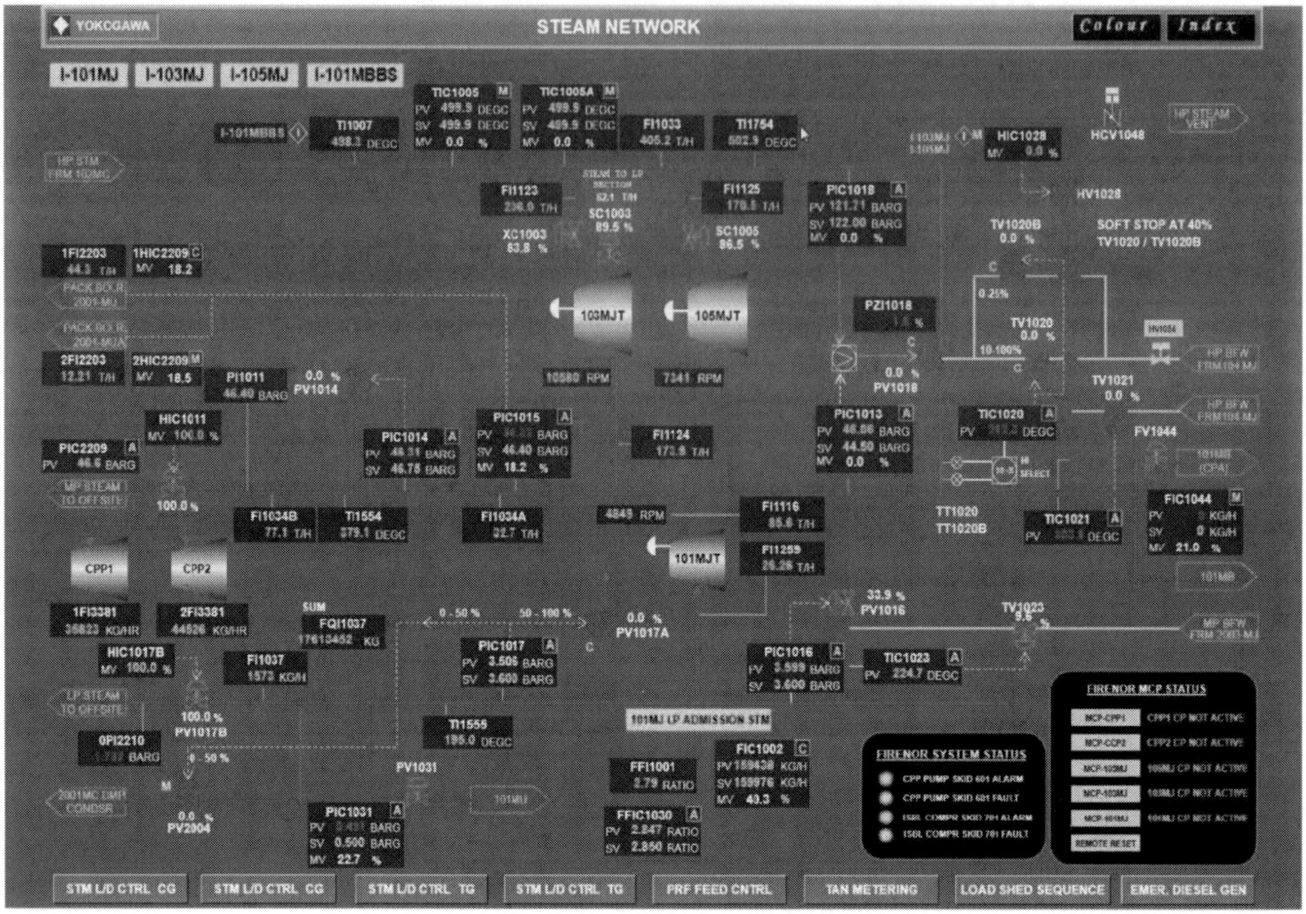

Figure 1. Steam System Overview

Design Intent and Current Operations

The utilities plant was designed to meet the demand of one single ammonia plant. Under normal operation, the package boilers were required to be on minimum turndown (15%). The boilers are kept running as a safeguard to be able to fire up to supply more steam in case of an ammonia plant upset.

In 2017, a new Technical grade Ammonium Nitrate (TAN) plant was commissioned. This new unit obtains its utilities from the ammonia plant including steam, electricity, nitrogen, plant air, desalinated water, demineralized water and nitrogen. With this new user, the operating philosophy changed so the package boilers increased steam production rates from 40 TPH to 80 TPH of MP Steam and the CPP's increased their power generation from 14MW to 18MW. The higher boiler and power demand reduced the original design redundancy (2 X100%) and has increased the risk of blackout from steam upsets.

Over the past four years the plant has experienced five reformer trips on low steam to carbon ratio and three black outs (loss of power from the CPP's). These events all occurred due steam system upsets which could not be managed by the control system and relied heavily on quick operator actions. The steam system dynamic study was revisited to enable a better understanding of the circumstances leading to the front end trips and blackouts during steam upsets, that normally should not lead to these trips.

Dynamic Steam Study

A steam system dynamic study involves developing a non-steady state simulation model of the steam system to predict the behavior of various parameters and control valve actions for a range of process upset scenarios.

The objectives of the study are summarized in the following list:

- Steam system behaviors under abnormal situations
- Steam system behavior with the TAN under steady operation
- Steam system behavior with start-up scenarios for both Ammonia plant
- Steam system behavior with start-up scenarios for the TAN plant
- Recommendations for better management of the steam system if issues are identified.
- Evaluate the start-up and emergency procedure capabilities of the plant by simulating the transient condition steam scenario for the select cases as defined. The result will include the steam balance for the final condition (in the form of Excel spreadsheets) and the rate curves for the critical parameters during the transient conditions.
- Review the tuning constants for the various controllers used in the steam system. The steam system control parameters can then be adjusted to make the system more stable during these upsets.

Scenarios Identified

The following steam upset scenarios were considered for the study:

Case 1: Loss of secondary reformer
Case 2: Loss of primary reformer
Case 3: Loss of 105-MJT
Case 4: Loss of 103-MJT
Case 5: Loss of 50T boiler
Case 6: Loss of 150T boiler
Case 7: Various TAN plant operating conditions
Case 8: Startup of 101-MJT
Case 9: Startup of 105-MJT
Case 10: Startup of 103-MJT

Study Basis

A dynamic model was created using proprietary software by KBR. The model included the installed equipment details for all key equipment in the steam system. Specific design data included the current PFD's and P&ID's, heat and material balance, steam balance, system volumes, exchanger duties, turbine efficiencies, control valve curves and datasheets, and controller tuning parameters.

The base ammonia model was first developed based on a rate of 2450 MTPD (2744 STPD). This basis was chosen as the highest rate where production could be stable and is where the plant is expected to operate in the future. An actual steam balance was provided by Yara for comparing the actual values to the simulated values. In review the base case model matched closely with the actual data during plant upsets - indicating the base case model was accurate and representative of the plant dynamics.

After the dynamic model was completed, preliminary cases were prepared. These cases were discussed in detail over a three-week site visit where subject matter experts from the plant were involved. Preliminary results and findings were discussed and agreed by the end of the site visit. These findings were then published in a draft report for review.

Time Constraints

Yara was in the middle of planning a turnaround (TAR) when this study was initiated. It was imperative that the study was completed within a reasonable time to ensure actions could be implemented in the July 2018 TAR. The study was initiated in April with completion in July, the TAR was planned in August, leaving one to two months to plan and implement the actions.

With a three month deadline, summarized in Table 1, the project team had to work diligently to develop the models, simulate the scenarios, review the results, discuss recommendations with the operations team and then finalize the TAR scope:

Milestone	Schedule
Kick-Off	Thurs Apr 6
Collect Data	Mon Apr 10
Model Update & Development	Fri Apr 21
Preliminary Critical Case testing	Fri Apr 28
Preliminary assessment of tuning and controls	Fri May 5
Critical Cases – at Site	Mon May 8 – Fri May 26
Less Critical Cases	Fri Jun 16
Preliminary Report	Fri Jun 29
Yara Review of Report	Fri Jun 30
Final Report	Fri July 7

Table 1. Steam Study Milestones

Dynamic Simulation Results

Main Findings

The study was completed well within the three-month timeframe, the configuration and reviews took about two months more to get it right before implementation at the end of August 2018.

The scenarios that likely lead to blackout and reformer trips are loss of process air, loss of synthesis gas flow and loss of the 150T package boiler. The process air compressor turbine trip and trips of any other turbines should not cause a blackout or reformer trip

The experience of blackouts on turbine trips were mainly due to the opening of the HP to MP letdown valve not configured to open enough. Process upsets from loss of flow with the turbines running are more likely to lead to a major steam upset and blackout because the turbines still consume steam and the steam production stops when the process flow stops. These cases are:

1. Loss of process air with air compressor online
2. Loss of synthesis gas flow with synthesis compressor online

Common findings

Some common issues were identified with all scenarios and are listed below:

HP to MP letdown valve, HV-1028, should open to match the MP demand, it only opens to a maximum of 75%. The steam letdown logic was originally configured so that the HP turbines remain online until the operator takes action to trip them or if they eventually tripped. The tripping of the 103MJT and 105MJT after 101MJ trip was implemented in 2007 and the steam letdown logic did not get updated to incorporate the change, this caused the HV-1028 to open a max of 75% when in some cases it should have been opened to 100%.

MP Boilers are slow to ramp up to compensate for the loss of MP steam from the HP letdown. This situation requires manual intervention for the operator to fire the boiler up quickly. An automatic action would improve this situation.

The boilers operate with one burner out of 2 when the load is below 40%. As such the ramp up of the boiler is configured to increase the rate to 40% until the second burner can be lit before ramping up further. This was all done by the boiler panel operator and takes significant time.

The steam to carbon trip set point was based on a design feed gas composition. Set at 2.3, this was close to the operating of 2.75, the actual gas supply was much leaner, and a lower set point is possible.

During all trips, the front-end pressure is high, and the temperature reduces quickly leading to condensation on the LTS catalyst. The LTS inlet and outlet MOV are too slow to protect the LTS catalyst from wetting.

Loss of Process Air to Secondary Reformer

A trip of the process air compressor turbine does not result in a reformer trip since the steam is not being used when the turbine is tripped. When the air is stopped with the compressor still online, the ammonia plant is expected to trip due to no HP steam production from the secondary reformer and the back end. The following can cause this scenario:

- The air compressor inlet MOV-1006 closes
- The air compressor anti-surge valve opens fully.

Recommendations

- Automate air compressor trip on loss of process air, in this case it is set at 80,000kg/h with a 30 second time delay. (The time delay is recommended to prevent spurious trips and give the operators time to act before steam balance becomes critical).
- Install air limiter on the process air flow controller to decrease air to gas set point to 2.65 if it reaches 2.9 or higher.

Loss of Synthesis Loop Circulation

If partial trip of 103-MJ is considered where converter will be isolated, and 103-MJ will keep on running on minimum governor speed with recycle valves open.

In this case, a large reduction in HP steam production occurs when steam generation in the back-end stops. The 103-MJT will keep on taking HP steam thereby reducing HP header pressure.

It is observed that the S/C ratio goes down to 2.3 in 2 minutes. It will be better to trip 103-MJT completely to maximize use of remaining HP steam to save MP header pressure.

Recommendations

To manually trip the synthetic compressor. It was difficult to determine an automatic action in this case, since there were many causes that cannot all be addressed with automatic action. The scenario was discussed with the experienced operators who preferred to stop the machine from the control room using the manual trip switch.

150T Boiler Trip

The 150T MP boiler is required to ramp up and supply the majority of MP steam during plant disturbances. The boiler should have a high reliability and availability, if the boiler trips, there is no redundancy and the ammonia plant will trip.
The ammonia plant will trip on low steam to carbon ratio within 10-15 seconds. A site wide blackout is expected, and both ammonia and nitrate plants will trip.

Recommendations

The plants can be saved if automatic load shedding of large steam and electricity consumers is initiated immediately upon a trip of the 150T boiler. Unanticipated problems like passing vents and inefficient pumps or motors will affect the success or failure of load shedding, but it can be compensated by incorporating additional load shedding to act as a safety margin.

- 7-8MW of power needs to be load shed immediately in case of 150T boiler trip.
- Power export to TAN plant approximately 3.5MW needs to be shed
- Auto ramp up of the 50T boiler to maximum rates
- Reducing plant rate will not help in this case
- Tripping the back end will not help keeping the front end online.

- Higher plant rates will result in a reduced load shedding requirement as more HP steam is produced (more steam export from ammonia plant) and the resulting steam demand from the 150T boiler is less.

Automatic Load Shedding

Manual load shedding within the 10-15 second window is not practical. The feasibility of Automatic load shedding was reviewed, the sequence of load shedding is just as important as the equipment involved (Fig 15).

Field adjustments have to be made at the same time. It is envisioned that this should not affect the overall outcome of the load shedding but will have an impact further down the timeline (more than 10 minutes) if these actions are not taken immediately.

- Trip one of the two running SCW pumps (2202-MJA/B/C), 2.4MW
- Start smaller emergency pump, 0.700MW.
- Trip the TAN plant power export, 3.5MW
- Trip BOG compressor,0.7MW
- Trip electricity export to external consumer, water corporation,0.7MW
- Change semi-lean pump from turbine drive to motor as motor uses less equivalent steam than the turbine, equivalent to 10 T/h steam savings
- Change 50T boiler from cascade to manual and ramp up to 70%.

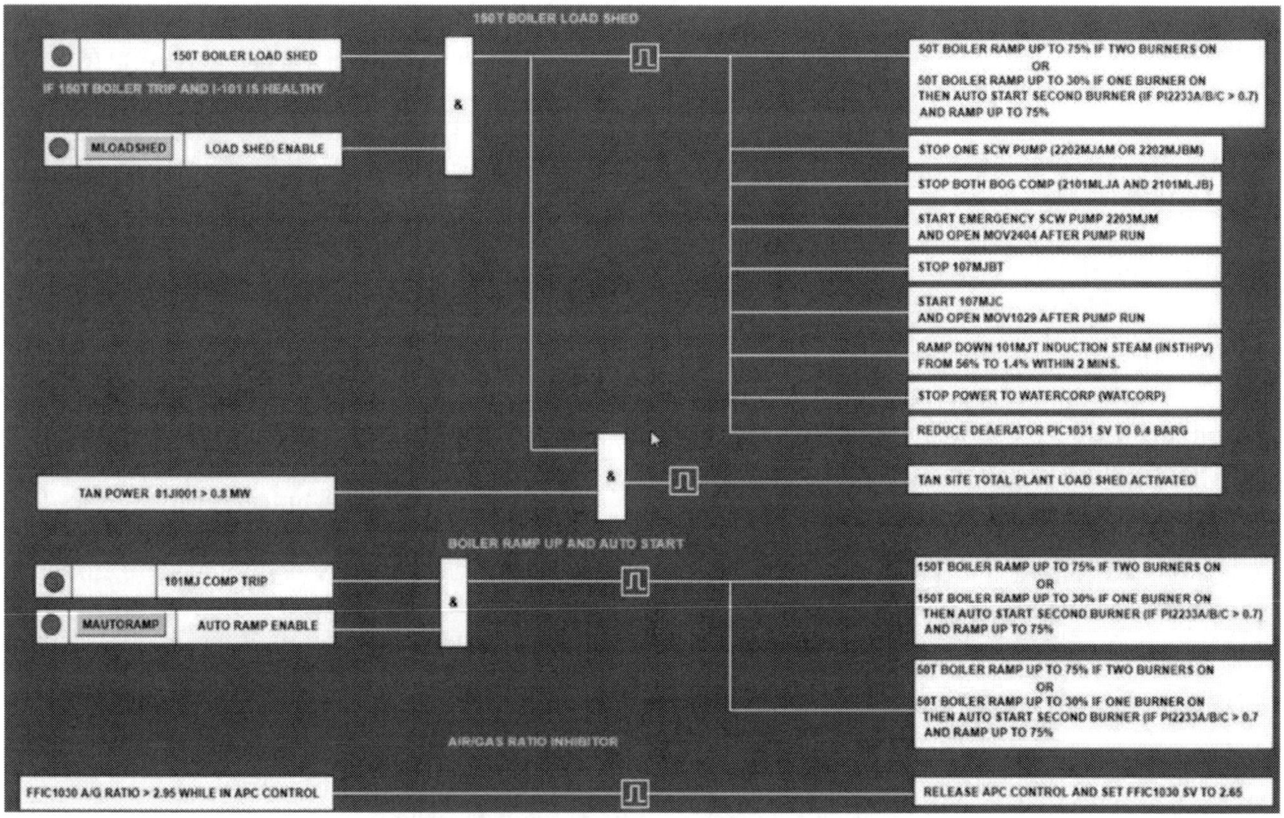

Figure 2. Automatic Load Shedding Logic

Automatic Boiler Ramping Up

Automating boiler ramping is a crucial action for providing additional steam following various plant upsets. Testing the ramp rate on the OTS is not realistic for boilers so the ramp up was tested during start-up for 10 and 20% increases. The increase in steam output was found to be significantly faster especially on the 20% increase, however additional tuning was completed to further speed up boiler ramp.

Tuning of the boiler fuel controller was required to achieve the maximum allowable ramp rate of 20% per minute. This was achieved in house using the help of advanced process control and the boiler manufacturer.

Safety Issues Identified

Loss of refrigeration compressor leads to warm gas to the molecular sieve and the synthesis loop pressure increases leading to possible water carryover and damage to the mol sieve, fouling of purifier.

The safe shutdown downstream of CO2 removal including the methanator is achieved by operator intervention, but in these cases, the panel operators are focusing on other issues.

These can be addressed by the following:

- Trip synthesis gas compressor to avoid over pressure of synthesis loop and heating of the refrigeration system.
- Trip methanator. As molecular sieve inlet is cooled by the refrigeration system, this will be heated and cause excess water to the molecular sieves. Tripping methanator will move the vent upstream and prevent this.

Challenges Faced

- Not all scenarios could be compared with plant data, in some scenarios the simulation data was assumed to be completely accurate.
- Some of the pumps will be changed over (sea cooling water and semi-lean aMDEA), however the discharge valves on these need to be opened automatically with these pumps starting. As it takes time to open these discharge valves the loss of SCW or semi-lean flow could be enough to trip the plant.
- SCW pump can automatically re-start after it is tripped, it will have to be put in manual after the trip.
- When transferring all power load to one CPP, any imbalance in frequency can cause both CPPs to trip.
- The emergency diesel generator will have to be automatically synchronized with the CPPs if it has to be started. This is difficult to implement and can cause significant damage to the CPPs if it is incorrectly configured.
- The plant rate controller is configured to automatically reduce feed gas flow when the steam is limited. The logic did not incorporate the air to gas ratio control as well. The separate air to gas control reacts after the plant rate controller reduces. In situations where the steam is lost and the gas flow is reduced, the air to gas ratio can quickly increase and this leads to a trip on air to gas ratio. This could not be resolved easily due to the complexity of the controllers, we anticipate the time delays and air limiter logic which was installed should allow for the controller to adjust in these cases.

List of Actions Implemented

- Update HP/MP letdown valve logic
- Change boilers to manual and ramp to 75% if Process air flow is low for > 30 seconds.
- Automate lighting of second burner during boiler ramp
- Automate air compressor trip on low process air for > 30 seconds

- Implement air limiter logic reducing air to gas set point to 2.65 if air to gas ratio increase to > 2.95.
- Load shed power from the TAN plant and non-critical consumers on 150T boiler trip
- Automate ramp of 50T boiler on 150T boiler trip
- Tune boiler fuel controller to achieve faster boiler ramp rates
- Trip synthesis gas compressor on refrigeration compressor trip
- Trip methanator following refrigeration compressor trip

Conclusions

The dynamic steam study identified low cost improvements to the safety and reliability of the plant. Simple cost-effective actions related to logic changes can improve the stability of the plant in upset scenarios and improve the overall reliability of the steam system. These changes could prevent some of the plant trips that have occurred in the past.

a steam system dynamic study should be conducted for all projects to ensure that the automatic actions are configured appropriately and that they meet the intent of the trips.

The study also identifies critical operator actions and the time to react which help determining the priority during upset conditions. This can also help with determining what needs automation and what doesn't. the information from the steam study can provide a basis for a cost benefit study to determine how much can be spent on automation as opposed to relying on operator action.

Startup scenarios should be considered when implementing automatic actions. When starting the plant, the load shedding can inadvertently activate and cause a major upset. These cases were identified when consulting with the production team and addressed by providing bypasses and instructions to remove the bypass when the plant was operating at normal rates.

Automatic load shedding is not as simple as setting up a trip and expecting an action. In some cases, the automatic actions have to be accompanied by field actions, e.g. throttling a pump discharge valve or opening a heat exchanger cooling water inlet valve to improve the flow distribution.

The load shedding actions could not be simulated offline as some equipment needed to be online.

References:

Wolfgang Jovanovic and Avinash Malhotra, "Construction and Commissioning of BFPL's 2200 MTPD – World's Largest Purifier™ Ammonia Plant", AIChE Ammonia Safety Symposium, September 2006

Installation of a New Flue Gas Tunnel Design Concept and Resulting Operation

Down-fired steam methane reformers utilize flue gas tunnels running the length of the radiant section floor to collect and transport radiant section flue gas to the convection section for heat recovery. Flue gas enters the tunnels through side wall openings which are distributed along the length of the tunnels in a pattern designed to achieve uniform flow of flue gas both in the downwards vertical direction and into the tunnel along the entire length of the tunnel. Conventional tunnels have been in use for several decades, but a new tunnel system manufactured by Blasch Precision Ceramics and designed and developed in conjunction with BD Energy Systems is now available and greatly improves upon the conventional design.

BD Energy Systems and Cherokee Nitrogen discuss the successful revamp project execution of installing the above discussed technology in an existing down-fired steam methane reformer along with operational results obtained from the reference Ammonia plant.

Joe Price
BD Energy Systems, LLC

Jason Hartsell
Cherokee Nitrogen, LLC

Introduction

Conventional tunnels incorporate physical features that constrain design efficacy resulting in non-uniform flue gas flow both along the length of the tunnels and among the tunnels. Non-uniform flue gas flow has been correlated to non-uniform catalyst tube convective heating, varied tube temperatures, and early tube failures due to localized overheating.

BD Energy Systems have developed a new patent pending Tunnel Optimal Performance (TOP) design method to achieve radiant flue gas flow uniformity and improved tube longevity/reliability. BD Energy Systems uses the Blasch Precision Ceramics designed StaBlox™ system, having mullite bonded high alumina refractory components that include unique features incorporating the TOP design within the StaBlox™ system and allowing the installation of the entire tunnel system without mortar utilizing interlocking features.

Previous articles have been written describing the design of the StaBlox™ tunnel system and the TOP designed orifice inserts engineered by BD Energy Systems that make up the new flue gas tunnel system that is the subject of this paper.[1] It is not the intention of this paper to restate all of the detailed information from the previous papers, but a summary of pertinent information is provided as a basis for further discussion.

Characteristics and Limitations of Conventional Tunnels

The following points list the inherent limitations of conventional tunnels discussed in previous papers.

- The openings in the side walls of conventional tunnels are rectangular in shape and are normally the size of a half brick. This also results in step-changes in flow area as opposed to a more fluid change in flow area.
- Conventional tunnels have buttresses to increase stability. This restricts the width of the tunnels and increases tunnel pressure drop with increased plant capacities due to the higher velocities inside the tunnels.
- The buttresses are also areas of the tunnels where flue gas flow openings cannot be supplied and create no-flow regions where flue gases must be diverted to adjacent openings resulting in increased flue gas flow to adjacent catalyst tubes.
- Conventional tunnels side walls also require expansion joints that create no-flow regions.
- Conventional tunnel slabs (covers) are simply laid on top of the tunnel wall and provide limited stability to the system. They are also a source of frequent failure due their mass.
- The mass of the tunnels is high and is a source of high temperature heat retention during cool down. The weight of the tunnel can also be detrimental to the tunnel's own stability due to deflection of the supporting floor structure.

Characteristics of the StaBlox™ Tunnels with TOP Design

- The openings in the side walls are circular and designed to accept an alumina orifice plate. There are over twenty different orifice sizes allowing fine tuning of flue gas flows.
- The StaBlox™ system does not require buttresses enabling the width of the tunnels to be maximized to reduce pressure drop.
- The system has its own floor component and all components are interlocking creating a very stable structure.
- Internal stabilizing bars are available for taller tunnels.
- There are no areas of no-flow.
- Thermal expansion allowances are designed into the components and installation. No discrete expansion joints are required.
- All components are hollow and light weight which decreases installation time, reduces mass by approximately 60%, and reduces heat retention.

Installation of New Tunnels

New TOP designed StaBlox™ tunnels were installed in 2018 at LSB's Cherokee Nitrogen facility in Cherokee, Alabama. The reformer has seven rows of 24 catalyst tubes and eight tunnels. The first step in installing the tunnels is leveling the radiant floor to receive the base pieces. Figure 1 shows the base pieces installed and ready for the remainder of the tunnel to be installed.

After the floor is leveled, installation of the base pieces and assembly of the tunnels proceeds at a much higher rate compared to conventional tunnels. The tunnels in this reformer were five feet (1525mm) high and forty feet (12,190mm) in length. Typically, three entire tunnels can be easily assembled in one 12-hour shift and at the Cherokee Nitrogen facility, one shift was able to complete four tunnels. Figures 2 and 3 depict a tunnel that is partially assembled and one that is completely assembled, respectively. Each of the tunnel components (base, bricks, and slabs) can be easily carried by one person.

Figure 1. Tunnel Base Plates Installed

Figure 2. Partially Assembled Tunnel

Figure 3. Completed Tunnel

For reference, Figure 4 shows an orifice insert installed in a tunnel side wall opening.

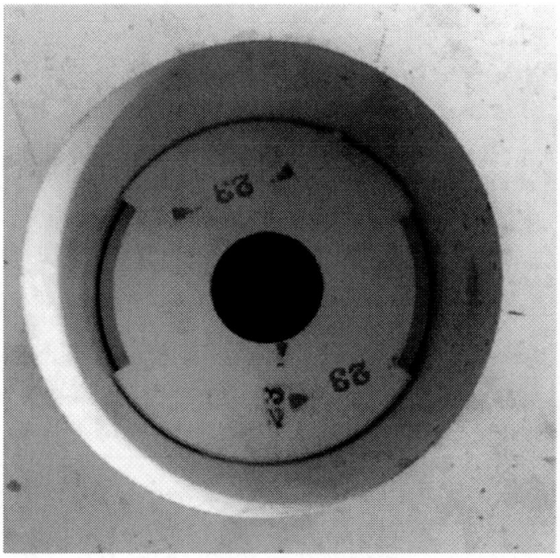

Figure 4. Tunnel Side Wall Opening with Orifice Plate Installed

Operational Data Analysis

As previously stated, the purpose of the flue gas tunnels is to assist the collection of the products of combustion in such a way as to create a uniform flow of flue gas in the vertical direction that results in uniform heating of the catalyst tubes. This goal implies that the flue gas must also enter the tunnels uniformly along the entire length of the tunnel otherwise crossflow occurs near the bottom of the radiant section as the flue gas follows the path of least resistance. The no-flow areas created by conventional tunnels create crossflows near the bottom of the catalyst tubes where the tube metal temperatures (TMTs) are the highest creating even higher temperatures compared to tubes located in front of no-flow areas.

Operating data were compared for periods of 2½ months before and 3½ months after the installation of the tunnels. Key points that were examined are catalyst tube metal temperatures and production rates.

Figures 5 and 6 show surface graphs by row and tube number of the TMTs before and after the installation of the tunnels, respectively. There is a noticeable improvement in both the temperature range and uniformity of the TMTs.

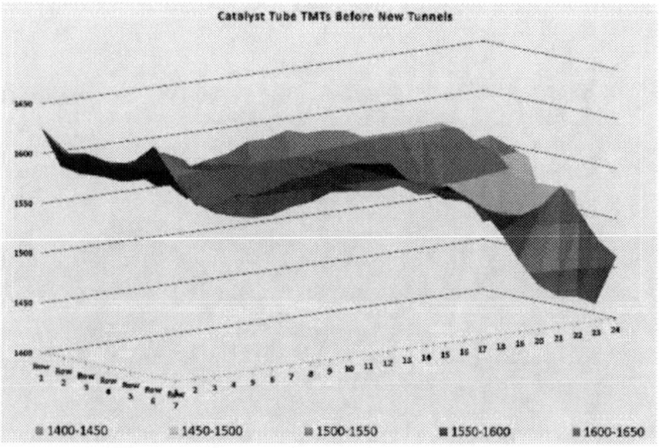

Figure 5. Catalyst TMTs Before New Tunnels

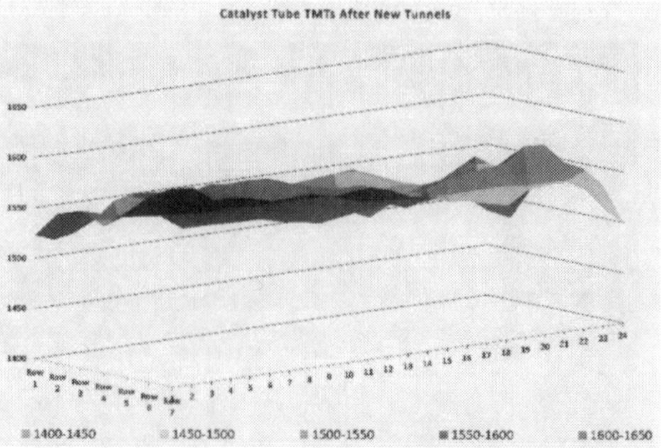

Figure 6. Catalyst TMTs After New Tunnels

Table 1 is a comparison of the TMT data.

	Avg TMT °F (°C)	TMT Std Dev °F (°C)	TMT Min to Max ΔT °F (°C)
Before	1556 (847)	42 (23.3)	222 (123)
After	1545 (841)	21 (11.7)	116 (64)

Table 1. Summary of TMT Data

It can be seen that not only did the average TMT decrease slightly but also the temperature range became narrower by approximately 50%.

The plant also experienced an increase in production of 3.3% after the installation of the new tunnels as well as an increase of 3.7% in the production/feed-rate ratio. These increases are a result of reduced methane slip from the front end of the plant due to more uniform heating of the catalyst tubes. More uniform heating of the catalyst tubes enables operators to safely increase the outlet process temperature while still maintaining the TMTs at safe levels.

The increase in production experienced at this plant is relatively high. It is estimated that larger plants would experience a smaller increase in the range of 1.3 to 1.8%.

Conclusion

The installation of the new tunnels had a positive effect on the catalyst tube metal temperatures causing the temperatures to be more uniform throughout the radiant section. This allowed an increase in production rates while maintaining the TMTs at a safe operating level.

References

1. Improving Down-fired SMR Flue Gas Flow Uniformity Using the New StaBlox™ Reformer Tunnel System, Joe Price, Tanner Howell, Joe Quintiliani, AIChE 2016, Nitrogen+Syngas 2016

Let's Be Safer by Making Process Safety Feel More Like Personal Safety

We all know and rightly deride manufacturing companies who apply the metrics of personal safety to Process Safety. The most egregious offenders are either very lucky or, increasingly likely to suffer a major process safety incident. But the driver for this is not as easily diverted as it seems. Personal Safety and, to an extent, lower tier Process Safety incidents are combatted by our senses - sight, smell, taste, hearing and touch. We jump out of the way of a dropped object, as we hear it clattering down and the cries of the dropper. We run in the opposite direction when we see and hear a fire. We react quickly when we touch a hot pipe. However, our major Process Safety risks, the like of which led to iconic incidents such as Bhopal, Flixborough, Texas City, Piper Alpha, are not aligned with easy detection and mitigation by our senses. This is because the hazards are kept in check by multiple independent barriers. It is only when those barriers are breached that our senses are alerted, by which time it is probably too late. This paper will examine the psychological and physiological distinctions between Personal or minor Process Safety incidents and Process Safety Disasters, why we respond differently to them and what steps we can take to be appropriately safe when we don't feel at all threatened in our workplaces.

Angus Keddie
Process Safety Matters Ltd

Introduction

Process Hazards is the insidious, evil twin of Personal Hazards. They are certainly related, but the Process doppelganger lurks unnoticed in the background and can strike devastatingly, often without warning. As they are related, perhaps we can try to understand their characters so that the aspects which enable us to keep ourselves and our team-mates safe can help improve Process Safety.

Personal and Process Safety

Personal Safety - Personal Safety generally applies to the individual at risk of harm to themselves or others on site due to scenarios he or she can anticipate and perceive, such that mitigative action can be taken ahead of or during an incident. Main categories include slips, trips and falls. The most common fatalities are caused by the 'fatal four': falls, being struck by an object, electrocutions, and being caught in between two objects. (Ref 1)

Process Safety - The CCPS [Chemical Center for Process Safety] defines Process Safety as: *'A disciplined framework for managing the integrity of hazardous operating systems and processes by applying good design principles engineering and operating practices. It deals with the prevention and control of incidents that have the potential to release hazardous materials or energy. Such incidents can cause toxic effects, fire or explosion and could ultimately result in serious injuries, property damage, lost production and environmental impact.'*

This distinction between the two is important because operating companies have traditionally been rather better at managing and rewarding the former than the latter. You need think no further than the number of times you have seen a banner proclaiming '1 million manhour without a Loss Time Incident (LTI)' or 'The Royal Society for the Prevention of Accidents (ROSPA) Award received for 6 years safe operation'. Personal Safety Risks are those which, for the most part, we can see, hear, smell, taste or feel. Alas, Process Safety Risks are 'senseless'...until it is too late.

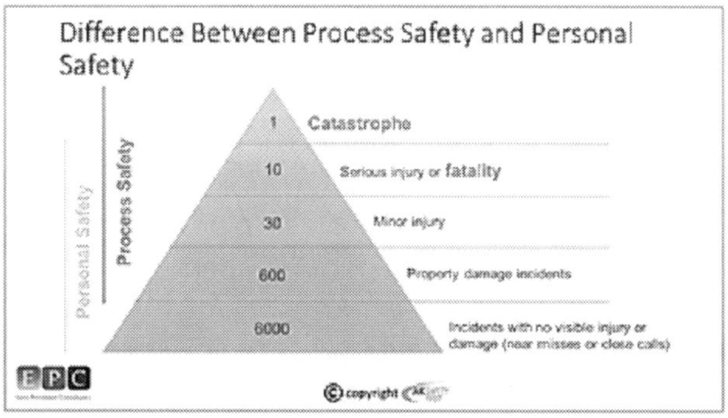

Figure 1. Process Safety vs Personal Safety

Figure 1 highlights the difference between process safety and personal safety.

Figure 2. The roof of the building on fire

<u>Example of a Process Safety Incident</u>

On September 29th, 2011, 4 hours after the start-up of the ammonia plant at GPN Grand-Quevilly, a hydrogen leak followed by a fire occurred on a 1-inch synthesis gas line on the discharge side of the synthesis gas compressor. The plant was tripped immediately, the plant personnel were evacuated to a safe location, and the fire was finally extinguished about 90 minutes later, after depressurization and nitrogen purging of the synthesis loop. The damages to the compressor room were considerable, but no one was injured and there was no release of toxic chemicals or pollutants outside the plant. The fire could be seen several miles away, and there was a large media impact locally (Ref 2

Figure 3. View of the west wall of the building in front of the compressor showing the origin of the fire

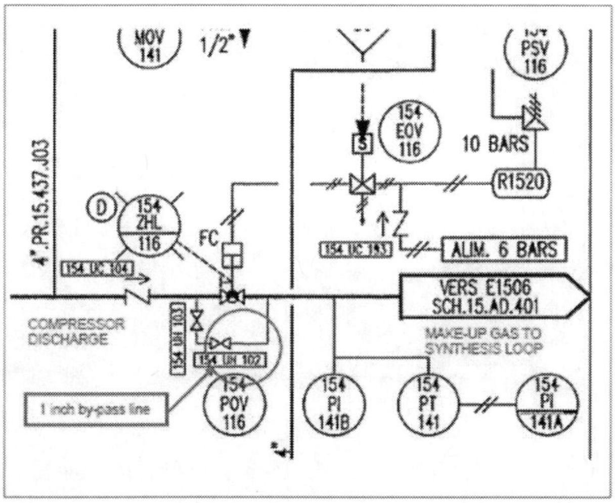

Figure 4. The root cause of the leakage was the failure of the yoke bushing on a 1-inch forged steel manual valve

Figure 2 and 3 show damage resulting from the fire. Figure 4 shows the location of the initiating leak.

This incident clearly meets the definition of a Process Safety Incident, where the cause and initial escalation cannot be sensed by the plant operators. Fortunately, upon detection, mitigation measures were activated to limit the damage and save lives.

Risk - A technical definition of risk is that it is the product of severity and likelihood. Mathematically, this means:

Risk = Severity of Ultimate Consequences X Likelihood of the Consequence Occurring

Interestingly, the risk associated with Personal and Process Safety can often be similar, which is because their severity/likelihood weightings are opposed (see Figure 5):

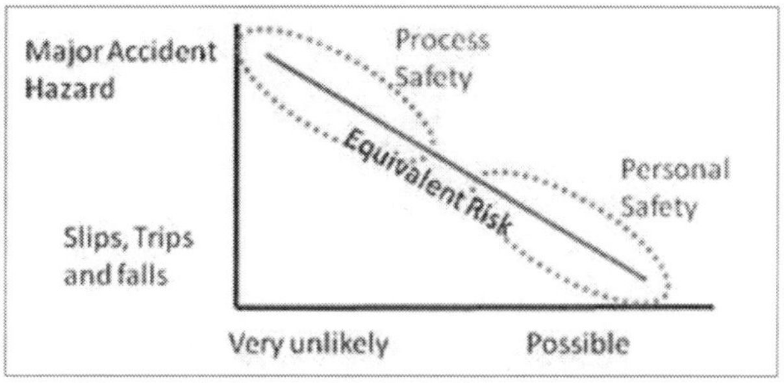

Figure 5. Same Risk, Different Outcome

Consequently, the cost (in all senses) of heeding rather than ignoring Process Safety can be extremely high:

Disaster	Fatalities / Injuries	Cost resulting from disaster	Prevention Cost	Prevention Intervention
Texas City	15 / 180	~$3,000 M	~$0.1 M	Installation of LAHH trip on Distillation Column
Deepwater Horizon	Nov-16	~$50,000 M	~$1 M	Investigation and probable replacement of safety barrier whose failure was evidenced by rubber in well fluid
Piper Alpha	167 / 35	~$3,000 M	~$0.1 M	Dismantling of ceiling between Condensate pump and its PSV
Bhopal	>3000 / >30000	~$1,000 M	~$0.05 M	Operation of Vent Scrubber System (which was not working at the time of the incident)

Figure 6. Costs of Heeding vs Ignoring Process Safety

It is, on average, 10,000+ times more expensive to recover from an accident than to have prevented it. In the case of the GPN Grand-Quevilly incident, the cost of the reconstruction was more than 16 million € ($20 million USD) including implementation of additional safety measures. The total length of the shutdown was 11 months. (Ref 2)

The Human Response to Fear

Fear is a feeling induced by perceived danger or threat that occurs in sentient organisms, which causes a change in metabolic and organ functions, ultimately resulting in a change in behavior, such as fleeing, fighting or freezing responses. Fear in human beings may occur in response to a certain stimulus occurring in the present, or in anticipation of a future threat perceived as a risk to body or life. (Ref 3) Physiologically, breathing rate increases, pupils dilate, heart rate follows suit, peripheral blood vessels (in the skin, for instance) constrict, central blood vessels around vital organs dilate to flood them with oxygen and nutrients, and muscles become tighter (causing hair to stand on end) and are pumped with blood. These changes make us more able to fight or flee the danger.

We can respond in this way because we have a brain, the main purpose of which is to be able to navigate our surroundings to spread our genes. To demonstrate this assertion, let me introduce you to the sea squirt (shown in Figure 7).

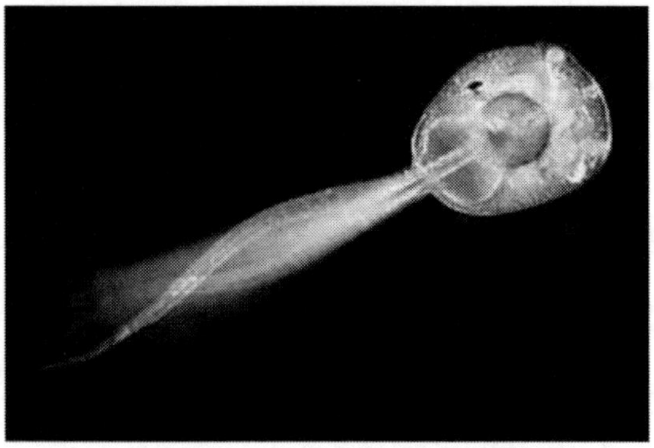

Figure 7. Juvenile Sea Squirt

The sea squirt (the clue is in the name) which develops a very simple brain, uses it to swim around to identify a suitable location upon which to perch – forever.

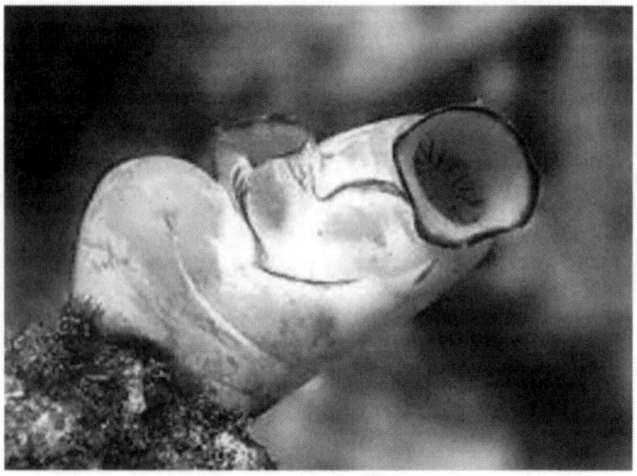

Figure 8. Adult Sea Squirt

Having attached itself to its rock, it morphs into a sedentary being, absorbing its brain in the process. Evolution bequeathed it with a brain, for the sole purpose of movement, in this case to find a suitable location to attach to and reproduce.

Of course, our brains with billions of neurons and trillions of interconnections are considerably more sophisticated than that of a pre-transformed sea squirt (which has a grand total of 177 neurons). They are, in fact, the most complex object in the known universe. Accordingly, as intelligent beings, we have access to multi-dimensional sensory interface between our brains and our environment. This is of significant advantage in minimizing personal safety risks in a site environment. This interface or, as it is better known, our senses, are generally accepted to be 5-pronged. In order of prominence, they are:

- Sight
- Touch
- Hearing
- Smell
- Taste

Another dimension is that, for most of our existence as a species, we have survived and then thrived in part because of our tribal sensibilities. Tribes were Us and everything else was Them.

Figure 9. A Tribe

Humans are social animals and ill-equipped to live on their own. (Ref 4)

Consequently, we have evolved discomfort at the prospect of separation from the tribe and an instinct to protect fellow members. This, also, has implications for Personal and Process Safety.

Personal Safety Benefits from Individual Senses and Tribal Attributes…

Tribal Attribute Barriers

According to a study by Robin Dunbar at the University of Liverpool, primate brain size is determined by social group size. (Ref 5) Dunbar's conclusion was that most human brains can really understand only an average of 150 individuals as fully developed, complex people (refer to Figure 10).

Figure 10. Dunbar's Number

Conveniently, personnel in operating facilities are generally lower than this number, creating a de-facto tribe. Individual shift teams are, of course, an order of magnitude smaller still and like any tribe, motivated to defend itself against perceived threats.
When that threat is of a Personal Safety kind, the team members who are the first to perceive danger will react for themselves, then, if time allows seamlessly segue into group protection with verbal warnings, gesturing or even physical interventions.

Individual Sense Barriers

When we are awake, our brain receives and can process around 2 GB/s – about the same a standard computer network (refer to Figure 11).

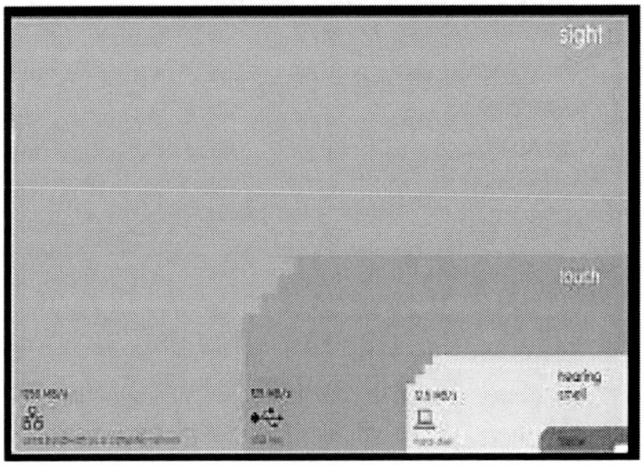

Figure 11. Brain Sense Processing Proportion

Ordinarily, our brain makes us aware of a mere 0.05% of the information it receives from the interface, which is, I guess, the amount needed for us to 'think' about stuff then act on it. However, it can rapidly react when it senses danger to the organism and engender a response. (Ref 6)

...while Process Safety Doesn't

Tribal Attributes inhibit appropriate responses

Attributes include:

Group Think
- Cultivated by innate discomfort of the possibility of being ostracized

Production Primacy
- If leaders emphasize profits and praise production uptime, without acknowledging the contribution of the underpinning good safety practice. Workers may then fear making a safety intervention for fear of blame for stopping profitable production.

Individual Senses act as Lagging (pre-fatal?) Indicators

Our physiological reaction to threats can lead to 3 types of outcome (Figure 12):
- Fight (engage with the perceived source of danger)
- Flight (escape from the perceived source of danger)
- Freeze (remain motionless, increasing the chances that the source doesn't perceive you or think you're dead)

Figure 12. Fight, Flight or Freeze

A significant and imminent Process Incident, such as a nearby high energy loss of containment of flammable or toxic fluids, would likely create dramatic indications of its occurrence, such as loud noise, visual eruption, pressure wave, powerful odors and acrid taste. In iconic Process Safety disasters, operator's physiological reaction to fear has led them to:

- Formosa Plastics – Fight. Running into the plant upon smelling the VCM release in order to close valves etc., which subsequently exploded and killed them.

Figure 13. Formosa Plastics Incident

- Texas City - Flight. Seeing a hydrocarbon release which caused 15 deaths, then fleeing on foot instead of driving away the pick-up truck which was subsequently identified as the source of ignition.

Figure 14. Texas City Incident

- Piper Alpha - Freeze. Operators, on adjacent platforms which controlled the gas flow, seeing and hearing the initial fire from adjacent platforms, froze (and didn't stop the flow) leading to an escalation which ultimately cost 167 lives.

Figure 15. Piper Alpha Incident

Tragically, in all 3 cases, the fear reaction of pivotal human actors actually made the final outcome worse…much worse.

Bridging the Gap: Ideas and Examples

Countering Groupthink

Ostracism is the exclusion from a tribe. For most of our evolution, ostracism equated to death. Cavewomen or Savannah Men don't fare very well on their own. And ostracism, was a possible punishment for standing up to the tribe, or tribal leadership.

Figure 16. What Ostracism Feels Like

Consequently, in general, we are often extremely uncomfortable at the prospect of standing up on our own against the group, for fear of ostracism (even though the modern consequences are much diminished) and that, insidiously, allows real threats to go unchecked and opportunities bypassed. As Robin William's character in Dead Poet's Society implores his charges to overcome *'the difficulty in maintaining your own beliefs in the face of others'*.

Ahead of the Texas City disaster, a Risk Assessment is said to have taken place about the siting of the Shutdown Contractor's Portacabins. The default supposedly was that the proposed location was safe and that the purpose of the meeting was to discuss and agree that notion. It wasn't challenged; the portacabins were sited; the subsequent explosion killed several contractors.

So how do we counter this in a Process Safety context. Extensive research has shown that an individual is significantly more likely to externalize and defend a counter position if he or she feels they are not the only one in the group to hold or even just voice it. One way that this can be achieved, for key Process Safety dialogues is to invite an appropriate 'devil's advocate' to the review, get them to propose and defend an alternate position (Figure 17). This may even be positioned as the starting point, which must have critiqued in place of the current default.

Figure 17. Devil's Advocate

The Power of Unplanned Alerts

I was carrying out a HAZID/HAZOP closeout for an Early Production Facility (see Figure 18) in Kurdistan. One of the actions was to validate the H2S alarm drill (H2S is a toxic gas, usually fatal at concentrations of 1000 ppm or 0.1%, which is often present in significant concentrations in well fluids). However, the day before the drill was due to take place, there was an actual alert. As you can imagine, the site crew all rapidly responded in the correct manner. In the debrief afterwards, one the team leaders noted that the response was much more dynamic than the recent planned drill and joked that apparently the alert was due to a faulty reading on the instrument. *'Maybe we should contrive some pretend alerts to make our guys pay more attention'* he said. And I thought...maybe we should.

Figure 18. Miran Site, Iraq

How to Become one of the Good Guys

As Process Safety Professionals, we're often seen as adversaries. We're the opposite of 'the bank that like to say yes'. We're the killjoys, the pedants, the cheerless gamekeepers. In a training session I ran recently, one of the delegates mused that his Process Safety department was located on the same floor as Internal Audit and Lawyers (archetypal 'No Guys'). But what if you took advantage of the fact that your entire genotype differs from everyone else on the planet by 0.06% (Every 1499 of each 1500 of your genes is identical to 7 billion other people). It then makes sense that how you would feel about and therefore react to a situation is, by and large, the same way as anyone else. Ta-dah! You're a mind reader and you didn't even know it.

Generally, we respond well to some behavior and poorly to others. We appreciate it when people are friendly and relaxed. We like it when we feel that our concerns are being listened to. We glow when our endeavors are praised ahead of our successes. However, we dislike being ordered aggressively to do things. We feel uncomfortable if information is stated rather than explained. We reject being made to feel part of the problem rather than part of the solution.

Leveraging these tenets in HAZOP reviews will optimize the outcome:

- Make the team comfortable as much of the time as possible
- Engage from the start
- Use team member names when addressing them
- Introduce and integrate new arrivals
- Give team members responsibilities
- Say when we'll finish and enable this

More Carrot less Stick

The psychology of adoption of appropriate Process Safety is from a company perspective that is driven by fear. Fear of paying a large fine; fear of having production halted while investigations take place; fear that a near miss could 'but for luck' have become a major incident. It takes a jolt for a company to realign its Process Safety dynamic - and once this happens, it often becomes the new norm. So, a big stick (or the threat of one) does appear to work at a corporate level.

However, the people who work in companies are differently driven and motivated. Sure, they'll respond in the short term to the big stick, to avoid pain and discomfort. However, over the long term, they will find ways to avoid it by taking a different path: perhaps secretly ignoring the directive and most likely tread a more hazardous path. So, a different approach is encouraged - less stick and more carrot. We need to find ways of making good Process Safety behavior more interesting, satisfying and psychologically rewarding. Let's celebrate the operator who prevents a near miss becoming an incident by a timely intervention. Let's put the fun into the Process Safety function. Let's tell the world when they ask us what we do 'we make incidents like Piper Alpha NOT happen' - making nothing happening for a long-time gratifying.

Perhaps a middle path could be distinguishing between different types of human error and responding to them differently.

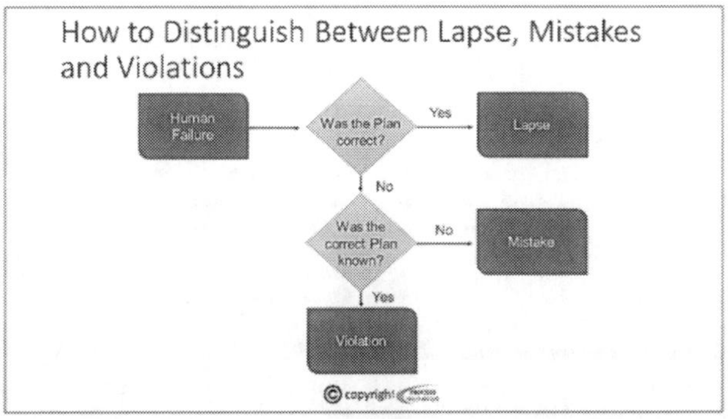

Figure 19. Lapses, Mistakes and Violations

Given that the majority of mistakes fall into the first two categories described in the title of Figure 19 above, I suggest the following responses:

- Slips – address task structure and methodology
- Mistakes – address training and feedback
- Violations – transparently punish

The Gamification of Process Safety

Figure 20. The Gamification of Process Safety

Yu-kai Chou is an authority on Gamification and Behavioral Design. He developed a Gamification Framework, Octalysis, which comprises 8 tenets to make any activity more captivating are:

- Epic Meaning and Calling
- Development and Accomplishment
- Empowerment of Creativity and Feedback

- Ownership and Possession
- Social Influence and Relatedness
- Scarcity and Impatience
- Unpredictability and Curiosity
- Loss and Avoidance

Octalysis Tenet	Description	PS Gamification Idea	Potential Participants
Epic Meaning and Calling	Participants feel that their PS mission is critical and larger than themselves alone	Computer Based Training (CBT) PS game with goal of keeping the hazard monster inside the pipe	Operators/ Front Line (FL) Supervisors
Development and Accomplishment	Generate short term gratification (e.g. milestone celebration) to encourage participants to endeavour with activity	Creating an animation of incident employee has prevented by intervention	Operators/FL Supervisors
Empowerment of Creativity and Feedback	Iterative and evolving mini-task combined with feedback	Macdermid Safety Cross program (prominently displayed cross with spaces for each day of the month (or week of the year) which is colour coded to indicate incident, no incident, action to avoid incident)	Operators/FL Supervisors
Ownership and Possession	Perhaps by creating a character which needs to be protected from PS incidents	CBT PS game with goal of keeping the hazard monster inside the pipe	Non-operating managers
Social Influence and Relatedness	Compare and contrast with the performance of other front-line teams. Create a PS league	3 monthly updates of where company sits of average and best of peers	Directors

Figure 21. A Process Safety Twist on Octalysis

We can apply Mr. Chou's findings to increase employee engagement with Process Safety with the examples in Figure 21 above:

Safety Third Anyone?

An idea promoted and applied by several key industry players is Chronic Unease, which is where front line operators are encouraged to be constantly vigilant to changes in their working environment during their working shift. Having to be alert for the whole working day seems extremely stressful and unsustainable. So, what can we do to steer a middle ground between abrogating personal responsibility for safety and being constantly alert? In 2009, Discovery Channel agreed to air a Mike Rowe one-hour special called Safety Third, where he talked candidly about mistakes he'd made on his series 'Dirty Jobs', and the unintended consequences of putting Safety First. He argued that many compulsory Safety programs discouraged personal responsibility in favor of Legal Compliance.

One way to engender this is to genuinely empower front line employees to 'call the safety shots' themselves. One company that applies this philosophy is RasGas (recently merged with Qatargas), a Qatar based LNG producer. Operators are encouraged to act on any discomfort they feel when on shift, to stop production if necessary. Managers are directed to fully support such decisions, while harshly penalizing any instances of violations (where the correct plan was known but another followed), however minor.

Vanguard Operators already 'Doing the Right Thing'

Many hazardous process operators have already adopted practices which reflect learning from Personal Risk Management. A selection is highlighted below:

Figure 22. RasGas Operator Vigilance

Figure 22 illustrates RasGas Operator Care. Empowering front line operators to 'Never be Wrong by Doing the Right Thing' – apologies to Mark Twain.

Figure 23. Eley Good Housekeeping

Figure 23 illustrates Eley Housekeeping. Eley, a Birmingham based explosives company with a clean and tidy plant. They may be amenable to offering a plant tour if requested by neighboring facilities.

Figure 24. Responding to a TTopco Alert

Figure 24 illustrates TaqTaq unplanned alert. Increased the likelihood that operators would respond correctly to any alert, whether simulated or real.

Figure 25. Bruce Webb of Neptune

Neptune's Bruce Webb (Figure 25 above), Operations VP from Neptune, a company recently formed from the upstream assets of Engie, has asks to be alerted when injury occurs on one of their facilities and follows up in person. (tribal support/ care/ tribal leader reverence).

Conclusions

As a species we have evolved to be able to prevent and mitigate personal safety risks on our sites, primarily due to our fear response. This same evolution has left us vulnerable to process safety risks which, although less common tend to be more severe. In some cases, it seems to actively conspire to worsen the outcome. However, by understanding and leveraging this response, we can develop new and more effective approaches to Process Safety Risks, saving many future lives as a result.

References

1. "OSHA Data & Statistics / Commonly Used Statistics". osha.gov. Retrieved 21 February 2017.
2. Ammonia Technical Manual 2013; Patrick Le Calvé GPN.
3. https://en.wikipedia.org/wiki/Fear
4. Isaacs, Harold Robert (1975). Idols of the Tribe: Group Identity and Political Change. Harvard University Press. p. 43. ISBN 978-0-674-44315-0.
5. Dunbar, Robin I. M. (2010). How many friends does one person need? Dunbar's number and other evolutionary quirks. London: Faber and Faber. ISBN 0-571-25342-3.
6. McCandless, David (2009). Information is Beautiful. London: William Collins. ISBN 9780007492893

Check Lists: The Low Hanging Fruit of Process Safety

One of the remaining 'low hanging fruit' in the realm of Process Safety is the rigorous application of appropriate checklists. In his book, 'The Checklist Manifesto' Atul Gawande argues that the preparation and adherence to hazardous activity checklists is a low cost, low effort route to reducing incidents. Working in the Health sector, he discovered numerous examples of checklist success, including the prevention of 8 deaths and 42 infections over 27 months by the adoption of a 5-point checklist in the Intensive Care Unit of a prestigious American Hospital.

Angus Keddie
Process Safety Matters

Dan Cojocaru
AmmoniaKnowHow.com

Introduction

There are powerful examples of the use of checklists in aviation, medicine and high-rise construction. This paper will explore how we can apply this principle to High Hazard Process Plants by:

- Examining the historical relationship between checklists, process safety incidents and their prevention.
- Demonstrating a correlation and postulate causation between appropriate use of checklists and a reduction in process safety outages.
- Describing best practice in the use of process safety checklists.
- Reviewing the evolution of checklists in related industries to determine which elements may be transferable.
- Contrasting and comparing current checklist tools.

An accident driving the checklist idea

On October 30, 1935 the US Army Air Corps held a flight competition for airplane manufacturers vying to build the military's next-generation long-range bombers.

A small crowd of army brass and manufacturing executives watched as the "flying fortress" of Boeing Model 299 test plane taxied onto the runway. The competitor was a much smaller airplane designed by Martin and Douglas.

Boeing's model was sleek and impressive, with a 103-foot wingspan and four engines jutting out from the wings, rather than the usual two.

Figure 1. Boeing Model 299

The plane roared down the tarmac, lifted off smoothly, and climbed sharply to three hundred feet. Then it stalled, turned on one wing, and crashed in a fiery explosion. Two of the five crew members died, including the test pilot. An investigation revealed that nothing mechanical had gone wrong. The crash had been due to "pilot error," the report said.

Figure 2. The Boeing Model 299 on fire after its crash at Wright Field on Oct. 30, 1935

Substantially more complex than previous aircraft, the new plane required the pilot to attend to the four engines, each with its own oil-fuel mix, the retractable landing gear, the wing flaps, electric trim tabs that needed adjustment to maintain stability at different airspeeds, and constant-speed propellers whose pitch had to be regulated with hydraulic controls, among other features.

While doing all this, the pilot had forgotten to release a new locking mechanism on the elevator and rudder controls. The Boeing model was deemed, as a newspaper put it, "too much airplane for one man to fly." The Army Air Corps declared Douglas's smaller design the winner.

So, a group of test pilots from Boeing got together and considered what to do. The test pilots made a simple list, brief, and to the point—short enough to fit on an index card, with step-by-step checks for takeoff, flight, landing, and taxiing. With the list, they check that the brakes are released, that the instruments are set, that the door and windows are closed, that the elevator controls are unlocked—dumb stuff. With the checklist in hand, the pilots went on to fly the Model 299 a total of 1.8 million miles without one accident. The army ultimately ordered almost thirteen thousand aircraft.

Historical relationship between checklists, Process Safety incidents and their prevention

Analyzing Fertilizer Industry Operational Risks Database (FIORDA) case studies, we notice that the energy and petrochemical industries still need to improve in learning lessons from incidents. This view is in part prompted by the reoccurrence of similar events, and also by anecdotal evidence of the difficulty of achieving long term changes in behavior and working processes when sharing lessons from accidents and incidents.

Previous research has indicated challenges at several stages in the Learning from Incidents (LFI) process including:

- Reluctance to report incidents due to fear of disciplinary action or the perception that reporting does not lead to any change
- Lack of human factors expertise in the analysis of incidents
- Lack of time and resources dedicated to helping people understand and make sense of disseminated lessons learnt across various disciplines and departments (e.g. operation, maintenance, QA/QC, project teams, etc.)
- Overload of recommendations, and the failure for all involved parties to agree on actions.
- Failure to check that implemented changes have addressed the underlying causes and reduced risk.

It is a common experience to hear the phrase, "We must learn lessons from this," following a major accident or following a more everyday event such as losing in a sporting competition. Indeed, this has become such a common phrase that one may feel that learning lessons is an automatic or natural process. In fact, the evidence from accidents and incidents indicates that it can be challenging for major hazard industries to learn effectively from such events.

If the only changes an organization is making are in response to Learning From Major Accidents (LFMA) rather than the broad range of potential events, as represented in the accident pyramid in Figure 1, this will typically lead to large disruptive changes following such major accident events in which risk will be reduced by large outlays in new safety-related equipment, with high associated capital expenditure (CAPEX) costs and reduced plant availability during implementation of such changes.

Over the longer term however, the memory of these low frequency events will be lost, and risk is likely to increase effectively unnoticed as the warning signs (weak signals) offered by incidents are not being effectively processed. Thus, solely adopting a LFMA process will lead ultimately to an increased frequency of major events, higher average risks, larger damage costs and more business interruptions.

An efficient LFI process will make use of the multiple opportunities for learning leading to a lower risk, more stable business environment as the organization makes small, optimizing adjustments in response to the learning from incidents. Here there is an opportunity to update the operator's checklists to look for reoccurrence of similar events and report them to upper channels for investigation and corrective actions.

In his book 'Thinking Fast and Slow', psychologist Daniel Kahneman states that estimation of the likelihood of occurrence of a major incident is higher than the actual number immediately following said incident, but then reduces to be below that number progressively over time. An analogy would be a hazard which recedes in your car's rear-view mirror as you drive away from it.

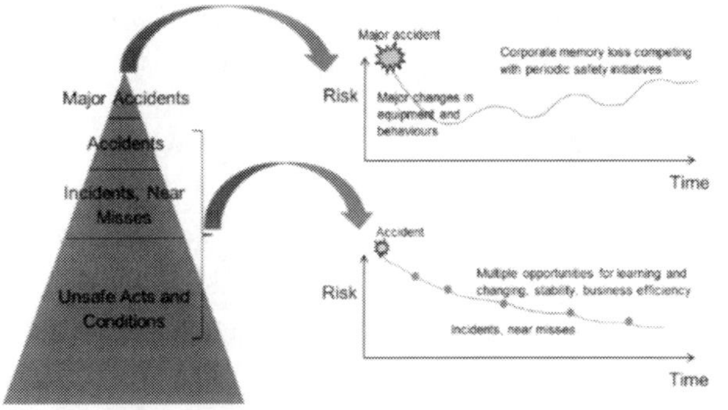

Figure 3. LFI vs LFMA

In the Fertilizer Industry Operational Risks Database, or FIORDA, we introduced a special classification called "Early Warning Signs" where we document the signs that preliminary may lead to a Major Accident or trigger near misses; e.g. frequent leaks, gradual increase of temperature or vibration, high pressure drop, abnormal color, etc. The purpose is to understand how Major Accidents and near misses can be prevented when understanding the process behavior in various conditions (start-up, shut-down, various types of plant upsets, abnormal downstream conditions, foaming, etc.)

Based on our industry records, a significant number of catastrophic incidents are occurring in primary reforming area during start-up. Explosion of primary reformer box, tubes damages, catalysis poisoning are constantly occurring as documented in multiple cases presented at this conference.

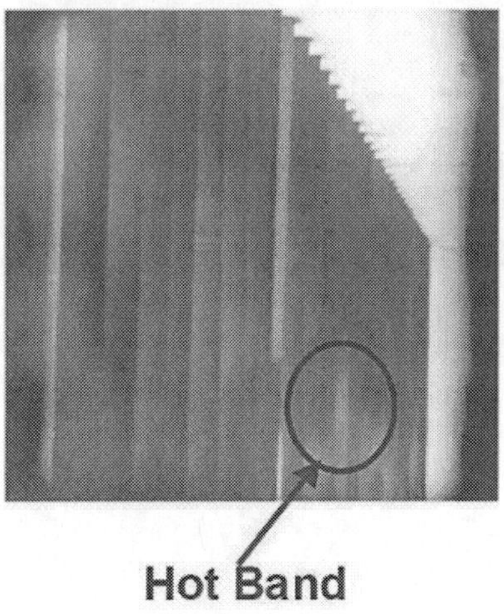

Figure 4. Hot bands in reformer tubes

Figure 5. Burner mis-alignment

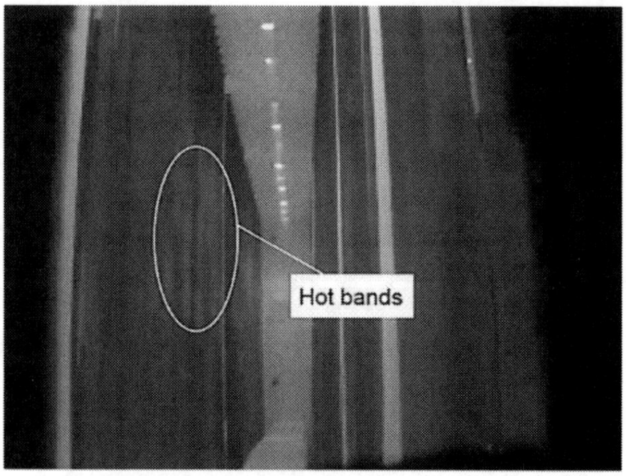

Figure 6. Hot bands in reformer tubes

Figure 7. Reformer tubes

Figure 8. Reformer after the blast

Figure 9. Reformer tubes failure

As the control system may be different from plant to plant, one common observation keeps coming up: the operator failed to make use of the check-list during the start-up or the start-up procedure does not include "how to" deal with a particular scenario. The operator either did not check the burners flame, tube color or fuel valves positions--all these contributing in the end to a very costly event. As mentioned by Johnson Matthey – at least once a year, a primary reformer suffers a complete or partial burn down and is caused by:

- Deviation from standard operating procedures in terms of operating pressure, fuel rates, etc.
- Increase in fuel gas calorific value.
- Low feed rates at too high firing rates.

Fourteen years ago, one of the conclusions was "It is unfortunately the case that with increasing frequency history is repeating itself and the hard-won learning of the last 50 years has been lost and now is being relearned." One of the recommendations was "checklists should be provided to cover unfamiliar tasks."

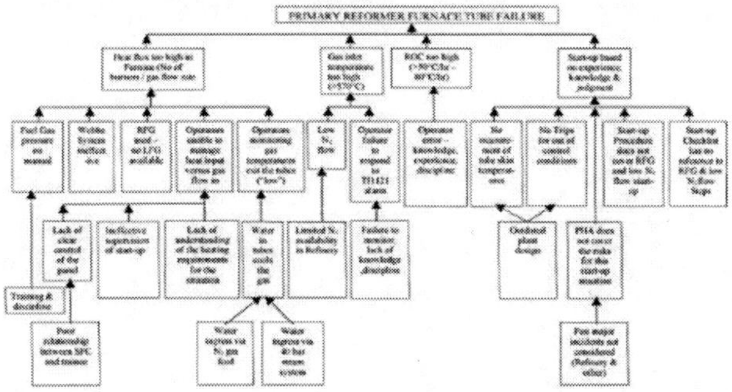

Figure 10. Example of Fault Tree analysis for Furnace Tube Failure

Companies aware of this trend already have developed Practices for Improvements in Reformer Performance and Reliability. A key aspect is the reformer performance monitoring training that includes a background information on reformer reliability management, recognition and troubleshooting of reformer performance issues, reformer tube damage mechanisms and inspection, fitness-for-service and remaining life of reformer tubes, and how to recognize, troubleshoot and prevent common reformer tube failures. Practical training on tube and reformer visual inspection during operator rounds is discussed in the classroom and then practiced in the field by participants using specifically prepared checklists.

Demonstrating a correlation and postulating causation between appropriate use of checklists and a reduction in PS outages

Safety efforts pay. This has been noticed 30 years ago by Norsk Hydro when they analyzed all the relevant incidents that have been recorded in the monthly Safety Report for the period 1986-1993.

The analysis shows that about 50% of the incidents could have been avoided through closer inspection of the plants in order to identify maintenance needs and to run more stringent checks on maintenance work (before, during and after activities). Measures taken during operations including use of possible identification method which could have prevented the incident (such as HAZOP, checklists, inspections, and so on) could have prevented 22% of the incidents, with Safe Job Analysis representing 12%.

The benefits can be seen clearly from the survey of expense saving Norsk Hydro has achieved since 1985 (see Table 1).

1984-1988	1990-1994
Fatal Accident Rate (per 100 million hours worked), FAR=3.9 in the period 1984-1988.	Fatal Accident Rate (per 100 million hours worked), FAR=0,35 in the period 1990-1994.
Average of 2 fatal accidents per year for Hydro employees.	Average of 1 fatal accident every 5 years for Hydro employees.
1,500-1,600 lost-time injuries.	365 lost-time injuries.
Average 15-17 days lost per injury.	Direct cost reduction alone amounts to USD 5.8 million a year.
USD 20 million in direct material costs as a result of fires and explosions, lost production not included.	USD 5.8 million in direct material costs as a result of fires and explosions.

Table 1. Norsk Hydro survey of expense results

Describing best practice in the use of Process Safety checklists

Nowadays, checklists are part of projects in the fertilizer industry from early stages of design until commissioning and operation.

Design

In the design phase, best industry practice makes use of lessons learned from other projects and include questions in the documentation review that address aspects like:

Have the reports of similar plants, citing accidents and the improvements made to counteract them, been reviewed and incorporated into the design?

Has the basic operating philosophy of the plant been reviewed and implemented, including the following:

1. *Normal/Alternative design cases?*
2. *Turndown?*
3. *Start-up?*
4. *Emergency cases?*
5. *Full or partial plant shutdown?*

Figure 11.

6. Is any material substitution made during the design stage been discussed and approved with the metallurgist and corrosion specialist?

For revamp cases

The checklist during design review shall include aspects like:

If the design is a new unit installed in an existing complex or a revamp of an existing unit, have the following been addressed:

1. *Is the overpressure protection philosophy for the new equipment compatible with the existing?*
2. *Is the design philosophy for new piping and equipment compatible with the existing or are additional overpressure protection devices required?*

Figure 12.

3. *Are the existing Utility/Auxiliary Systems compatible with the new equipment and is the capacity of these existing systems sufficient to meet the additional demand from the new unit?*
4. *Are materials of construction for the new/revamped unit compatible with the existing complex?*

Based on previous incidents, other questions to be included in the project checklist during the design phase are:

1. *Have the criteria for avoiding fatigue failures from acoustic and flow induced vibration in piping subject to high velocity and/or large pressure drops been checked? If flow or acoustic vibration is considered a possible problem, have the proper support guidelines been specified?*
2. *If air or oxygen can enter the process, has the possibility of deflagration been reviewed and the system designed accordingly?*
3. *Where a runaway reaction is possible, has this risk been adequately addressed in establishing the design? (e.g., materials of construction, relieving loads, trips and interlocks, etc.)*
4. *Has overfilling of equipment been considered?*
5. *Have all potential problems (e.g., corrosion, erosion, reactivity, improper mixing, vibration, etc.) caused by injection points been addressed? How?*
6. *Are all potential problems (e.g., corrosion, erosion, reactivity, improper mixing, vibration, etc.) included and discussed in the HAZOP studies?*

Pre-commissioning

Pre-commissioning procedures shall include checklists with questions related to equipment testing:

Figure 13.

1. *When equipment is planned to be included in the piping system during testing, is it ensured that piping test pressure is compatible with equipment design pressure?*
2. *For pneumatic testing, has maximum stored energy been evaluated?*
3. *Has the procedure for pressure testing existing lines been agreed with EPC Contractor (this should be part of tie-in coordination meeting)?*
4. *Has the procedure for pressure testing been determined for systems containing high pressure, high temperature piping welded to vessels (e.g. high-pressure steam systems)?*
5. *Are there any national or local regulations prohibiting/limiting pneumatic testing?*

Commissioning

In order to avoid any safety risk, a Pre-Start-up Safety Review (PSSR) shall be performed using a certain formal procedure. It is important that at this stage of the project, the HAZOP actions have been closed, all P&IDs comments are included in the final revision and updated to as-built status.

The checklist shall address the Mechanical Completion status, Instrumentation and Control loops check and ESD function checks.

As per previous experience, the fact that trip logic checks and PSSRs have detected incorrect conditions shows that they are needed. The tools should not only rest with the construction company, but with the technology provider who should apply them because the technology provider:

- knows where in (pre-)commissioning that risks could arise
- is the originator of the reference documents such as trip and interlock descriptions

Operation

Over time, plants undergo modification as a result of continuous improvement programs or revamp projects. New equipment is installed along with piping modification and new control loops. As the projects are handed over to operation, new SOPs (Standard Operating Procedures) should be developed together with new checklists for operators. However, often due to lack of time and resources, these are documents that are often remained left to be updated. In extreme cases, working instructions are passed over verbally from a senior operator to a junior and so on. We have seen cases where an SOP and checklist have been developed, proved efficient and have been presented officially during a training program, but then was abandoned by operators after one month since "the Senior Operator knows better".

Updating the P&IDs and SOPs together with checklists are very often becoming actions and recommendations in the HAZOP studies.

Reviewing the evolution of checklists in related industries to determine which elements may be transferable

Even the most expert among us can gain from searching out the patterns of mistakes and failures and putting a few checks in place.

In medicine, a study performed to identify the impact of using the checklist showed that the rate of major complications for surgical patients in eight hospitals fell by 36 percent after introduction of the checklist. Deaths fell 47 percent. Infections fell by almost half. The number of patients having to return to the operating room after their original operations because of bleeding or other technical problems fell by one-fourth. Overall, in this group of nearly 4,000 patients, 435 would have been expected to develop serious complications based on earlier observation data. But instead just 277 did. Using the checklist had spared more than 150 people from harm and 27 of them from death.

Tom Wolfe's book, "The Right Stuff," tells the story of our first astronauts and charts the demise of the maverick test-pilot-culture of the 1950s. It was a culture defined by how unbelievably dangerous the job was. Test pilots strapped themselves into machines of barely controlled power and complexity, and a quarter of them were killed on the job. The pilots had to have focus, daring, wits, and an ability to improvise—the right stuff. But as knowledge of how to control the risks of flying accumulated—as checklists and flight simulators became more prevalent and sophisticated—the danger diminished, values of safety and conscientiousness prevailed, and the rock star status of the test pilots was gone.

Figure 14.

Nowadays, before the pilots started the plane engines at the gate, they adhered to a strict discipline - the kind most other professions avoided. They ran through their checklists. They made sure they'd introduce themselves to each other and the cabin crew. They did a short briefing, discussing the plan for the flight, potential concerns, and how they'd handle troubles if they ran into them. And by adhering to this discipline—by taking just those few short minutes—they not only made sure the plane was fit to travel but also transformed themselves from individuals into a team, one systematically prepared to handle whatever came their way.

There is one element that we can learn from aviators and that is discipline

We don't know what we don't know

.. there are known knowns; ... We also know there are known unknowns; . . . But there are also unknown unknowns the ones we don't know we don't know. - Donald Rumsfeld

The "known knowns" are those issues that pose a hazard, have been identified, and have probably been addressed effectively including via a checklist used during design or operation.

The "known unknowns" are the focus of most hazards analyses. For example, a process may involve the transfer of chemicals from a high-pressure vessel to another vessel that operates at lower pressure. The designers of the system may not have considered the possibility that the pressure gradient could reverse, i.e., that the pressure in the first vessel could drop to a value less than that in the second vessel, thus creating the possibility of reverse flow. Such a scenario is a valid topic for the hazards analysis team to discuss. This scenario is a "known unknown" - it may not have been considered, but it is part and parcel of a normal hazard analysis.

It is the "unknown unknowns" that the operation and engineering team and particularly the team leader needs to give special consideration to. These situations require creative and imaginative thinking. In this context, another Rumsfeld quotation: "absence of evidence is not the evidence of absence" should be considered. Getting team members to "think the unthinkable" is one of a team leader's biggest challenges. First, we all know the, "I've never seen it happen, therefore it can't happen" syndrome. Second, these low probability scenarios usually involve the simultaneous occurrence of contingent events (which is why they occur so rarely). Once more, team members typically have trouble accepting and understanding unlikely combinations of events. To help overcome this block, the leader may choose to describe a number of real accidents that occurred elsewhere to show how "weird" they were - yet they happened.

Conclusion – Human Factor

Are the checklists the magic bullet for all our reliability and operational issues? The simple answer is No.

As much as we would like to take credit as safeguard for our operation procedures and "operators are trained to deal with this situation" the reality proves otherwise. During the stressful conditions of plant trips, sometimes on night shifts, overloaded with alarms and radio calls, even well-trained operators can make mistakes despite having the best documented checklist in hand.

Research indicates that for a complicated non-routine task, like restarting a plant to original process state, the error rate is very high at 0.1. For routine tasks where care is required, a simple task like resetting a valve after some related work has a physical operation error of 0.01 (equivalent with 1 error in 100 operation). However, there are of course many other scenarios which make errors much more likely, acting as a multiplier factor to original error probability. For example:

- if operator is unfamiliar with a situation that is potentially important, but which is either novel or occurs only infrequently, the nominal error probability multiplies significantly.
- If there is a shortage of time for error detection and correction, the probability of making mistakes also grows. For general high stress situation, the error rate is 0.25 meaning that 1 in 4 maneuvers may go wrong.

But, as Norsk Hydro demonstrated, intelligent application of checklists can be a life saver. Applying the latest development in automated systems and training their operators, including simulating non-routine tasks, considering previous industry incidents, the nominal error probability can be significantly reduced.

The safest future involves trained and confident operators applying dynamic checklists for significant non-routine operations.

References:

1. The Checklist Manifesto – Atul Gawande
2. PSM Lessons from a Major Reformer Furnace Failure - RDC Prior, SH Excellence CC, Johannesburg, South Africa, IChemE HAZARD Symposium Series No 159
3. Implementation of Best Practices for Improvements in Reformer Performance and Reliability - Giuseppe Franceschini Process Owner Inspection, Yara International, James R. Widri, Manager, Advanced Engineering, Quest Integrity, LLC, AICHE Ammonia Safety Symposium 2015
4. Dr David J. Smith, 'Reliability and Maintainability and Risk'
5. Process Risk and Reliability Management 2010, Ian Sutton.
6. Safe Start-up of Ammonia / Urea Plants under Challenging Circumstances - Klaus Noelker ThyssenKrupp Industrial Solutions AG, AICHE Ammonia Safety Symposium 2014
7. 99 Questions to Ask during a Design Review Session, AmmoniaKnowHow.com
8. Common Problems on Primary Reformers - Bill Cotton and Peter Broadhurst, Johnson Matthey Catalysts, AICHE Ammonia Safety Symposium 2004
9. Guidance on Learning from Incidents, Accidents and Events Edward Smith, Principal Consultant (DNV GL), Richard Roels, Senior Consultant (DNV GL), DNV GL, IChemE HAZARD Symposium Series No 160
10. Commissioning Handbook; Plant Design and Operations and Operation Procedures book by Ian Sutton
11. Safety Management System: Safety Performance Indicators and Results, Willy Bjerke Norsk Hydro, N-0240 Oslo, Norway, AICHE Ammonia Safety Symposium 1997
12. Catastrophic Failure of Reformer Tubes at Courtright Ammonia Plant, Bhaskar Rani Terra Industries Courtright, Ontario, Canada, AICHE Ammonia Safety Symposium 2006
13. Primary Reformer - Firebox Explosion New BMS and HAZOP Actions overview - Aurelien Flamme YARA, Tertre, Sammy Van Den Broeck YARA, Sluiskil, Wim Versteele YARA, YPO

The extension of tubes' life on a Top fired Primary Reformer: A Greek case study

This paper describes the case study of New Karvali Fertilizers s.a. (NKF) industrial complex, which operates a small-scale ammonia plant using Natural Gas (NG) as raw material. The NKF's choice to operate the convectional top fired primary reformer with steam to carbon (s/c) ratio 3.4 mole/mole, higher than the conventional ones, is described and the results of this choice are presented. Among others, the interesting case of failure-free reformer tubes after 191,000 hours of operation with only one catalyst change in between, which is almost the double expected lifetime of the tubes, according to the manufacturer's data sheet, is analyzed. In this time-period of operation, as well as during the several shut downs of the plant, all the appropriate non-destructive and destructive tests were executed without important findings, the catalyst was from time to time screened and only once fully replaced. Finally, the conclusions of surveys and performance reports are outlined, aiming in reaching a new world record on conventional primary reformer tubes and catalyst operating life.

Dr Theodora Kouloura
Chief Operation Officer, New Karvali Fertilizers s.a., Greece

Emmanuel Siopis
Deputy Production Manager, New Karvali Fertilizers s.a., Greece

Introduction

The operational life of NKF's primary reformer tubes and catalyst was extended longer than the usual one. This result was based on one side to the ratio S/C 3.4 mole/mole, higher than the conventional one and on the other side to the process followed on screening and only partial replacement of the existing catalyst.

Background

NKF is the only nitrogenous fertilizer producer in Greece. Production facilities are located in Northern Greece and includes several production units in operation as an ammonia, a nitric acid and two sulphuric acid units and an AN/CAN fertilizer production unit with a capacity of 300,000 MTY. Additionally, different utilities as demineralised water, cooling water, power generation from waste heat (21MW), 15,000 MT atmospheric ammonia tank, 1,800 MT semi-refrigerated ammonia sphere as well as off-site facilities (bagging and palletising units, storage tanks for acids and bulk storages for sulphur and fertilizers) are involved.

More specifically, the ammonia plant conceived to exhaust small gas field at maximum possible plant design capacity-energy efficiency, unprecedented in early 80's for small capacity Ammonia units. Licensor (Park Process) and general contractor for EPC project was KTI –Zoetermeer /Holland, today Technip.

The ammonia plant was started up in 1988 with a nominal capacity of 400 MTD. The Plant ceased operation in late 1992 due to exhaustion of the local field reserves and was mothballed for a long shut down. The plant resumed operation in early 1998 after construction of Greek national NG grid providing the plant with imported Russian natural gas (NG) (94.0% CH_4).

In 1999 plant was revamped for capacity increase and improved energy efficiency. Plant capacity is maintained stable between 470-480 MTD up to date.

System Configuration

The main characteristics of this ammonia plant are the following:

- Natural gas is used only as primary reformer feed
- No secondary reformer, LTS and methanator
- Very quick start up process
- Purification of hydrogen by PSA mol sieves. High purity syngas
- No energy consuming CO_2/ residual CO/CH_4-removal. Zero synloop purging
- Primary reformer burners designed to use low calorific purge gas from PSA unit and enriched air from nitrogen generator
- Steam generated from process waste heat to drive power generator covering all motive electrical power requirements
- Synloop reciprocating electrical motor driven compressors due to relatively small plant capacity and high loop pressure
- Air cooled steam turbine condenser
- Practically nil purge gas from synloop
- Superheated steam generated from process waste heat at high pressure 45 bar_g and 440°C.

Furthermore, the plant incorporates many innovative features first applied in an ammonia plant design:

- Innovative flexible thin tube-sheet process gas cooler with *external* low temperature bypass, 28 years without failure.
- Enriched air from nitrogen generator (35% O_2) used as reformer combustion air
- First application of CASALE axial-radial ammonia converter.
- First application of UOP PSA technology for hydrogen purification by innovative N_2 purge and recompression. High H_2 recovery minimum 93% and savings in N_2 compression energy.
- First application TSA technology for direct H_2S, COS and mercaptans removal by innovative low energy cost regenerative mol-sieves. Last traces of sulfur components are converted to H_2S and removed in Zink Oxide vessels.
- Innovative thin tube sheet, low alloy synloop WHB. High specific steam production of 1.35.

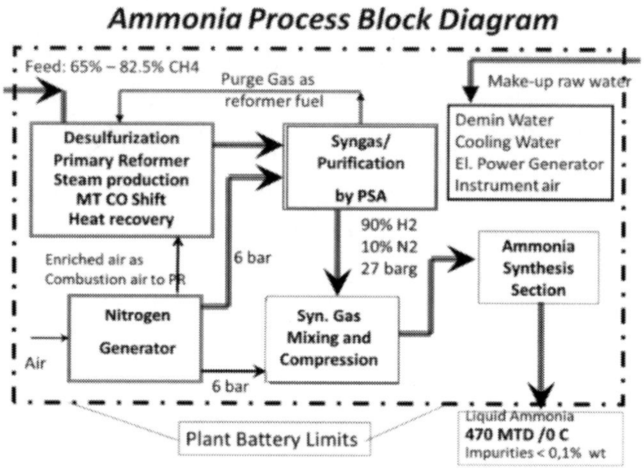

Figure 1. Ammonia unit block diagram

As it is shown in figure 1, the plant was designed for two feedstock qualities. One associated gas from off-shore Prinos sour oil field with only 69.6-76.7% CH_4, Feedstock A and one sweet lean gas with 80.6-84.7% CH_4 from South Kavala off-shore gas field, Feedstock B. Due to the declining of the Prinos gas, South Kavala gas had to be added to make up the required gas quantity. Gradually the composition of the supplied gas ended up becoming Feedstock B until the exhaustion of the South Kavala field by the end of 1992.

As the plant resumed production in early 1998 with 94-98% CH_4 sulfur free Russian gas, the plant was revamped to increase capacity up to 480 MTD.

Desulfurization. As the Prinos field associated gas contained 200 ppm sulfur compounds of which 160-170 ppm methyl/ethyl mercaptans, 20-40 ppm COS and 10 ppm H_2S, the unit was equipped with an innovative efficient desulfurization sub-unit.

Instead of the conventional energy consumption, the NG passes first at ambient temperature over a 4-bed fully automatic TSA (Temperature Swing Adsorption) mol-sieve desulfurizer where all the H_2S, most of the mercaptans and COS was adsorbed. The last traces of mercaptans and COS are converted to H_2S over a small COMOX catalyst bed at 320^0 C after hydrogenation. This remaining H_2S is absorbed by ZnO guard-bed. The NG stream used for mol sieve regeneration is then used as fuel in the reformer. It is obvious that, since Russian NG was introduced in 1998, the 4 TSA desulfurizers are staying idle and being bypassed!

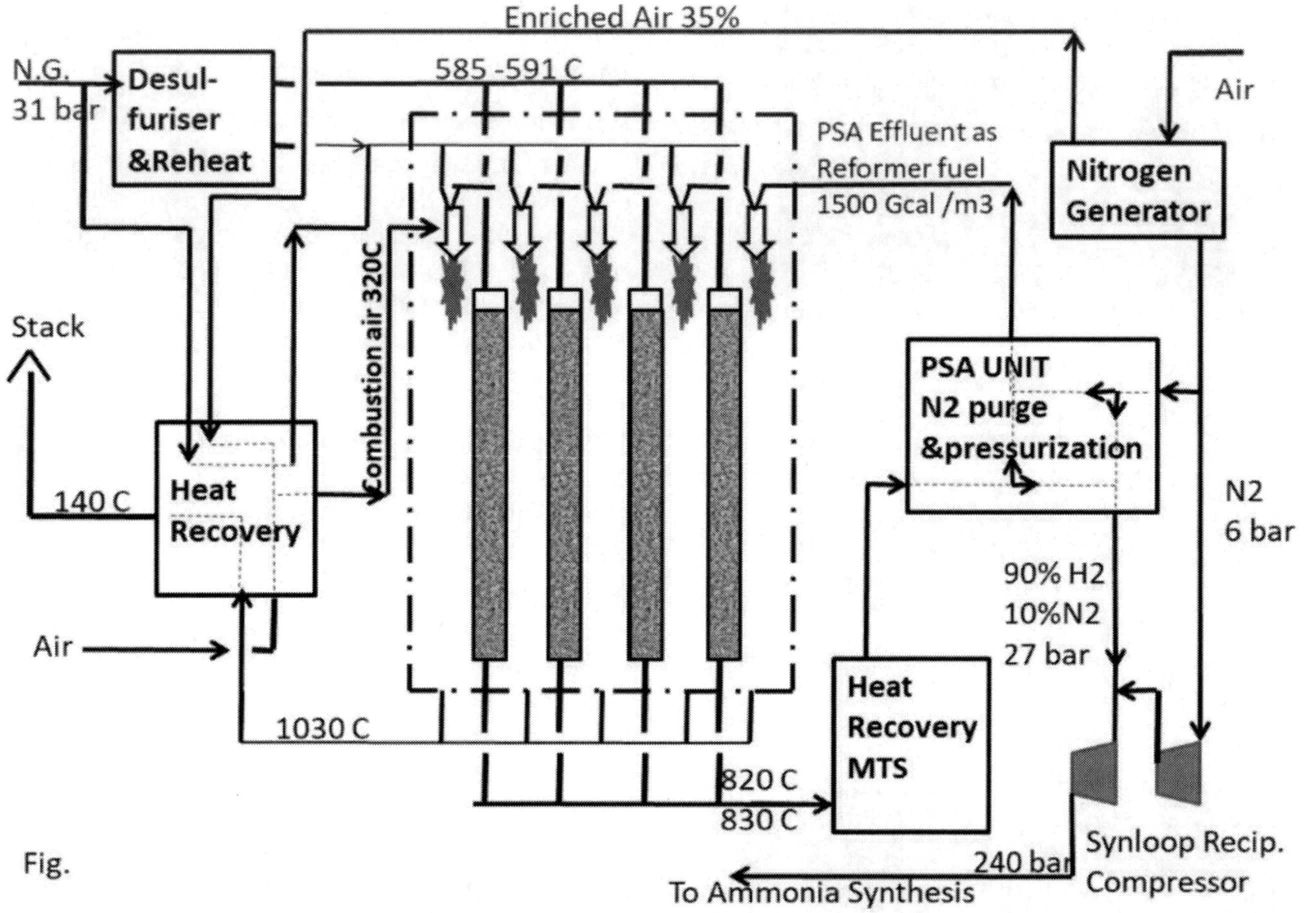

Figure 2. Simplified PFD at nominal plant capacity

Primary reformer. Top fired conventional primary reformer KTI design as per Figure 3.

Tubes. 136 tubes (25Cr 35 Ni+Nb- S&C G4852), ID/OD 100/122.8 mm, placed in 4 rows, catalyst loaded height 11.62m, tube pitch 260mm, lane spacing 2,200 mm.

Catalyst: 12.2 m^3/10,509 Kg Synetix 57-4, up to May 2015, then JM 57-4Q.

Burners: 50 John Zink low NO_x burners in 5 rows able to operate with NG or with low calorific value PSA effluent, or with mixture of both as per figure 4. Capacity of the outer rows is about 50-60% of middle rows burners.

Figure 3. Tubes and burners (inside view)

Figure 4. Burners using NG and purge gas

Interventions and Inspections

During July 2000, new catalyst, ICI 57-4, was charged in all 136 tubes after 81,000 hours in operation. Five years later, in December 2005 all 136 tubes (120,000 hours in operation) were inspected over the complete tube length. Eddy current, tube expansion, metallographic –tensile-strength –hardness examination was executed. Tube C 12 was replaced. Only in the tube-bottom was an expansion of 2%, found by eddy current. By metallographic examination the real expansion revealed to be 4%.

Furthermore, in October 2008 all 136 tubes (140,600 hours in operation) were inspected (eddy current, tube expansion). Although OD growth rate had not exceeded 1.4% and only small indication of cracks observed, 24 tubes were replaced. 19 out of the 24 tubes were from the two middle rows. For top and bottom parts of tubes B-19, B-22, C-9, C-14, with expansion up to 1.5 - 1.9% metallographic examination, tensile strength, hardness has been performed. Screened catalyst plus about 20% new catalyst was charged in the replaced tubes.

In the next shut down, during May 2011, all 136 tubes (112 tubes 163,000 hours in operation) were inspected again. 42 tubes in middle rows and 4 in one outer row with OD growth rate between 2.0 and 2.4%. Many hot bands observed were replaced. Screened catalyst plus 20% new catalyst was charged in the replaced tubes only. The old catalyst was load on the bottom and the new one on the top.

Finally, in May 2015 last 65 tubes were replaced (182,000 hours with plant at nominal capacity plus 9,000 hours in the last year with plant in abnormal condition at reduced capacity), but without tube failure incident. New catalyst JM 57-4Q was charged in the all 136 tubes. The history of replacement is illustrated in Figure 5.

Samples from four tubes were selected based on hot spots, that had been detected with a pyrometer during the last year operation due to the deterioration of the catalyst. Tubes were removed from C29, C30, D9 and D24 position. These tubes were temporarily exposed to temperatures between 940°C and 965°C. Normally the tubes are exposed to metal (tube skin) temperatures of 880°C (top part) and 905°C (bottom part) and inner pressure of 31 bar. The condition of the examined samples show that the tube material is widely damaged due to creep.

By metallographic examination a slight decrease of number of cavities can be noticed in comparison to the samples previously examined. The results of hardness measurement and tensile test are slightly higher (worse) compared to the previous results.

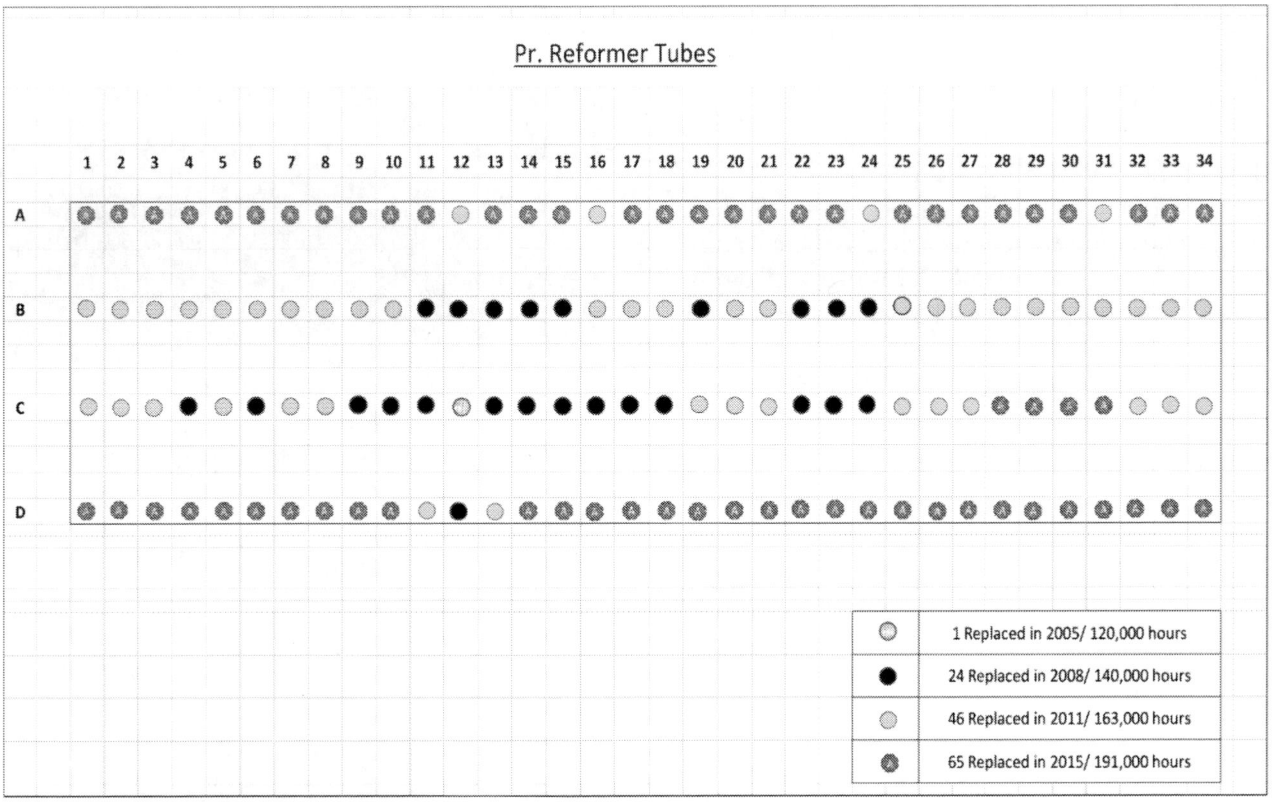

Figure 5. Arrangement of tubes including marking tube replacement data

Reformer tubes monitoring. The trend of:

a) the calculated CH_4 Approach to Equilibrium (ATE),
b) the uniform pressure drop of loaded catalyst
c) the Tube Wall Temperature (TWT) and
d) the balanced reformer firing

is important to be followed during routine plant operation, in order to optimise the catalyst's performance, as well as to provide early detection of conditions that could lead to catalyst or tube failure.

A. ATE An increased ATE or Methane Slip (MS) from the reformer, mainly due to the natural or exceptional catalyst activity loss, results not only in increased energy consumption, but also on hotter tube wall temperatures.

As this plant has no secondary reformer to compensate MS increase, the unique action left is increasing reformer firing. It is well known that even a slight increase in the tube wall temperatures will have a serious impact on the tube life. A TWT increase of as little as 10^0 C may result in up to a 30% shortened life-time of the reformer tubes.

When trying to alleviate the effect of increased ATE or MS by operating the primary reformer at higher exit temperature and higher S/C ratio, this is done almost certainly at the cost of increased fuel consumption and at the cost of reformer tube life, running them with hotter wall temperatures.

NKF decided to operate under constant exit temperature of process gas at 820°C and constant S/C 3.4 mole/mole. Under this operation conditions at the initial catalyst loading the MS was 6.7% and the ATE 1.5. Following the diagram in figure 6, although the methane slip was stable for almost four years, the ATE was increased.

It is important to be notified that in every shutdown of the reformer or after sudden trips of the ammonia plant due to several causes, which had many hours to be restored, NKF was proceeded on steaming of the reformer's catalyst.

The goal was to operate the reformer under conditions that result in the lowest possible tube wall temperatures (TWTs).

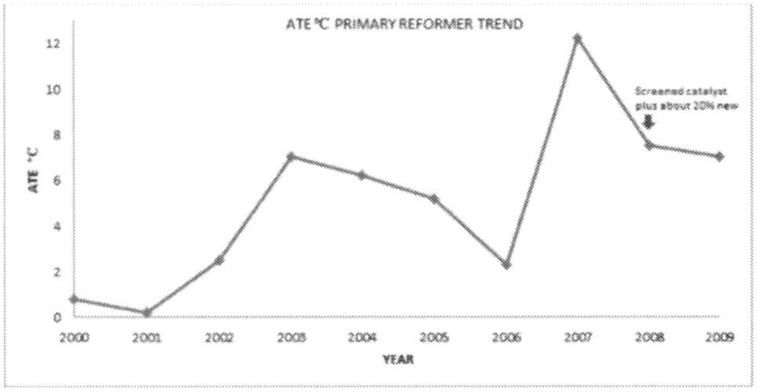

Figure 6. The trend of ATE

The reformer evaluation reports carried out explain the ATE trend given above. As already mentioned, the catalyst has undergone frequent reversible deactivations (carbon formation, oxidation) and subsequent regenerations.
Also, it is visible that after the catalyst screening and the charge of the new catalyst both the MS and ATE have been improved

B. Catalyst loading. The reformer being one of the most intensive and time-consuming reactors during catalyst loading, it is obvious that to replace and load tubes, is imposing severe penalties in case that the tubes are loaded improperly.

At the catalyst loading in all 136 tubes in year 2000 and at partial tube replacements in 2008 and 2011, the new tubes were loaded properly (UNIDENSE method), but non-replaced tubes remained enclosed, loaded with old catalyst having served for about 60,000 hours (close to 8 years) up to 15 years.

Figure 7. Catalyst before and after screening

Almost 20% of the catalyst, served 11 years in a reformer tube was broken (Figure 7). Broken pieces certainly are increasing the Dp and reducing the gas flow through. Sintering due to high TWT and mechanical degradation causing pellet breakage, are irreversibly affecting catalyst activation and Pressure Drop.

C. Tube Wall Temperature (TWT) Measurement. Various non-contact infrared methods are used to measure reformer tube temperatures. However, these measurements rely on the accurate determination of tube surface emissivity, and require a method of compensating for reflections from the hotter surrounding furnace environment and from hot flue gas. Often routinely used tube skin temperatures measurement may be significantly in error. Most infrared thermometers require the operator to estimate the target emissivity to obtain the true target temperature.

NKF executed combined portable infrared measurement with gold cup thermometer and pulse laser technology with narrow bandwidth single-color infrared thermometers to automatically determine emissivity values and accurately correct for true target temperature. The gap in the results of the two different ways of measurements was relatively small and gave a room of confidence for the real TWT.

D. Balanced Reformer Firing. As it is shown in figure 5, 61 tubes out of 68 from the outer rows A and D were replaced in May 2015, when only 4 out of 68 from the middle rows were replaced in the same period. This is indicative for a poorly balanced reformer firing, where the middle rows B and C got a higher heat flux /higher TWT to the detriment of the expected life of the tubes in these two rows (see figures 8 and 9).

Figure 8. Tubes from row A top (2011)

Figure 9. Tubes from row B top (2011)

In year 2008 tubes were for 140,600 and the catalyst for 60,000 hours in operation. It is evident that due to the Dp difference non-uniform flow rates were established between tubes.

That in turn had affected to these tubes to operate hotter than tubes with lower Dp. This is manifested with wall temperature differences up to 80^0C recorded for the Primary Reformer tubes.

In order to be sure that no failure of tubes would be occurred, the process of steaming and reactivating the catalyst was followed in every time opportunity.

For sure, there are other conventional primary reformers where HP 45 tubes operated for 17-21 years without failure, but it is not recorded up to now cases with the same catalyst in the tubes for 15 years.

Unfortunately, samples of discharged catalyst in 2005, 2008, 2011, and 2015 have never sent for examination of its residual activity.

Thermal cycling effect to reformer tubes

In the figure 10, the numbers of reformer trips, shut downs, hot and cold start-ups from 1985 to 2015 are given. Shut downs, hot and cold start-ups are generating very high through-wall stresses. The cumulative effect of several thermal cycles is very damaging and results in accelerated creep cracking. The impact of shutdown and start up on the tube life is very important. The impact of transients during trips and hot start up is also quite important. The reformer tubes have had in average 2.2 cold start-ups per year and 2.6 transients.

The new tubes of the same material and with upgraded JM catalyst 57-4Q are expected to get about 1 cold start-up and 1 transient per year in average. Based on the previous experience, and improving the recorded process, NKF assumes that the life cycle of both, Primary Reformer tubes and catalyst, could be extended more than the conventional ones.

TRIP REFORMER

MONTH	1985	1986	1987	1988	1989	1990	1991	1992	1993	1998	1999	2000	2001	2002	2003	2004	2005	2006	2007	2008	2009	2010	2011	2012	2013	2014	2015	TOTAL
IAN		4	1		1								1		1							1						
FEBR			3+1			1		1					1															
MAR			4	1		2												1				1						
APR										2		1		1							1							
MAI				1	1					5										1								
JUN			1		1												1											
JUL					1				1					1			2+1	2			1							
AUG		1		1				1	1	1							2+2			1								
SEPT		1	1+1									1																
OCT		1								1									1		1							
NOV		2																1										
DEC	4		1	1																			1					
TOTAL	4	9	13	4	4	3	0	2	2	9	0	2	2	2	1	0	8	4	1	2	3	2	1	0	0	0	0	78
COLD STARTS				14																								
TRIP OUTSIDE CAUSES				14																								

SHUTDOWN REFORMER

MONTH	1985	1986	1987	1988	1989	1990	1991	1992	1993	1998	1999	2000	2001	2002	2003	2004	2005	2006	2007	2008	2009	2010	2011	2012	2013	2014	2015	TOTAL
IAN	1										1														1			
FEBR					2																			1				
MAR		1			1																							
APR					1	1		1	1	1		1	1	1														
MAI							1	1	1														1				1	
JUN						3			1			1														1	1	
JUL						4			1					1							1							
AUG				2		2							1											1				
SEPT		2	2	1	1				1																			
OCT			1			1							1					1							1			
NOV			3		1		2										1			1								
DEC			1		3	1																						
TOTAL	1	3	7	3	9	12	3	2	5	1	1	2	3	2	0	0	1	1	0	1	1	0	1	2	2	1	2	66
COLD STARTS			66																									
OUTSIDE CAUSES			5																									

Figure 10. The number of reformer trips, shut downs, hot and cold start-ups from 1985 to 201

Conclusions

- ➢ NKF's ammonia plant as well as its primary reformer have a specific design oriented to low NG specific consumption and use of low calorific value purge gas
- ➢ NKF's ammonia plant operates under specific conditions in terms of feedstock synthesis as well as zero synloop purging.
- ➢ Under these conditions, the life cycle of the reformer's catalyst is affected from several parameters and could be extended using screening process and addition of a small portion of new catalyst.
- ➢ The life cycle of primary reformer's tubes could be extended using an increased ratio of S/C up to 3.4 mole/mole and proceeding on steaming and regenerating of the catalyst from time to time.

Acknowledgements

We would like to express our gratitude to all those who gave us the possibility to complete this paper. We want to thank:

1. New Karvali Fertilizers s.a. President and the Board of Directors, who give us the opportunity to attend the conference
2. AIChE, who invited us to attend the 64th Annual Safety in Ammonia Plants and Related Facilities Symposium

Safe Solution for Pre-Reduced Ammonia Synthesis Catalyst Replacement

For more than 70 years, Topsoe has played a leading role in the evolution of the ammonia industry by continuously improving its process-related know-how of the chemical industry and by developing of state-of-the-art catalysts. Among these, the pre-reduced version of Topsoe ammonia synthesis catalysts (KMR 111 and KM1R) possesses a synergetic combination of operational reliability (selectivity, activity, stability, resistance, robustness, simple and fast activation, etc.) and logistics strength (compliance with environmental and transportation legislations and the low risks associated with storage, installation and activation, etc.) that makes them the quintessential example of Topsoe's contribution to a safer and more reliable ammonia industry.

This paper addresses two fundamental aspects associated with the replacement of an ammonia synthesis catalyst charge: The inherent safety hazards of installing a pre-reduced catalyst and the plant operation at transient conditions such initial start-up.

Carlos Axel Diaz-Tovar and Bo Hartvigsen
Haldor Topsoe A/S, Denmark

Introduction

Since the introduction of the first side-fired steam reformer in 1957, the company founded by Haldor Topsoe has played a leading role in the evolution of the ammonia industry.

Through the development of the radial flow ammonia synthesis converter in 1966 to the groundbreaking SynCOR™ technology in 2017 [[**1**], Topsoe has focused on the continuous improvement of its process-related know-how of the ammonia industry.

Alongside the improvement of its know-how, Topsoe has made substantial research efforts in developing state-of-the-art catalysts since the first Topsoe ammonia synthesis catalyst charge was produced in 1947.

These research efforts are aimed at having catalysts with such characteristics that ensure:

a) Operational reliability (selectivity, activity, stability, resistance, robustness, simple and fast activation, etc.)
b) Logistics strength (compliance with environmental and transportation legislations, low risks associated with storage, installation, activation, etc.)

The synergetic combination of operational reliability and logistics strength makes the KMR 111 and KM1R catalysts the quintessential example of Topsoe's contribution to a safer and more reliable ammonia industry.

This kind of contribution is key when the long operational lifetime (in some cases more than 30 years) of a Topsoe ammonia synthesis catalyst makes the installation and activation of the new charge (i.e. replacement) possibly a once-in-a-lifetime experience for most of the personnel in an ammonia plant.

The unloading, installation/loading and activation/startup need to be meticulously planned in order to avoid any incidents, which could result in serious personnel injuries and/or equipment damage, while minimizing the plants downtime. Consequently, the purpose of this paper is to address two fundamental aspects associated with the replacement of the catalyst charge:

1. The inherent safety hazards of installing a pre-reduced catalyst and
2. The plant operation at transient conditions such as start-ups

In general, any catalyst installation is associated with many hazardous activities. Among this, the confined space entry in a long and narrow reactor with limited egress is one of the most challenging from a safety standpoint; especially when the catalyst to be loaded is in its pre-reduced state as it is liable to self-heat in contact with air and without an energy supply. Furthermore, when this type of catalyst is not passivated to withstand the weather and work conditions found in a typical loading, the installation normally takes place under an inert atmosphere.

Consequently, having an ammonia synthesis catalyst in its pre-reduced state that is stable and that can be installed in normal atmosphere is, from a safety and an economic stand point, highly valuable for any ammonia plant owner.

Although installing a pre-reduced catalyst has the drawbacks previously mentioned, they are significantly outweighed by the incentive of reducing 2 to 4 days the time that the plant is operating in transient conditions, i.e. activation/start-up of the ammonia synthesis catalyst.

The benefits that the pre-reduced version of Topsoe KM catalysts have brought to the ammonia industry are tangible in more than 1200 charges of ammonia synthesis catalyst that have been successfully manufactured, installed, and activated. The vast majority of which have been with our recommended combination of pre-reduced and unreduced catalysts (Figure 1).

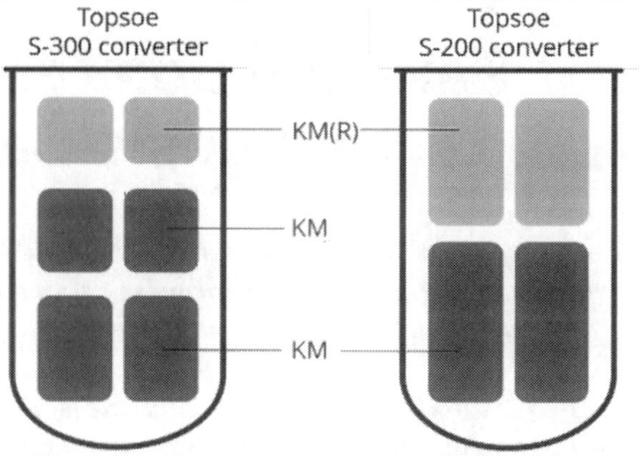

Figure 1. Topsoe's recommended loading scheme for ammonia synthesis converters. (R) denotes pre-reduced catalyst.

KMR 111 & KM1R: The path of highly stable pre-reduced ammonia synthesis catalysts

To produce a catalyst that offers both operational reliability and logistics strength, the path (Figure 2) from production to normal operation has to be carefully planned.

Clearly, the quality of the raw materials, the optimal amount of the promotors as well as the manufacturing procedure (i.e. the production step) are the key factors that determine the final quality of an ammonia synthesis catalyst. However, the scope of this paper is not to focus on the production but to highlight the steps that allow the production of a highly active pre-reduced KM catalyst that is safe to handle, transport, install and activate.

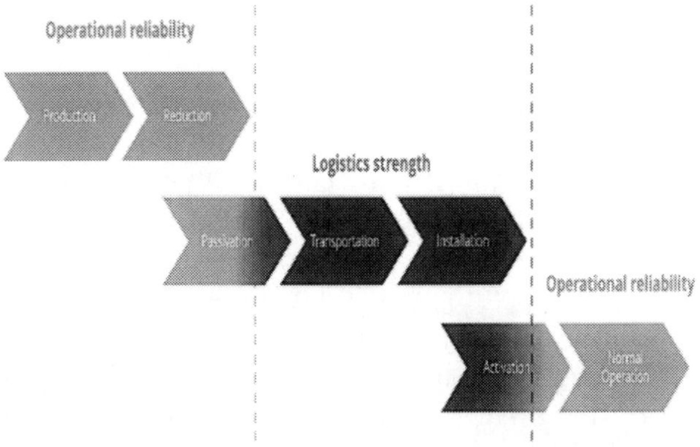

Figure 2. The path of producing, installing and activating Topsoe pre-reduced ammonia synthesis catalysts

In the following sections, each one these steps will be discussed.

Reduction

The intrinsic catalyst activity depends on the particle size in which the catalyst is reduced from iron oxide to elemental iron and, of course, on the reduction procedure [2].

<u>*The effect of water*</u>
Water and high temperatures are promoters of hydrothermal sintering. This phenomenon is characterized by the growth of the iron crystallites. This growth has a negative impact on the catalyst activity as it decreases the available superficial area and could also block or collapse pores, making the crystallites inaccessible for the reactants.

Additionally, water inhibits the ammonia synthesis reaction; which means that higher temperatures are required for reaction to take place.

<u>*Particle size*</u>
The larger the particle size, the longer it takes to complete reduction of a given particle and the more damage is therefore inflicted to the following catalyst layer due to water released in the reduction of this particle.

<u>*Reduction procedure*</u>
As it is well known for other metal catalysts made by the reduction of their oxides, the method of reduction influences the properties of ammonia synthesis catalysts.

The reduction procedure should ensure that the pore system formed as a result of the removal of the oxygen ions increases the superficial area as close as possible to the maximum theoretical one. Furthermore, the characteristics of the pore structure that develops during the reduction depend strongly on the heating rate and the local gas composition (mainly H_2O) to which the catalyst is exposed.

A controlled heating rate is relevant as it prevents the reduction from developing rapidly, leading to high water vapor concentrations in the gas, and because the crystalline structure of the already reduced catalyst is prone to be altered, reconstructed and accelerated by water vapor and it is in detriment of the final catalyst activity.

It is therefore of outmost importance to limit the amount of water vapor in contact with the catalyst and this can be achieved with a high space velocity. Although, an initial space velocity of 3,000-5,000 V/V/H is difficult to achieve in an industrial installation, it can be done in the manufacturing facilities of the KM catalysts.

A pre-reduced catalyst charge will then not only ensure a higher activity of the catalyst but also that the heat generated by the ammonia synthesis reaction will allow the plant to increase the space velocity and to keep it as high as possible throughout the activation of the charge. This will lead to a more controlled heating rate and a lower vapor water concentration in the exposed catalyst.

In Figure 3, when the reaction is fast, the reduction would occur in a narrow zone that moves progressively from the outer surface to the unreacted dense center (left); when the reaction is inhibited, e.g. due to the presence of high concentrations of H_2O, there would be an uneven profile of reduction degrees throughout the KM1 granules (right).

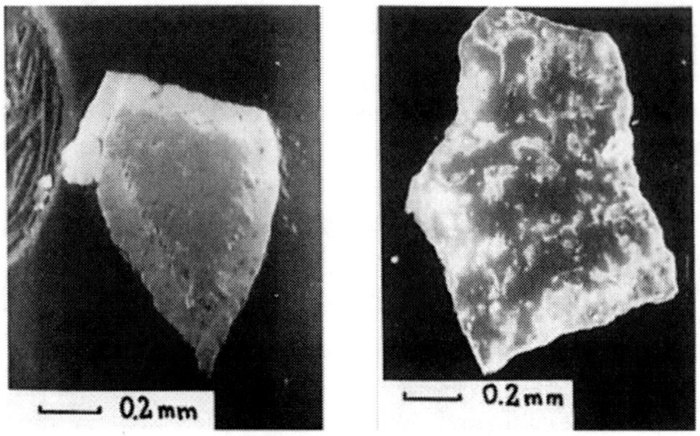

Figure 3. Fast reaction (left) vs inhibited reaction (right). Image obtained from [3]

Passivation

A reduced ammonia catalyst is strongly self-heating in contact with air even at ambient temperature. Consequently, the pre-reduced catalysts used in the industry have been given a passivation treatment upon reduction so that they can be transported and installed.

The passivation treatment is carefully conducted considering that:

- Skin oxidation has an influence on the pore structure and the surface area, since voids may be closed off and the mouth of the pores shrunk reduced
- It is mandatory that KM-catalysts comply with the international transport regulations UN 2881 Metal Catalyst, Dry (Iron), Class 4.2, Packing Group II; which are part of the document "Recommendations on the Transportation of Dangerous Goods, Model Regulations" [4]

Additionally, the state-of-the-art passivation treatment given at manufacturing facilities provides Topsoe's latest development within ammonia synthesis catalysts, the KMR 111, and the industry-leading KM1R catalyst, with sufficient stability so that they can be loaded in atmospheric air.

Installation / loading

The loading of the catalyst is usually overlapped with the mechanical installation of the beds, inter-bed heat exchangers, among others, and it is associated with many high-risk activities such as: handling and exposure to hazardous substances, working at heights, and the most critical one, confined space entry in a long and narrow reactor with limited egress.

According to the Catalyst Loading Hierarchy (see Figure 4) issued by the European Catalyst Manufacturers Association [5], avoidance of reactor entry is the preferred option. However, if reactor entry is required, particularly in the case of inert/nitrogen atmosphere (or other potentially hazardous residual or incidental substances), an appropriate breathing apparatus such as an integral helmet and appropriate air supply is normally required.

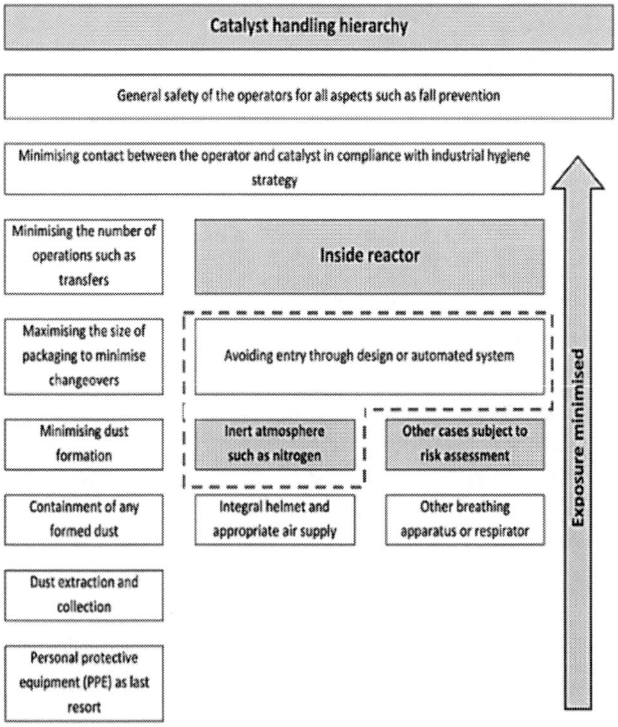

Figure 4. Catalyst handling hierarchy. Image obtained from [5]

Reactor inert entry is a high-risk operation and strict procedures are required including, in the case of European countries, Europe Conformity (CE) approved specialized equipment, supervision and training required (see Figure 5).

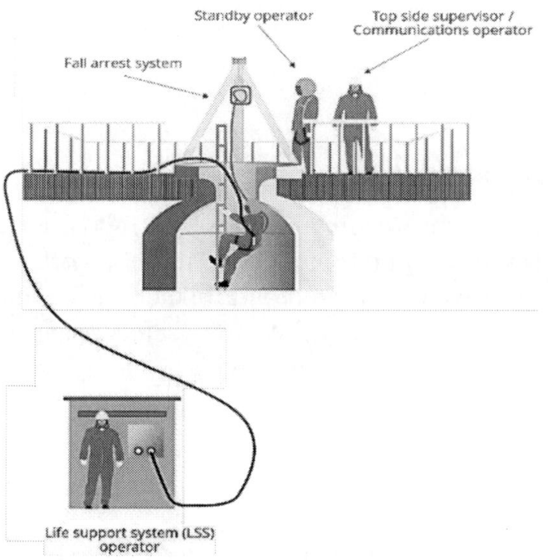

Figure 5. Example of life support system (LSS) organization for inert entry on the top of the reactor. Image taken from [5]

When considering that over the past decade, at least two full charges of pre-reduced catalyst and over 10 of the pre-reduced/unreduced combination have been installed in atmospheric air (Figure 6) every year worldwide without any incidents, then one can conclude that Topsoe pre-reduced KM catalysts possess a unique feature that allows customers to not only reduce the loading time, but also to avoid the additional safety risks and hazards associated with inert entry.

Additionally, from an economic perspective the expenses of loading a pre-reduced catalyst in inert atmosphere can be at least 25% higher than loadings done in atmospheric air. These expenses are not minor when one takes into consideration that the cost of loading in nitrogen atmosphere a standard size 80 m^3 ammonia synthesis converter is approx. $200,000 (additional cost of one bed in a split charge is approx. $50,000).

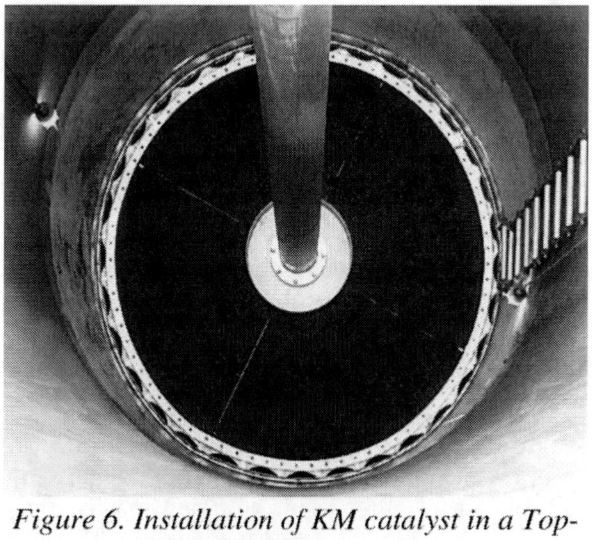

Figure 6. Installation of KM catalyst in a Topsoe radial flow converter

Activation / Start-up

The protective layer added during the passivation treatment must be removed again before the catalyst regains its activity for ammonia synthesis.

Evidently, the amount of water formed during the re-reduction of the catalyst is much lower than the one formed during the reduction of the oxidized catalyst. Figure 7. shows a comparison of the expected amount of water to be formed (in short tons) during the activation of three different charges of ammonia synthesis catalyst with different percentages of pre-reduced catalyst.

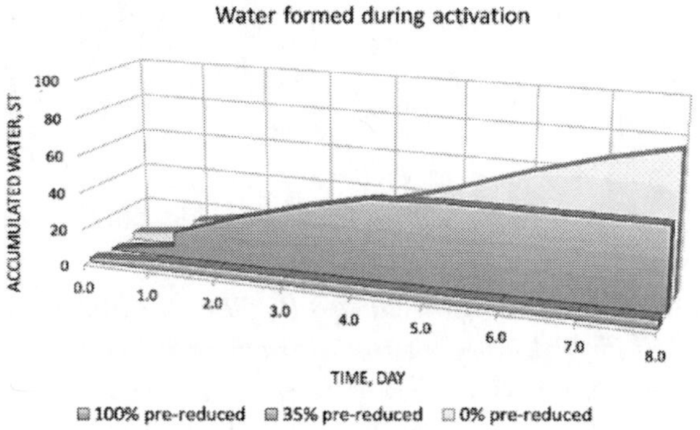

Figure 7. Expected amount of water formed during the activation of three 90m³ catalyst charges with different content of pre-reduced catalyst (0, 35 and 100%)

While it is expected that the charge with only pre-reduced catalyst stops forming water as early as 36 hr. after the activation had started, the full charge of unreduced catalyst will continue forming water during the first eight to ten days after the activation commenced.

However beneficial a much lower amount of water produced during the activation of the catalyst charge, the biggest profit from using pre-reduced catalyst is the possibility of reaching design capacity within few days.

Figure 8 shows the expected production of ammonia during the first ten days after the activation of the three different catalyst charges commenced. As it can be seen, there is a significant increase in the production rate (2-fold) when moving from a full unreduced charge to a split charge of pre-reduced (first bed) and unreduced (lower beds) catalysts.

However, this same step up increase is not seen when moving from that same split charge to a full charge of pre-reduced catalyst. Consequently, Topsoe's recommended loading scheme is the split charge of pre-reduced (first bed) and unreduced (lower beds) catalysts.

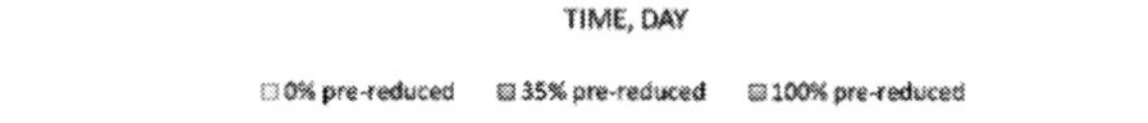

Figure 8. Estimated production of ammonia during the activation of three 90m³ catalyst charges with different content of pre-reduced catalyst (0, 30 and 100%)

Industrial experience

Proven stability

The use of vibrating roads is a common practice for increasing the loading density of an ammonia synthesis catalyst charge in horizontal converters. Although, the suggested type for this kind of jobs is the pneumatically-driven one, the electrically-driven are also commonly used.

Background

During the loading of a KM1R/KM1 catalyst charge in a horizontal converter at a world-scale North American ammonia plant (see Figure 9), the electric vibrators became a safety hazard after being in operation for four days.

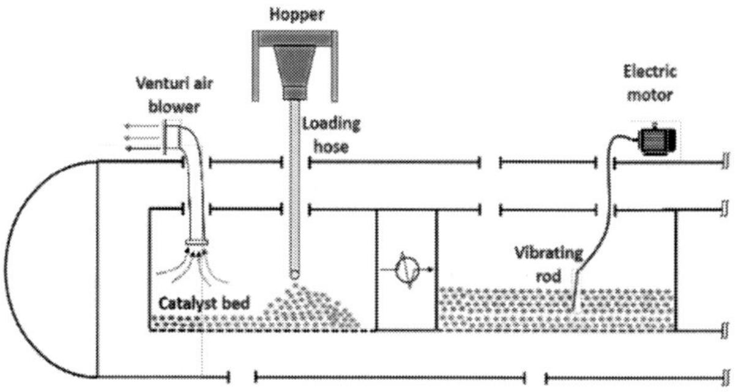

Figure 9. Typical set-up for loading a horizontal ammonia synthesis converter by means of vibrating rods

Shortly after the day shift of day five commenced, the vibration of the pre-reduced catalyst in the first bed was taking place when a loading crew member expressed his concern about the vibrating rods getting extremely hot.

The loading crew member mentioned that he had observed "smoke" coming out of the vibrating rods and the catalyst. The foreman of the loading crew measured the temperature in the catalyst and observed that it had reached approx. 90°C.

In order to ensure the safety of all the personnel as well as the integrity of the catalyst and the converter, the vibration of the catalyst was stopped, and the catalyst temperature was closely monitored in order to ensure that there was no risk of a temperature runaway.

Troubleshooting

After testing of all the available vibrating rods, it was concluded that the rods were indicating electrical malfunction and, consequently, the heat generated by them was not being efficiently dissipated. This hypothesis was confirmed during the testing when the temperature of the rods reached 60-70°C in just a few seconds.

Figure 10. Photo of Topsoe KMR 111 catalyst.

Corrective action

The loading procedure was modified to account for the fast temperature increase in the rods and the maximum allowable temperature for them was set to be 50°C. Furthermore, the loading crew was reminded to inform immediately when they see outstanding taking place.

Lesson learned

Vibrating rods were originally designed for removal of air bubbles trapped in concrete slump. Their usage in loading of ammonia synthesis catalyst, besides the enormous benefit of increasing the loading density, has a major drawback, the low heat capacity of the catalyst compared to that of the concrete slump.

This drawback is reflected in the difficulty of dissipating adequately the heat generated by the rods; especially when they are electrically driven. Although this is not a concern when vibrating an oxidized catalyst, the heat generated by the rod could have created the conditions needed for the pre-reduced catalyst to react with the oxygen in the air. Fortunately, the stability of the KM catalyst prevented an undesired outcome.

This incident confirmed the unmatched properties of Topsoe pre-reduced KM catalysts that makes them unique in the market. These properties are:

1. The catalysts can be loaded in atmospheric air
2. The stability of the catalyst is such that, even with a temperature of approx. 90°C, the catalyst did not react with the oxygen in the air

Faster activation

Plants can experience a head start after installing a split charge of Topsoe ammonia synthesis catalysts KM1R and KM1.

Background

In the mid 1990's, a world-scale ammonia plant in Northern Africa installed its first Topsoe S-200 ammonia synthesis converter (pressure shell and basket) with a charge of KM1 catalyst. The converter and catalyst were in successful operation for more than 20 years.

When the basket and the catalyst reached the end of their operating lifetime, the basket was replaced with the same Topsoe S-200 solution. However, the catalyst combination was changed to the recommended loading scheme of KM1R/KM1.

Activation

Error! Reference source not found. and Figure *12* show the development of the temperature profiles across the converter during the activation of the KM1 and KM1R/KM1 catalyst charges.

As it can be seen in **Error! Reference source not found.**, the temperature gap between the inlet and outlet measurements in bed 1 (blue and green solid lines, filled with un-reduced KM1 catalyst) remained constant during the first 60 hr. of the activation. This period evidently corresponds to the reduction of the KM1 catalyst in the first bed.

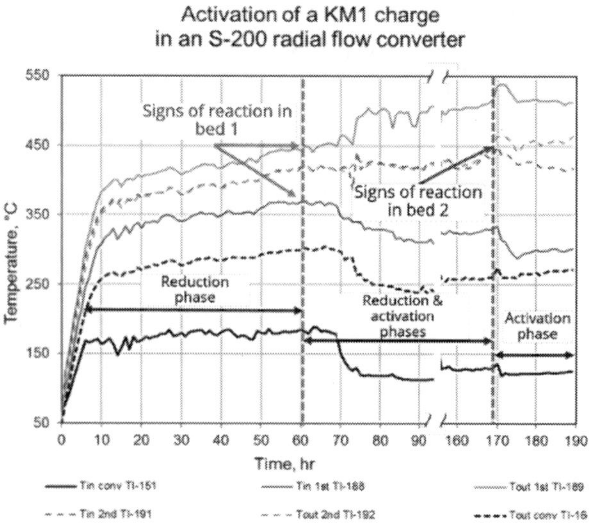

Figure 11. Activation of a KM1 catalyst charge in a Topsoe S-200 ammonia synthesis converter

As the reaction developed in the first bed, the inlet temperature to the second bed was brought up to and maintained at approx. 420°C. Excluding the downtime, the reduction phase of the KM1 catalyst in the second bed lasted approx. 60 hrs.

The benefit of installing KM1R catalyst in the first bed became evident to the customer when the first signs of the ammonia synthesis reaction were seen with an inlet temperature to the first bed of 320°C and only 25 hr. into the activation procedure (see Figure 12).

Furthermore, the head start given by the pre-reduced catalyst allowed the KM1 catalyst in the second bed to have a reduction time which was approx. 15 hrs. less than in the previous activation.

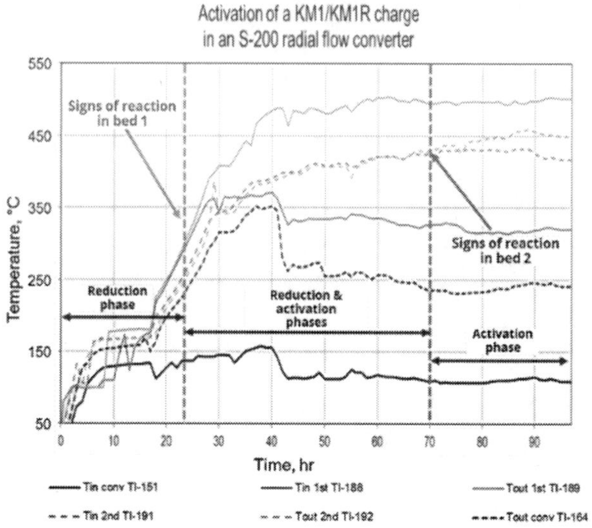

Figure 12. Activation of a KM1/KM1R catalyst charge in a Topsoe S-200 ammonia synthesis converter

Tangible benefits

When two different ammonia synthesis catalyst charges (KM1 and KM1R/KM1) are activated under the same conditions (plant, synthesis loop, converter type, and start-up heater to mention some of them), the comparison of the reduction and activation times between charges is much more reliable and realistic.

The split charge of KM catalysts installed and activated in the mid 2010's was reduced and produced ammonia in both beds only 70 hours after the activation process started. During the same period, only the first signs of reaction were seen in the first bed during the activation of the previous charge.

By reducing the activation time from approx. 120 to 70 hrs., the plant owner saw direct benefits in terms of more production and shorter operation in transient mode by installing the pre-reduced catalyst in the first bed.

Conclusion

Nowadays, a catalyst is state-of-the art when it not only delivers superior operational reliability but also logistics strength. These two concepts are of significant relevance when the catalyst could reach an operational lifetime as long as 30 years.

This is the case of the pre-reduced Topsoe ammonia synthesis catalysts KMR 111 and KM1R. Two products that are manufactured to deliver high activity during normal operation as well as to minimize the risks associated with handling and loading, while ensuring a shorter operation of the plant in the transient condition that an initial start-up represents.

These top-of-the market qualities make the KMR 111 and KM1R catalysts the quintessential example of Topsoe's contribution to a safer and more reliable ammonia industry.

References

[1] P. J. Dahl, C. Speth, A. Krøll Jensen, M. Symreng, M. K. Hoffmann, P. A. Han, S. E. Nielsen, New ATR Based Ammonia Process, Ammonia Technical Manual, 2017, pp. 161-172.
[2] A. Nielsen, An Investigation on Promoted Iron Catalysts for the Synthesis of Ammonia, Jul. Gjellerups Forlag, 1967.
[3] A. Reizer, A. Baranski, The topochemistry of the reduction of an iron catalyst for ammonia synthesis, Appl. Catal. 9, (1984), 343
[4] UN Recommendations on the Transport of Dangerous Goods. Model Regulations, UN, ECE, 2017
[5] European Catalyst Manufacturers Association, Catalyst handling best practice guide, ECMA1004q (2018)

AUTHOR INDEX

AIChE
120 Wall Street, FL 23
New York, NY 10005-4020

ISBN 979-8-3313-0306-8

CURRAN ASSOCIATES INC.
proceedings.com